Library of
Davidson College

1990 Year Book of Developmental Biology

Joel M. Schindler

CRC PRESS, INC.
Boca Raton, Florida

Year Book is a federally registered trademark of Year Book Medical Publishers, Inc. and is used pursuant to a license from Year Book Medical Publishers, Inc.

This book represents information obtained from authentic and highly regarded sources. Reprinted material is quoted with permission, and sources are indicated. A wide variety of references is listed. Every reasonable effort has been made to give reliable data and information, but the author and the publisher cannot assume responsibility for the validity of all materials or for the consequences of their use.

All rights reserved. This book, or any parts thereof, may not be reproduced in any form without written consent from the publisher.

Direct all inquiries to CRC Press, Inc., 2000 Corporate Blvd., N.W., Boca Raton, Florida, 33431.

©1990 by CRC Press, Inc.

International Standard Book Number 0-8493-3304-0
International Standard Serial Number 1042-0932

Printed in the United States

THE EDITOR

Joel Schindler received his B.Sc. degree in biology from the Hebrew University in Jerusalem, Israel in 1973 and his M.Sc. degree in biochemistry from the same institution in 1975. The following year, he returned to the United States with his doctoral mentor, Professor Maurice Sussman, to complete his doctoral studies. He was awarded his Ph.D. from the University of Pittsburgh in 1978.

From 1978 to 1981, Dr. Schindler was a postdoctoral research fellow at the Roche Institute of Molecular Biology in Nutley, New Jersey. During this time period, Dr. Schindler's research efforts focused on the study to investigate changes in gene expression during the peri-implantation period of mouse development. In addition, he was involved in a series of studies aimed at unraveling the mechanism of action of retinoids in inducing murine embryonal carcinoma cell differentiation.

Following his tenure at Roche, Dr. Schindler became an Assistant and subsequently, an Associate Professor in the Department of Anatomy and Cell Biology at the University of Cincinnati College of Medicine in Cincinnati, Ohio. In addition, he was a member of the Graduate Program in Developmental Biology at the Institute for Developmental Research, Children's Hospital Research Foundation, Cincinnati. Dr. Schindler participated in several team-taught courses to both graduate and medical students and was primarily responsible for the areas of cell differentiation and early embryo development. His research efforts remained focused on the regulation of gene expression during cell differentiation and specifically included defining the role of polyamines in regulating the differentiation of both murine and human embryonal carcinoma cells.

In 1987, Dr. Schindler accepted a position in the Genetics and Teratology Branch at the National Institute of Child Health and Human Development (NICHD), in Bethesda, Maryland. His current responsibilities include developing and overseeing NICHD-supported projects in the areas of basic developmental genetics and early embryo development. His unique position allows Dr. Schindler to closely monitor current progress and publications in the field of developmental biology.

Dr. Schindler has received several fellowships and awards; has been a Visiting Fellow at Macquarie University, New South Wales, Australia; and has served as both an editorial and grant reviewer for numerous journals and funding institutions. He is a member of the Society for Developmental Biology, Sigma Xi, the American Society of Cell Biology, the American Association for the Advancement of Science and the New York Academy of Sciences. He is the author or coauthor of scientific reports in numerous journals, books and symposia volumes. This volume is Dr. Schindler's second editorial venture with CRC.

EDITORIAL BOARD

Stephen Alexander, Ph.D.
Associate Professor, Division of Biological Sciences, University of Missouri, Columbia

Robert C. Angerer, Ph.D.
Professor, Department of Biology, University of Rochester, Rochester, New York

Marianne Bronner-Fraser, Ph.D.
Associate Professor, Department of Developmental and Cell Biology, Developmental Biology Center, University of California, Irvine

Joseph G. Culotti, Ph.D.
Associate Professor, Department of Medical Genetics, University of Toronto, Ontario, Canada

Marcelo Jacobs-Lorena, Ph.D.
Associate Professor, Department of Genetics, Case Western Reserve University, Cleveland, Ohio

Terry Magnuson, Ph.D.
Associate Professor, Department of Genetics, Case Western Reserve University, Cleveland, Ohio

Thomas D. Sargent, Ph.D.
Senior Staff Fellow, Laboratory of Molecular Genetics, National Institute of Child Health and Human Development, National Institutes of Health, Bethesda, Maryland

Joel M. Schindler, Ph.D.
Health Scientist Administrator/Program Officer, Genetics and Teratology Branch, National Institute of Child Health and Human Development, National Institutes of Health, Bethesda, Maryland; Adjunct Associate Professor, Department of Anatomy and Cell Biology, University of Cincinnati College of Medicine, Cincinnati, Ohio

E. Charles Snow, Ph.D.
Associate Professor, Department of Microbiology and Immunology, University of Kentucky Medical Center, Lexington, Kentucky

Gary M. Wessel, Ph.D.
Research Associate, Department of Biochemistry and Molecular Biology, University of Texas, M. D. Anderson Cancer Center, Houston, Texas

JOURNALS REPRESENTED

Cell
Development
Developmental Biology
Differentiation
EMBO Journal
European Journal of Immunology
Genes and Development
Genetics
Journal of Cell Biology
Journal of Experimental Medicine
Journal of Immunology
Journal of Neuroscience
Molecular and Cellular Biology
Nature
Neuron
Proceedings of the National Academy of Sciences of the United States of America
Science

TABLE OF CONTENTS

1
Developmental Genetics 1

2
Developmental Gene Expression 53

3
Developmental Cell Biology 81

4
Maternal Controls, Cytoplasmic Determinants, and
Imprinting 117

5
Cell Interactions 143

6
Cell Lineage and Developmental Fate 181

7
Cytodifferentiation — Cell- and Tissue-Specific Gene Expression
and Maintenance 207

8
Homeobox Genes 245

9
Morphogenesis and Pattern Formation 279

Author Index 299

Subject Index 307

INTRODUCTION

Progress in developmental biology continued at a dizzying pace during the past year. More details about questions that are central to the field were uncovered. New and improved technological advances helped facilitate progress. The complexity of detail that underlies even the simplest organisms was further underscored. If a recurring theme were to emerge from all the detail, it would be that a sense of consistency has begun to materialize that reflects basic similarities among varying developmental systems.

Our increased knowledge of the molecular and cellular detail that directs development supports a theme which suggests that there are certain bioactive molecules that are universally expressed and appear to be essential for all normal development to proceed. This emerging concept of "universality" suggests that certain developmentally active molecules are related among different species and share functional roles. In addition, the same molecules can perform different functions at different times in the same organism. Thus, a potentially limited number of "molecular families" could be identified that play major roles in the direction of many fundamental developmental processes. Three such families of molecules that have received particular attention this year are growth factors, homeobox-containing proteins, and retinoids.

The relationship between proto-oncogenes and growth factors continues to be a central focus of investigation. Molecular analysis indicates that many proto-oncogenes encode proteins that function as growth factors or growth factor receptors. Families of such growth factors continue to expand and incorporate new members as they are discovered. Such families include members from a broad range of organisms, extending from frog to mouse and man. Such phylogenetic conservation suggests that these types of molecules play important and necessarily conserved roles in development. In addition, certain growth factors seem to play multiple roles throughout ontogeny. Thus, a single class of molecule can play a pivotal role in development in multiple species and at multiple times. The regulation of their function should continue to fascinate and excite the field.

The function and continued identification of the increasing number of homeobox-containing proteins remains an important focus of investigation. Phylogenetic conservation again suggests that such proteins play important roles in directing developmental events. As DNA binding proteins, the mechanistic detail of the mode of action of many homeobox-containing proteins will clearly have an impact on our understanding of how proteins and nucleic acids interact, how that interaction is regulated, and what the consequences of such interactions could be. In addition, families of DNA binding protein genes have emerged that are similar but not identical to homeobox-containing genes, again suggesting that such classes of molecules have central roles in multiple developmental processes. Continued dissection of the regulatory circuitry that coordinates the expression of these gene classes will certainly provide new insight into our understanding of the underlying genetic regulation of development.

Retinoic acid and its related class of compounds, known collectively as retinoids, also play a unique developmental role. This role includes the function of retinoic acid as a morphogen during amphibian and avian development and its role in maintaining various differentiated phenotypes in mammalian development. The molecular dissection of its mechanism of action indicates that several components exist, including cytoplasmic binding proteins and nuclear receptors. Furthermore, there are multiple nuclear receptors, all of which are related to a superfamily of such receptors, that collectively function to mediate the effects of small molecules on gene expression. Thus, retinoids are an excellent example of how small molecules, not encoded by genes, can exert enormous developmental influence. They remind us that not only gene products and genetic regulation but also other classes of molecules can play major roles in directing developmental events and that their mechanistic modes of action can be sophisticated and complex.

The most far-reaching technological advancement that has been applied to the investigation of developmental events has been the use of the polymerase chain reaction (PCR). This powerful technique has already had an impact on the ability to generate reagents, select mutants, and investigate the molecular structure of developmentally relevant genes. The rapid incorporation of this new technology into the "standard operating procedures" of the field indicates the extent to which new technologies can quickly and effectively have an impact upon the field and expedite progress.

The area of mammalian developmental biology has, in particular, experienced much of the most exciting progress during the past year. This is to a large extent due to improvements in certain evolving technologies that have facilitated rapid expansion in the field. Our ability to direct genetic alterations through homologous recombination, culture embryonic stem cells more efficiently, and generate chimeric and transgenic animals have all contributed to our increased knowledge in this arena. The use of these technologies together will clearly enhance our ability to generate specific animal models in order to investigate defined developmental anomalies.

Assigning articles to chapters in this book remains difficult as distinctions between categories within developmental biology continue to blur. This breakdown in distinctions speaks well of both our increased knowledge about various developmental events and our increased appreciation for their complexity. As we understand more of the detail that explains a certain developmental event, we recognize that it involves several previously distinct biological events. While certain important components may emerge as "universal" throughout much of development, their own functions in different organisms and at different times underscore the difficulty of relying on categories to define the discipline. The more we learn, the less distinct it becomes.

Finally, plant development has been included in this volume to a limited degree. Its relative incorporation is not a comment on its importance but rather the editor's limited understanding of it, and the extent of its representation will be expanded next year. We hope that this volume again presents the breadth and depth of developmental biology as a discipline and continues to excite and surprise you.

Developmental Genetics 1

INTRODUCTION

Experimental manipulation aimed at understanding the underlying genetic basis for developmental events remains the central theme for most investigations in developmental genetics. Such manipulation can either be classic, as is the case in standard genetic crosses, or molecular, as is the case in transfections. Ultimately both can be used together.

By continuing to study the abnormal, we can learn much about what normally should occur during development. Therefore, the ease with which we can select and isolate interesting mutants can greatly facilitate our progress. The characterization of mutant phenotypes allows us to identify some of the specific components that constitute the genetic circuitry that underlies development. Subsequent genetic crosses can uncover how different genetic lesions interact and thus provide the detail necessary to fill in pieces of the cascade of events that direct development.

The articles in this chapter focus on various aspects of using mutant phenotypes to understand development. Several demonstrate different ways to generate new mutants and use novel selectable markers to identify modified genotypes. Transposable elements are discussed as insertional mutagens and their mechanism of transposition is explored. The ability to regulate such transposition can greatly enhance the general utility of using transposable elements as molecular markers. The polymerase chain reaction is discussed as a means of amplifying developmentally interesting genes and analyzing their structure. Embryonic stem (ES) cells are further exploited as excellent vehicles to introduce exogenous DNA into the mouse germ line through chimeric mice. Gene targeting is explored and transgenic mice that can be considered as experimental models for specific deficits are presented. Anti-sense RNA experiments are described as a means of disrupting normal development, genetic rearrangements are explored, and the "macromanipulation" of DNA through chromosome microdissection and artificial chromosomes is presented. Classic and molecular approaches are wed. Improvements in our ability to artificially manipulate genomes in order to address specific issues of development is beautifully detailed and intellectually quite satisfying.

Hygromycin Resistance as a Selectable Marker in *Dictyostelium discoideum*

T. T. Egelhoff, S. S. Brown, D. J. Manstein, and J. A. Spudich
Mol. Cell. Biol., 9, 1965—1968, 1989

Dictyostelium discoideum is useful in biological study, especially since a DNA-mediated transformation system for this organism has been established. Because only a single selectable marker, resistant to the neomycin derivative G418, exists for *D. discoideum*, however, experimental possibilities using this organism are limited. The work presented here was designed to create a second selectable marker usable for DNA transformation in *D. discoideum*.

Two plasmids derived from pUC119 were created in which the promoter and the first 8 codons of the actin 15 gene of *D. discoideum* were fused to the third codon of the hygromycin resistance gene (hygromycin phosphotransferase, *hph*) of *Escherichia coli*. When the integrating vector, which also contained the gene encoding resistance to G418, was introduced into the axenic cell line Ax4 under selective conditions, transformed colonies appeared. Transformation was confirmed by Southern blot analysis, showing variable plasmid copy number ranging from a few to about 200. Selection for G418 resistance resulted in transformants that were resistant to hygromycin, although selection for hygromycin resistance did not reproducibly result in transformants that were resistant to G418. To circumvent this problem, the high-copy number extrachromosomal plasmid containing the hygromycin resistance cartridge was used for transformations; this resulted in very efficient transformation and hygromycin-resistant colonies. Southern blot analysis confirmed the high copy number, extrachromosomal presence of the plasma in the transformant lines.

Sometimes colonies spontaneously resistant to hygromycin appeared. For this reason, the use of control transformations is recommended, as is confirmation using Southern blotting. The development of this *D. discoideum* transformation vector widens the types of experiments possible using this organism.

♦This report of the construction of a transforming vector based on hygromycin resistance opens up many new avenues of experimentation on the development of *Dictyostelium*. Previously, the introduction of DNA into cells of *Dictyostelium* was limited to vectors conferring G418 resistance. Thus, secondary transformation was impossible. The vector, pDE109, is an extrachromosomal vector which can be used to transform genes, via hygromycin selection, into cells already transformed by other G418 resistance based vectors. Therefore, *in vitro* mutagenized genes now can be cloned into gene-disrupted strains, and the interactions of cloned regulatory elements can be tested. The paper also reports an integrating vector based on hygromycin selection, but indicates that transformation was unpredictable. *Stephen Alexander*

High-Frequency Switching in *Dictyostelium*
B. Kraft, D. Steinbrech, M. Yang, and D. R. Soll
Dev. Biol., 130, 198—208, 1988

High-frequency switching of phenotypes is known to occur in *Salmonella,* trypanosomes, and yeast. Recently, several high-frequency switching systems have been identified in the dimorphic yeast *Candida albicans,* with these identifications based on the ability to plate large numbers of colonies for subsequent determination of phenotype. Recent work from this laboratory has shown that *Dictyostelium discoideum* can generate timer mutant variants at a very high frequency. The present investigation aimed at characterizing such high-frequency phenotypic changes in *D. discoideum,* and comparing such changes to switching in *C. albicans.*

The AX3, clone RC3, and the previously described developmental mutant FM1, when plated to low density, yielded variant colonies spontaneously at frequencies of about 10^{-2}. High frequencies of sectoring also occurred. Low doses of ultraviolet radiation that resulted in 5 to 20% lethality stimulated variant colony formation fivefold and also resulted in increased sectoring frequencies. Both aberrations in morphogenesis and changes in developmental timing appeared to undergo switching. While switched phenotypes were reproducible and heritable, they also exhibited high spontaneous frequencies of interconvertibility between variant phenotypes and high spontaneous frequencies of reversion to the wild-type phenotype (Figure 1-2).

These variant phenotypes in *D. discoideum* exhibit many characteristics of the switching system of *C. albicans.* While the mechanism of switching in *D. discoideum* is unknown, it may be due to a high-frequency reversible transposition system, although several other mechanisms cannot be ruled out. Work in progress involves elucidating the molecular mechanism of switching in *Dictyostelium.*

Transfer RNA Genes: Landmarks for Integration of Mobile Genetic Elements in *Dictyostelium discoideum*
R. Marschalek, T. Brechner, E. Amon-Böhm, and T. Dingermann
Science, 244, 1493—1496, 1989

Transfer RNA (tRNA) genes function to encode tRNAs, but also have other functions. In *Saccharomyces cerevisiae,* tRNA genes are often asso-

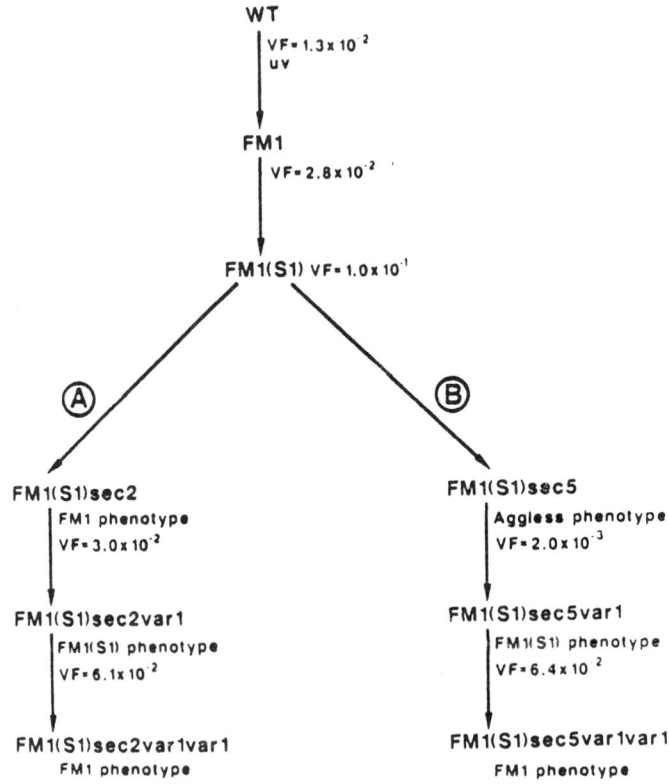

FIGURE 1-2. Two switching sequences exhibiting sequential reversibility. WT, wild type; FM1, original fast mutant 1; FM1(S1), slow mutant 1 isolated from the original FM1; sec, variant sector; var, variant colony; VF, frequency of varaiant colony formation; uv, ultraviolet treatment. Note that the sequence of switch phenotypes (A) of FM1(S1)sec2car1 is WT → FM1 → FM1(S1) → FM1 → FM1(S1) → FM1, and the sequence of switch phenotypes (B) of FM1(S1)sec5var1var1 is WT → FM1 → FM1(S1) → aggless → FM1(S1) → FM1. (From Kraft et al., *Dev. Biol.*, 130, 1988. ©Academic Press.)

ciated with transposons, although the reason for this association is unclear. The work described here involved a detailed characterization of 24 genomic fragments from the axenic *Dictyostelium discoideum* strain AX2, containing different tRNA genes.

Sequence analysis of these genes showed regions of extreme similarity occurring at conserved positions relative to the tRNA genes. The first element, DRE1, was composed of a core of 199 bp, with nucleotides 1 to 72 as a direct terminal repeat. This element was found singly, or in two to four tandem copies. Two major classes of DRE1 were found. The second element, DRE2, was always found separated by 39 dA-dT base pairs from DRE1. DRE2 included an open reading frame of more the 939 nucleotides. DRE1 and DRE2 together may function as a unit of a composite element. Both were extremely unstable even when cloned in *Escherichia coli rec A*⁻ strains. The third element, Tdd3, had been previously described, but in this study it was always found about 100 nucleotides downstream from mature tRNA coding regions. In related strains, Tdd has been found in different locations, causing a 9- to 10-bp duplication of the target site DNA. DRE1/DRE2 also appears mobile.

These findings suggest that tRNA genes in *D. discoideum* are preferential targets for mobile genetic elements. These genes are also preferential targets for the sigma and tau elements of yeast. It is not clear whether the association with mobile genetic elements might disrupt tRNA gene expression in either organism.

♦ *Dictyostelium* is a popular experimental material for studies on development and molecular cytology. Mutants play a large role on these studies. However, it is often difficult to identify the gene responsible for the phenotype. Clearly, a system for transposon tagging would be extremely useful. These two reports suggest that such a system may be forthcoming.

Kraft et al. extend on earlier studies showing that *Dictyostelium* is capable of rapidly and heritably switching phenotypes. They now show that this phenomenon is rapid (10^{-2} to 10^{-3}), reversible, and capable of generating many different phenotypes. Significantly, switching is inducible by low doses of UV light. The system is very similar to that seen in the yeast *Candida albicans* where it is thought to be involved in pathogenicity. If the underlying mechanism is due to a transposable element, and the element can be identified, this would provide a powerful tool for molecular genetic analysis in this organism.

Marschalek et al. have identified three repetitive elements with transposon-like structures (terminal direct repeats) associated with tRNA genes. Two of these elements are newly identified and one, Tdd3, has been previously identified. All are consistently found at certain distances from the tRNA genes, although there is no obvious sequence specificity of the integration site. Comparison of the same tRNA genes in different strains shows that some have these putative transposons while others do not. Whether these differences in unrelated and highly mutagenized strains are due to movement of the sequences, and whether this movement can be controlled, remains to be seen. *Stephen Alexander*

Signal Transduction in *Dictyostelium fgd* A Mutants With a Defective Interaction Between Surface cAMP Receptors and a GTP-Binding Regulatory Protein
F. Kesbeke, B. E. Snaar-Jagalska, and P. J. M. Van Haastert

Because of its small genome size, *Dictyostelium discoideum* is an excellent organism in which to study signal transduction in chemosensory mutants. One such group of mutants comprises the *fgd* A complementation group, previously isolated by Coukell and colleagues. Mutants in this group fail to respond to cAMP with chemotactic reactions nor with the induction of EDTA-resistant contact sites; they do, however, possess cell surface cAMP receptors. The experiments reported here were designed to characterize these mutants both biochemically and functionally.

The *fgd* A mutants had both high- and low-affinity surface receptors for cAMP, but they were missing the B^{ss} form. Cyclic AMP induced down regulation and covalent modification of these receptors, but did not induce chemotaxis nor activation of adenylate nor of guanylate cyclase. In isolated membranes, the inhibition of cAMP binding by GTPγS and GDPβS were reduced. Both basal and cAMP-stimulated high-affinity GTPase activity were reduced in these mutants. In membranes from these mutants, GTP-mediated stimulation and inhibition of adenylate cyclase were normal.

These findings suggest that transmembrane signal transdution is defective in *fgd* A mutants. It is likely that this defect lies in the interaction between surface cAMP receptors and a particular G-protein. While defects in the cAMP receptor, a specific G-protein, or an unknown component required to activate a G-protein would all explain the characteristics of these mutants, it may be that their primary defect lies in the G-protein that mediates the receptor stimulation of the phosphatidylinositol cycle. If so, that transduction pathway is essential for chemotaxis and the stimulation of adenylate cyclase in *D. discoideum*.

♦ This study presents a thorough biochemical and physiological analysis of a group of previously isolated chemosensory mutants termed "frigid". These mutants fall into five complementation groups. Three of these appear to be simple "program mutants" that fail to enter the developmental program while the A and C complementation groups appear to be defective in signal transduction. The latter is useful in the molecular dissection of the signal transduction process in *Dictyostelium* which is involved in both cell aggregation and subsequent cell differentiation.

The data presented in this paper indicate that these mutants are suitable for subsequent studies of the cAMP signal transduction mechanism. cAMP receptors are present and were down regulated and covalently modified by cAMP. However, no cellular responses were induced by cAMP such as activation of the adenylate or guanylate cyclases or chemotaxis. Several properties associated with G-proteins are defective in the mutants although the different mutants exhibit quantitative differences in the phenotypes. Overall, the data show that the signal transduction defect in these strains lies somewhere between the cAMP receptor and a specific G-protein. The authors indicate that a G-protein subunit of 40-kDa (detected with a Gα-common antiserum) is absent in the "frigid" strains. These mutants should continue to be useful in working out the molecular details of cAMP signal transduction during dvelopment of *Dictyostelium*. Stephen Alexander

Genome Linking With Yeast Artificial Chromosomes
A. Coulson, R. Waterston, J. Kiff, J. Sulston, and Y. Kohara
Nature, 335, 184—186, 1988   1-5

A physical map of the genome of *Caenorhabditis elegans* would be useful to facilitate the molecular cloning of interesting loci, but currently, such mapping has been limited to a linking of cosmid clones into clusters or contigs. This technique seemed to have stalled when 90 to 95% of the genome had been cloned into 17,500 cosmids in about 700 contigs, and when the linking clones needed proved difficult or impossible to find. The present work was designed to use the yeast artificial chromosome (YAC) vector to map these cosmids.

Large (50 to 1000 kilobase) genomic fragments from *C. elegans* were introduced into YAC vectors that provide centromeric, telomeric, and selective functions. These YACs were hybridized with probes from the previous *C. elegans* library of Lorist cosmids. This approach was facilitated both by the fact that little or no homology exists between YAC and Lorist vectors and that the nematode genome consists of relatively few repeat sequences.

Grids of YAC clones were probed with cosmids from the ends of contigs, permitting walks. Cosmid clone grids were also hybridized with individual YAC probes. In 7 months and 1000 probings, the number of contigs representing the genome was halved. Only about one third of the contig joinings based on hybridization have been confirmed with cosmid overlaps, but the remainder, which require confirmation, appear genuine by a variety of criteria.

Use of YAC maps will not likely supplant cosmid and lambda clone maps, because smaller clones are required for certain techniques. It is not certain whether this technique will be useful in mapping more complex genomes that contain a higher proportion of repeated sequences. Fingerprinting procedures may need to be developed in lieu of the hybridization techniques used here.

♦This paper represents a technical advance which should greatly facilitate the physical mapping of genomes, such as that of *C. elegans* and eventually the human genome. YACs (or yeast artifical chromosomes), which are much larger than cosmid or phage clones, can be used to fill the gaps in physical maps (based on the analysis of cosmid or phage clones) by hybridizing to clones at the ends of contigs (islands of overlapping cosmids). Correlating the physical map with the genetic map of genetically amenable organisms, such as *C. elegans*, has many potential benefits, not the least of which is that it will facilitate the cloning of any gene that has been precisely mapped genetically. This means that the molecular nature of defects in developmental mutants can be determined rapidly and efficiently. This should help to elucidate the molecuar mechanisms underlying a large number of developmental paradigms. *Joseph G. Culotti*

Identification and Purification of a *Drosophila* Protein That Binds to the Terminal 31-Base-Pair Inverted Repeats of the *P* Transposable Element
D. C. Rio and G. M. Rubin
Proc. Natl. Acad. Sci. U.S.A., 85, 8929—8933, 1988

1-6

P elements are transposable elements found in *Drosophila melanogaster* that exhibit highly regulated, tisssue-specific transposition. P elements have 31 base-pair inverted terminal repeats and their insertion results in duplications of 8 base pairs of target DNA. Host-cell-encoded functions may be required for P elements transposition, but a direct identification of these functions is difficult in the absence of an *in vitro* P element transposition system. The present work was an attempt to identify *Drosophila* proteins involved in P element transposition using DNase I footprinting and nuclear extracts of *Drosophila* K_c tissue culture cells.

A nuclear extract fraction known to contain a number of DNA binding proteins was assayed for binding to labeled P element DNA probes using DNase I footprint protection. A binding activity was identified that specifically protected a region of the 31-base-pair terminal inverted repeats. Irrespective of the duplicated target DNA sequence, this binding activity was found on inverted repeats at both the 5' and 3' ends. Ultraviolet photochemical crosslinking experiments resulted in the specific covalent binding of a 65- to 70-kDa polypeptide to the 31-base-pair repeat DNA. Purification of the polypeptide that binds to the inverted repeat, based on site-specific DNA affinity chromatography, resulted in the identification of a 66-kDa P element binding protein. This protein is not the same as a protein of similar molecular mass encoded by the P element.

This DNA footprinting study resulted in the identification of a protein of molecular mass 66,000 that binds to the terminal 31-base pair inverted repeats of the P transposable element. This protein may be a host factor involved in transposition of P elements. By analogy to various prokaryotic proteins required for mobility of transposable elements, the 66-kDa protein might interact with the P transposase during the transposition reaction.

♦ Drosophila transposable P-elements gained enormous importance when it was discovered that they can mediate integration of genetic material into the fly's germ line. However, little is known about the mechanisms that mediate transposition.

There are two requirements for P-mediated transposition to occur: the presence of an active transposase and the presence of defined terminal DNA sequences flanking the DNA to be transposed. The transposase is coded by the intact 2.9 kb P-element. Although transcription occurs in all cells, regulation at the level of splicing allows transposase to be produced only in the germ line. The 31-bp inverted repeats that define the extrme 5'- and 3'-ends of the P-elements were known to be required but little was known about additional sequence requirements.

The complexity of the sequences required for transposition suggest that multiple host factors are involved (the coding capacity of the P-element itself is rather limited). The identification of such factors and the characterization of the corresponding genes will certainly be the focus of forthcoming research. The paper by Rio and Rubin represents the first step in this direction. They have purified a 66 kDa protein from cultured Drosophila cells that recognizes a 16 bp sequence at the end of the 31-bp repeat. This protein may well represent a host factor essential for transposition. It is reasonable to expect that additional host proteins exist that recognize other sequence elements within the region required for transposition. The further characterization of these proteins may lead to novel findings concerning not only host cell functions but also P-element transposition. *Marcelo Jacobs-Lorena*

Mobility of *P* Elements in Drosophilids and Nondrosophilids
D. A. O'Brochta and A. M. Handler
Proc. Natl. Acad. Sci. U.S.A., 85, 6052—6056, 1988

P elements are naturally occurring, highly mobile transposable elements, first found in *Drosophila melanogaster,* that have been employed as efficient gene vectors. It has been suggested that P-elements are capable of transposition in a wide variety of species and hence would be useful as general gene vectors in various insect and noninsect systems. Tests of the latter possibility have been inconclusive. The present work was based on the premise that the usefulness of P-elements as general gene vectors is limited by the mobility of P-elements in various taxa. In these experiments, the phylogenetic limits of P-element mobility were investigated using a P-element mobility assay.

Embryos of seven drosophilid and two nondrosophilid tephritid species were injected with a P-element-excision indicator plasmid and a transposase-producing helper plasmid. After incubation and heat shock, the plasmids were recovered and transformed into *Escherichia coli*. If precise or nearly precise excision of P element from the plasmid had occurred *in vivo*, Lac Z^+ (blue) colonies were detected on indicator plates.

In all drosophilid species tested, including *Chymomyza proncemis* and *Zaprionis tuberculatus*, P element excision was detected. The frequency of excision was dependent on the relatedness of the species to *D. melanogaster*. In contrast, no excision was detected in the related dipterans *Anastrepha suspensa* and *Toxotrepana curvicauda*.

These findings indicate that P element mobility *in vivo* is phylogenetically restricted. It is likely that the *in vivo* excision assay used here results in more direct assessment of P-element mobilization than methods that rely on genetic selection or germ-line transformation. The identification of species that do not support P element transposition should prove useful in the study of the mechanisms and regulation of transposition and excision.

♦ The development of P-elements as an efficient vector for introduction of genetic material into the Drosophila germ line has allowed significant progress to be made in the understanding of many aspects of insect biology. Earlier studies demonstrated that P-elements can also be mobilized into species other than *Drosophila melanogaster*, such as *Drosophila simulans* and the more distantly related *Drosophila hawaiiensis*. However, attempts of P-element transformation into non-drosophilids have so far been unsuccessful. Although a single germ line integration event has been obtained in the mosquito *Anopheles gambiae*, this event was not due to P-element transposition.

Achieving germ line transformation in other species, such as is insects of medical or economic importance, is an important goal. The assays available so far for P-element activity in living insects have been laborious, cumbersome, and relied on drug selection schemes that had not previously been tested in those insects. O'Brochta and Handler have adapted a very sensitive and efficient assay originally developed by Rio et al. to measure P-element funcions in living insect embryos. As illustrated in the figure, two plasmid DNAs are coinjected into the embryo. One codes for a functional P transposase. The other contains an ampicillin-resistance gene and a target β-galactosidase sequence interrupted by the insertion of a defective P-element. Successful excision of the P-element insert by the transposase regenerates β-galactosidase activity. After incubation of the injected embryos, plasmid DNA is extracted and used to transform *E. coli*. The bacteria are then tested on ampicillin plates containing an indicator of β-galactosidase activity. Excision of the P-element is assayed by measuring the number of β-galactosidase-positive colonies.

The experiments revealed that transposase-dependent P-element excision occurs in embryos of several species belonging to the *Drosophila genus* but not in the related Tephritidae (order: Diptera). This specificity of P-element mobilization could be due to a number of reasons, including the absence of required host-encoded factors (see report on the papers by Mullins et al. and by Rio and Rubin in this volume) or the presence of inhibitory factors in the host cytosplasm. The results of this paper are significant not only because they provide additional evidence for genus-specificity of P-element mobility, but also because they establish a relatively simple assay that should prove useful in further studies. *Marcelo Jacobs-Lorena*

A Copia-Like Transposable Element Family in *Arabidopsis thaliana*
D. F. Voytas and F. M. Ausubel
Nature, 336, 242—244, 1988

Arabidopsis thaliana is a popular subject of plant molecular genetics studies, due in part to its short generation time and small genome size. The work described here was an attempt to identify endogenous transposable elements from this plant, in order to facilitate future genetic analysis. To do this, a screen of restriction fragment length polymorphisms was performed on 16 geographical races (ecotypes) of *A. thaliana*, based on the rationale that insertional polymorphisms might be found that might have arisen by DNA transposition.

An insertional polymorphism was found in the ecotype Kas-1. When this insertion was used as a hybridization probe on Southern blots, it was found to be present in one to three copies in most ecotypes studied, and to be a member of an interspersed family of sequences named Ta1. A 5.2-kilobase (kb) Ta1-1 element from the Landsberg ecotype (La-0) was fully sequenced and was found to be similar in structure to retrotransposons and integrated retroviral proviruses.

The Ta1 element sequenced had a large internal domain of 41,900 base pairs (bp) bounded by 514-bp long terminal direct repeats (LTR) ending in 4-bp inverted repeats. A 5-bp direct repeat, due to duplicated target sequences, flanked the insertion of the sequenced Ta1 element. An internal 4.2-kb region contained a single open reading frame similar to that of retrotransposons and retroviruses, composed of segments that appear to include an RNA-binding domain, a protease domain, and a reverse transcriptase domain.

This transposable element resembles the *Drosophila melanogaster* retrotransposon in a number of features that suggest they are more related to one another than to other transposable retrotransposons and retroviruses. Apparently, retrotransposons are common to both plants and animals.

◆ The first suggestion that transposable genetic elements existed at all was the result of studies on maize. Since then, such genetic elements, most notably P-elements in *Drosophila*, have been discovered in many organisms and used successfully to facilitate genetic analysis in them. Voytas and Ausubel report the discovery of a new family of transposable elements, called Tal elements, in the conifer *Arabidopsis thaliana*. Since this organism has become the subject of intense molecular investigation, this discovery could have enormous implications for the entire field of plant molecular genetics. The structural homology between Tal elements and *copia* elements suggests that such elements play a role in observed genetic variability within species and that they may share a common ancestral origin. *Joel M. Schindler*

Antisense RNA Inhibition of Polygalacturonase Gene Expression in Transgenic Tomatoes
C. J. S. Smith, C. F. Watson, J. Ray, C. R. Bird, P. C. Morris, W. Schuch, and D. Grierson
Nature, 334, 724—726, 1988   1-9

Antisense RNA regulates gene expression naturally in bacteria, and has been used experimentally to reduce gene expression in several animal and plant species. The work described here was designed to achieve stable inhibition of the expression of the polygalacturonase (PG) gene, a developmentally regulated endogenous gene, in stably transformed tomatoes. The PG gene seems to exist in only a single copy in tomato, and its gene has been cloned. PG and its mRNA are synthesized when fruit ripen; the enzyme acts in fruit softening by partially solubilizing the pectin of the cell walls.

Constitutive expression of PG antisense RNA was achieved in tomato plants, after transformation with a vector in which a 50-bp untranslated region and the translation start site of the PG gene was fused, in inverted orientation, to the cauliflower mosaic virus 35S RNA promoter and the 3' end of the nopaline synthase gene. Northern analysis showed that transformed plants expressed the PG antisense gene stably, in a polyadenylated form, with at least three transcripts including one which appeared full-length. Expression was constitutive in leaves and fruit of some transformants; progeny of transformed plants also showed the presence of the antisense gene by Southern analysis. PG protein and PG enzyme activity was reduced up to 90% in ripening fruit from transformants.

Although antisense PG expression might have ultimately resulted in fruit with lengthened shelf-life, no differences in fruit softening as measured by compressibility were found between transformants and control fruit. Further studies are in progress to measure the degree of pectin depolymerization during ripening of fruit from transformed and normal plants.

♦As is the case in animal developmental biology, progress in plant developmental biology is linked to technological advances that can have a positive impact on the field. Such progress has taken place in parallel with studies on traditional animal systems and now allows plant developmental biology to be explored in great detail. The paper by Smith et al. incorporates several technological advances that are also used in animal studies to address a basic question in plant developmental genetics. The authors blend transgenic technology with antisense technology to investigate the developmentally regulated expression of the polygalacturonase gene. They successfully demonstrate that such genetic manipulation is feasible and that this approach is a viable strategy to investigate general questions related to plant development, gene expression, or breeding. *Joel M. Schindler*

The SCID-hu Mouse: Murine Model for the Analysis of Human Hematolymphoid Differentiation and Function
J. M. McCune, R. Namikawa, H. Kaneshima, L. D. Shultz, M. Lieberman, and I. L. Weissman
Nature, 241, 1632—1639, 1988   1-10

Study of the human immune system and of human hematopoietic cells has been limited by the lack of a suitable animal model. The present work was designed to generate such a model, employing mice with a genetically determined severe combined immunodeficiency. These SCID mice usually die young of opportunistic infections; although they cannot mount successful cellular or humoral immune responses to foreign antigens and lack functional T and B cells, probably due to defects in antigen receptor gene rearrangement, the rest of their immune and hematopoietic systems, including the thymic stroma, is intact. The experiments described here involved the production of the hematochimeric SCID-hu mouse, formed by the engraftment of human fetal liver, thymus, and lymph node into the SCID mouse.

Control mice died young, unless they were maintainted on tri-

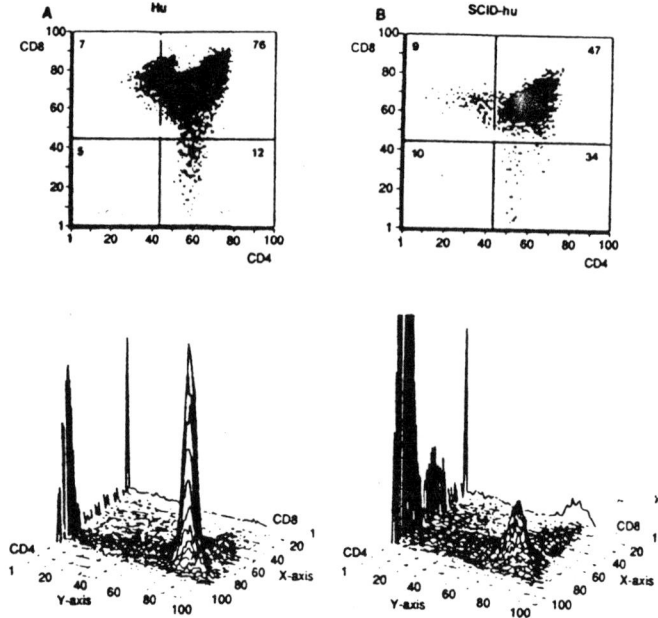

FIGURE 1-10. FACS analysis of SCID-hu thymocytes. Two-color profiles of age-matched human thymus are shown on the left (A); those of the SCID-hu thymus are on the right (B). Subpopulations that are double-negative, single-positive, and double-positive for CD4 and CD8 (and their respective percentage of total thymocytes) are shown in the corresponding quadrants of the top graphs. FITC-Leu 3a and PE-Leu 2a were used as described by the supplier (Becton Dickinson, Mountain View, CA). (From McCune et al., *Science*, 241, 1632—1639, 1988. With permission.)

methoprim-sulfoxasole. Hematochimeras implanted with human fetal thymus and human fetal liver cells remained healthy without prophylactic antibiotics. Implanted human fetal thymus grafts showed microscopic anatomy comparable to that of its normal human counterpart. Double-color analysis by fluorescein-activated cell sorting showed the presence of CD4 and CD8 thymocytes in similar proportions to those found in normal human fetal thymus (Figure 1-10). These thymocytes were of human origin. No human thymocytes were found in the host thymus. Human T cells were found only transiently in the engrafted thymus, and were found later and transiently in the peripheral circulation. Some SCID-hu mice that had been engrafted with human fetal liver cells, human fetal thymus, and human fetal lymph node produced human IgG at concentrations about 10% that found in normal human serum.

These findings indicate that a murine model system has been created in which human T and B cell lineages differentiate. Although such immune system reconstitution was partial and transient, with no evidence of physiological functioning of human lymphoid cells, it may be useful for analysis of human retroviral diseases and their potential therapies.

Transfer of a Functional Human Immune System to Mice with Severe Combined Immunodeficiency
D. E. Mosier, R. J. Gulizia, S. M. Baird, and D. B. Wilson
Nature, 335, 256—259, 1988

In an effort to develop an improved animal model of the human immune system, mice with severe combined immunodeficiency (SCID) were injected with human peripheral blood leukocytes (PBL). The survival and functioning of these cells within the SCID mice were investigated.

Initial experiments demonstrated that xenogeneic graft-versus-host disease was not a significant problem, intraperitoneal injection was optimal, and that the number of PBL transferred and the EBV status of these cells influenced survival. Human PBL could successfully be established in SCID mice with 10×10^6 PBL from EBV-positive donors or 50×10^6 cells from EBV-negative donors.

After injection of human PBL into SCID mice recipients, there was a rapid increase in human immunoglobulin secretion. The transferred human lymphocytes proved capable of responding to antigen. Flow cytometry demonstrated that 2 weeks after transfer, SCID spleen were composed of 80 to 92% human lymphoid cells. Although normal lymphocytes survived and expanded, the relative DB90-011 proportion of lymphocyte subsets was altered in some SCID mice as compared to the normal human immune system.

These results indicate that the transfer of human PBL into SCID mice may provide a useful animal model of the functioning of the human immune system and its response to pathogens.

♦A drawback in most attempts to study human tissues directly is the obvious limitations imposed upon such experiments. This has made the comprehensive analysis of the human immune system almost impossible. This has become more sharply into focus in the wake of such mammoth social problems as AIDS. It is important for scientists interested in the human immune system to have at their disposal approaches beyond *in vitro* mitogenic stimulation of human peripheral blood leukocytes. In the past year, two laboratories have published accounts of the engraftment of human hematopoietic cells or peripheral blood leukocytes into C.B.-17 *scid/scid* mice, which display a severe combined immunodeficiency syndrome and lack the ability to mount either a cell-mediated or humoral immune response (Bosma et al., *Nature,* 301, 527, 1983). Both of these reports indicate that such mice demonstrate both human cellular and humoral responses. This advance will allow a more careful analysis of the human immune repertoire, a system to follow the purification and subsequent propagation of human hematopoietic cells, and possibly a better experimental approach for the further characterization of human diseases, such as AIDS. *Charles Snow*

The *scid* Defect Affects the Final Step of the Immunoglobulin VDJ Recombinase Mechanism
B. A. Malynn, T. K. Blackwell, G. M. Fulop, G. A. Rathbun,
A. J. W. Furley, P. Ferrier, L. B. Heinke, R. A. Phillips,
G. D. Yancopoulos, and F. W. Alt
Cell, 54, 453—460, 1988   1-12

Complete variable immunoglobulin regions are assembled through DNA rearrangements mediated by conserved recognition elements that appear to be targets of a site-specific recombination system (VDJ recombinase) (Figure 1-12). SCID mice do not produce mature, functional lymphocytes, due to aberrant variable region rearrangements that frequently lead to large deletions. To elucidate the recombinase defect in SCID mice, A-MuLV transformed cell lines were derived from SCID and normal mice. The molecular nature of the variable region rearrangements in the SCID mice cells were determined and compared to those of nomal mice.

Ongoing heavy chain gene rearrangement was detected in the transformed cells from both normal and SCID mice. Therefore, the VDJ recombinase appears to be active in the SCID mice. However, as normal IG heavy chain mRNA was never expressed from the SCID mice cells, all rearrangements appeared to be aberrant. The nucleotide sequence of the junction fragments was determined from subcloned A-MuLV transformed SCID lines. Normal ligation at both ends was never detected in SCID rearrangements and large deletions were frequently observed. Secondary rearrangements occurred in the SCID cells and were also aberrant.

Therefore, SCID pre-B cells have an active, but defective VDJ recombinase. The most likely defect is impaired joining of ends generated by site-specific endonucleolytic activity. As unrepaired double-strand breaks, such as those due to a site-specific endonuclease, are lethal to cells, illegitimate recombination events presumably mediate the aberrant joining in surviving SCID pre-B cells (Figure 1-12).

♦ It has long been thought that the defect responsible for severe combined immune deficiency (*scid*) might reside in the absence or malfunctioning of the VDJ recombinase (Schuler et al., *Cell,* 46, 963, 1986). The basis for

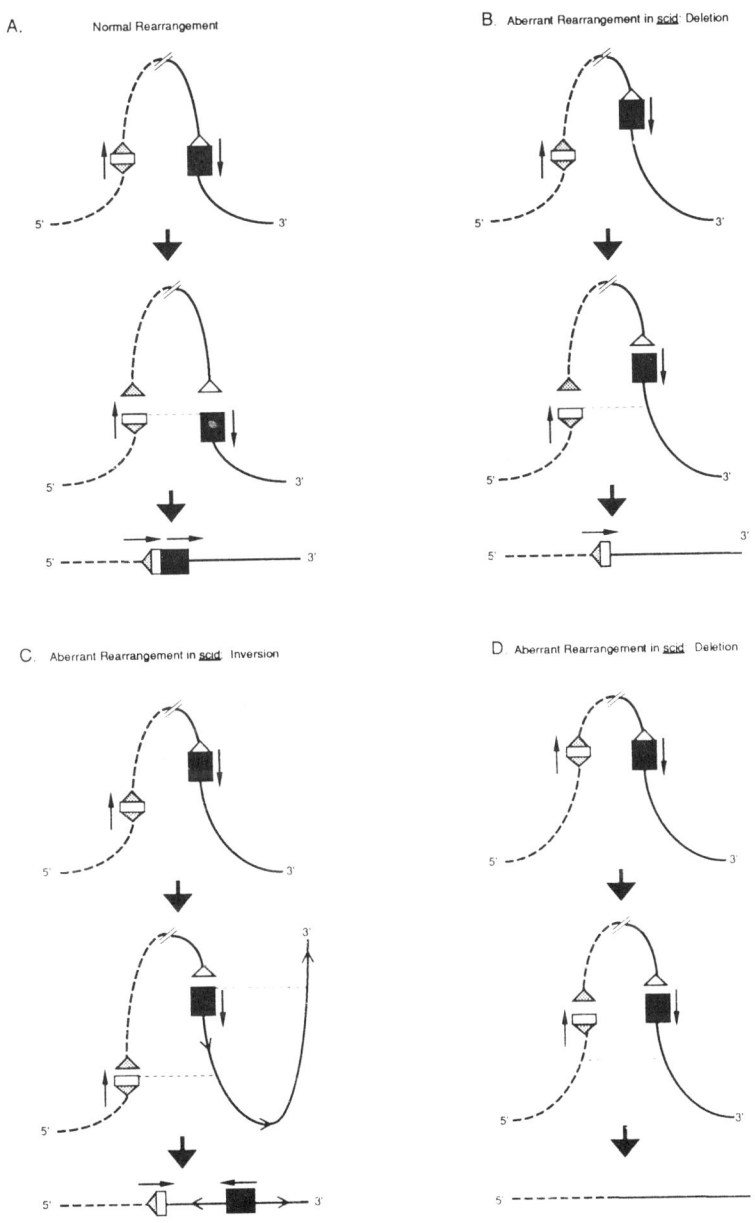

FIGURE 1-12. Model for VDJ recombinase-mediated rearrangements. The germ-line Ig locus is represented by a D coding segment (open box) flanked by a 12 bp spacer signal element (shaded triangles) and positioned at an undefined distance upstream of a J_{H} coding segment (black box) flanked by a 23-bp signal element (open triangles). Arrows indicate the transcriptional orientation. A proposed normal joining mechanism is shown in panel A (adapted from Alt and Baltimore, 1982). Illegitimate recombination events proposed to resolve the defective joining process in *scid* pre-B cells may result in deletion of one (B) or both (D) coding segments or inversion of a chromosomal fragment (C). See text for further details. (From Malynn et al., *Cell,* 54, 453, 1988. ©Cell Press.)

this point of view was that both B and T cells from such animals displayed highly aberrant abilities to successfully rearranged the appropriate VDJ regions. Not only is this topic important in terms of the disease process itself, but systems designed to study the *scid* defect provide opportunities to further our understanding of the rearrangement process. Very little is currently known about this vital process and a paper out this year provided details to this puzzle. Apparently, this defect is not at the early endonucleolytic scisson mediated by the recombinases, but rather at the later step involving the ligation of the rearranged strands. *Charles Snow*

V Gene Rearrangement Is Required to Fully Activate the Hypermutation Mechanism in B Cells
J. Roes, K. Huppi, K. Rajewsky, and F. Sablitzky
J. Immunol., 142, 1022—1026, 1989   1-13

The V regions of secondary response antibodies are encoded by V regions that usually contain somatic mutations due to hypermutation during clonal expansion. The extent of somatic mutation in nonproductively rearranged L or H chain loci was investigated in five hybridoma cell lines, which express heavily mutated antibodies.

Sequence analysis revealed that somatic point mutations occur equally in nonproductively and productively rearranged loci, if a V gene is rearranged. In nonproductive H chain loci with a DJ-H rearrangement, the frequency of somatic mutations is 0.2%, while that of VDJ-H loci expressed by the same cells is 2.5% (Figure 1-13).

The hypermutation mechanism of B cells appears to be targeted to V genes rearranged to the J locus. It may require juxtaposed sequences from both V and J regions in order to function optimally.

Early Onset of Somatic Mutation in Immunoglobulin V-H Genes during the Primary Immune Response
N. S. Levy, U. V. Malpiero, S. G. Lebecque, and P. J. Gearhart
J. Exp. Med., 169, 2007—2019, 1989   1-14

The mechanism of hypermutation of the rearranged V gene region is not well understood. In order to understand this somatic mutation process, the dynamics of *in vivo* mutation have been studied to define a population of B cells undergoing mutation.

Mice were immunized. Splenic mRNA was prepared on day 5, 7, and 13 and annealed to gamma constant region primers to produce cDNAs encoding V-H genes. Mutations were identified through sequencing of 103 cDNA clones corresponding to 18 different V-H genes.

On day 5 after immunization, there was a low level of mutation. Muta-

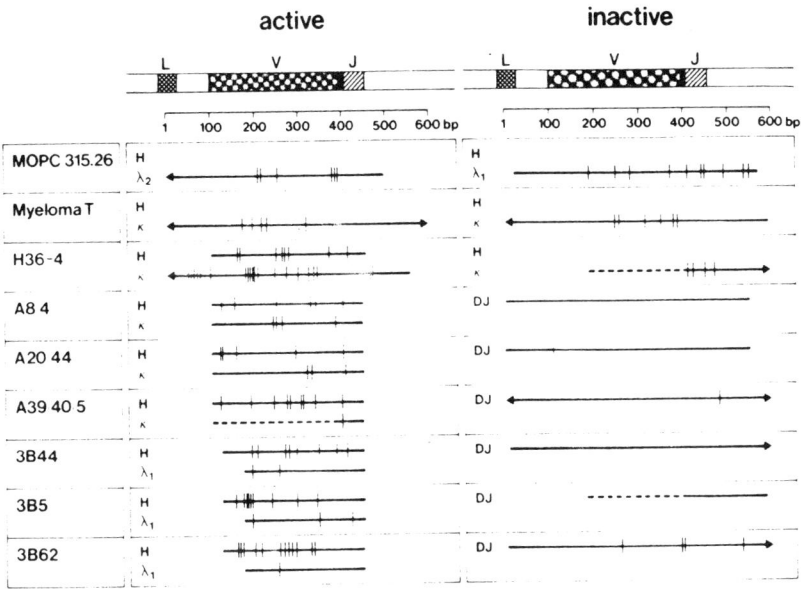

FIGURE 1-13. Schematic presentation of productively and nonproductively rearranged heavy and L chain loci analyzed from cell lines MOPC315-26 (14, 18, 19); myeloma T (14); H36-4 (8) A8/4 and A20/44 (10, 20); A39/40/5 (3); and 3B44, 3B5, and 3B62 (6). Ig loci are aligned according to their rearranged J segment. Somatic point mutations in a region of maximally 600-bp length are indicated by bars. Dashes represent regions of the nucleotide sequences where somatic point mutations could not be determined because corresponding germ-line sequences were not available. Arrows indicate that the sequence extends either at the 5' or 3' end. (From Roes et al., *J. Immunol.*, 142, 1022—1026, 1989. ©American Association of Immunologists.)

tions were detected on day 7 at a rate of 10^{-3} mutations per nucleotide per generation. On day 13 the number of mutations had not increased above the day 7 level.

These data are consistent with brief activation of hypermutation mech--anisms at the end of the first week after immunization. Subsequently, selection for antigen binding appears to preserve the beneficial mutations.

♦ The continued expansion of the repertoire not only centers on the rearrangement of VDJ region, but also upon mutations within the variable regions. This nongermline contribution to the repertoire has long intrigued B cell biologists, and two reports advanced our understanding of this hypermutational event. The first report examined the potential involvement of the recombination event itself to the appearance of somatic mutation.

This statistical analysis of sequenced heavy and light chain variable regions exhibiting DJ vs. complete VDJ rearrangements, suggests that a successfully rearranged segment may participate in the initiation of or the formation of sequences necessary for the functioning of a hypermutation mechanisms. The second paper provided evidence for a temporal relationship operative in the control of the hypermutation mechanism.

This paper indicates that following the administration of specific antigen there are periods of intense somatic mutation. *Charles Snow*

Clonal Anergy Induced in Mature V-Beta-6+ T Lymphocytes on Immunizing *Mls*-1-b Mice with *Mls*-1-a Expressing Cells
H.-G. Rammensee, R. Kroschewski, and B. Frangoulis
Nature, 339, 541—544, 1989   1-15

Although tolerance to specific antigens can be induced in adult animals, the mechanism is not understood. In an effort to investigate the mechanism of tolerance in adult animals, adult *Mls*-1-b mice were specifically made tolerant to *Mls*-1-a antigens. The majority of murine T-helper cells that express the V-beta-6 T-cell receptor gene segment are specific for *Mls*-1-a antigens. Therefore, an anti-V-beta-6 MAb was used to follow these *Mls*-1-a-specific T cells in tolerant mice.

The frequency of V-beta-6+ T cells within the lymph node was examined in untreated and tolerant mice. There was no correlation between number of V-beta-6+ cells and tolerance. Therefore, the proliferation of these cells in response to *Mls* and H-2 antigens was investigated in a mixed lympocyte reaction. Although the levels of IL-2 receptor expression were the same in cultures from both untreated and tolerant mice, only untreated cultures produced IL-2 and proliferated.

The induction of *in vivo* tolerance to specific antigen in mature mice is described in this paper. Tolerance appears not to be due to clonal deletion, but to clonal anergy. Such anergic clones do not express IL-2 or proliferate in response to antigen.

♦Finally the very first direct evidence for clonal anergy was recently presented. These anergic T cells still express the TCR and the interleukin 2 receptor, but apparently do not produce the interleukin 2 needed to maintain the proliferative response, a result quite similar to an experimental model already published for T cell unresponsiveness (Jenkins and Schwartz, *J. Exp. Med.,* 165, 302, 1987). These reports have really provided an exciting year in this area of T cell ontogeny. *Charles Snow*

Mouse Embryonic Stem Cells and Reporter Constructs to Detect Developmentally Regulated Genes
A. Gossler, A. L. Joyner, J. Rossant, and W. C. Skarnes

Science, 244, 463—465, 1989

Mouse embryonic stem cells (ES) remain pluripotent and can be cultured, manipulated, and then induced to form chimeric embryos. The *lacZ* gene linked to a weak promoter can be integrated and used as a reporter gene to detect *cis*-acting elements. In an effort to increase understanding of the genes involved in early mammalian development, such *lacZ* constructs were introduced into ES cells and these cells were used to construct chimeric mice as an assay for early developmental gene activity.

Two *lacZ* constructs were used in these experiments. In the "enhancer trap" construct, *lacZ* was fused in frame to a minimal promoter, while in the "gene trap" construct, *lacZ* lacking a promoter and transcription initiation site was placed just 3' to a splice acceptor site. In the gene trap construct, a splicing event can generate expression of a *lacZ* fusion protein. These constructs were introduced into the ES cell line D3 via electroporation.

After enhancer trap construct transfer to D3 cells, 60 independent cell lines were established, of which 6 expressed *lacZ*. After gene trap construct transfer, 8 independent cell lines expressing *lacZ* were established. Chimeric embryos were produced from these cell lines. Identical patterns of *lacZ* expression were detected in all embryos derived from the same transformed lines. The reporter constructs were transmitted to progeny and expression of *lacZ* followed the same pattern in these progeny.

Of the 14 lines expressing *lacZ*, 3 demonstrated well-defined spatially and temporally restricted patterns of expression in chimeric embryos. The most interesting line, D3-6-28, was expressed exclusively in the posterior neuropore at day 8.5 and then extended further anterior in the spinal cord during development.

These experiments have demonstrated successful introduction of reporter *lacZ* constructs into ES cells and incorporation of these cells into chimeric embryos. Such cells were successfully used to detect and mutate mouse genes expressed during early development. Expression of *lacZ* was used to assess the pattern of the expression of these genes both temporally and spatially within the embryo. This approach can be used to identify novel genes involved in the regulation of early mammalian development.

♦ An important aspect of studying any developmental system is to be able to produce mutations that affect specific processes or stages of embryogenesis. Although mutations can be easily induced in mouse by chemicals, X-rays, or insertions of transgenes or retroviruses, efficient methods do not exist to allow one to do a stage- or tissue-specific screen. There are a few instances where targeted mutagenesis of particular genes with selectable phenotypes has been achieved (for example *Hprt*), or saturation mutagenesis screens of specific regions of the genome have been conducted (for example the proximal portion of mouse chromosome 17). The above report by Gossler and colleagues showed that introduction of a *lacZ* reporter construct that is either promoterless (gene trap), or contains a minimal promoter that provides a TATA box and translation initiation codon but lacks an enhancer (enhancer trap), can be used to detect cis-acting elements in the mouse genome. The important point to stress is that *in vivo* expression can be analyzed directly in chimeric embryos without the need for generating transgenic mouse lines. Once an interesting stage- or tissue-specific pattern of expression is observed, a line of mice can be derived from the remaining cells that are kept in culture or are cryopreserved. Thus, it is now possible to conduct stage- or tissue-specific screens in an efficient manner. For those integration events that fall within the endogenous gene, it should be possible to clone the preinsertion site, allowing one to conduct a molecular genetic analysis of the interrupted gene. Although the investigators claim that selection for expression of the reporter construct in embryonic stem cells is a valid approach for identifying genes expressed in a stage- or tissue-specific manner in the embryo, it is important to devise methods for identifying those stem cells where the reporter construct is not expressed. If not, some very important genes may be missed. *Terry Magnuson*

Disruption of Proto-Oncogene *int-2* in Mouse Embryo-Derived Stem Cells: A General Strategy for Targeting Mutations to Non-Selectable Genes
S. L. Mansour, K. R. Thomas, and M. R. Capecchi
Nature, 336, 348—352, 1988   1-17

This paper describes a general method for isolating ES cells containing targeted mutations in any gene for which a cloned fragment exists and intron-exon boundaries are known. A positive selection for incorporation of the vector is combined with a negative selection against random integration to enrich for cells containing the targeted mutation. This procedure was used to isolate ES cells containing insertions in the *int-2* gene.

The replacement vector contains 10- to 15-kilobase (kb) homologous to the target gene with a neomycin resistance gene (neo-r) inserted into an exon and an HSV-thymidine kinase (HSV-tk) gene adjacent to the homologous sequences. The neo-r gene both disrupts the coding sequence and is a selectable marker. When replacement of the endogenous gene by the vector occurs by homologous recombination, the HSV-tk gene will not be transferred (Figure 1a). Cell lines in which the targeting event occurred will be target-neo-r HSV-tk-, while random integrants will be target+ neo-r HSV-tk+ (Figure 1b). Therefore, by selecting for cells contain-

a **Gene Targeting**

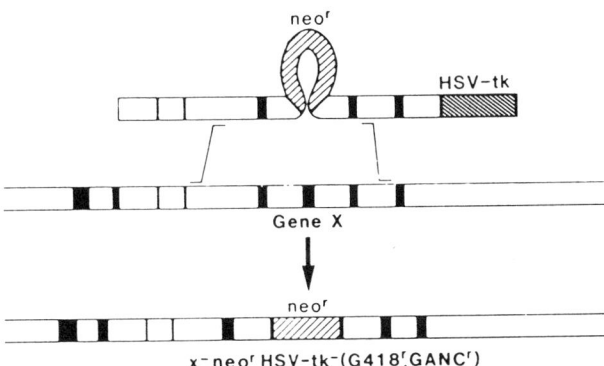

b **Random Integration**

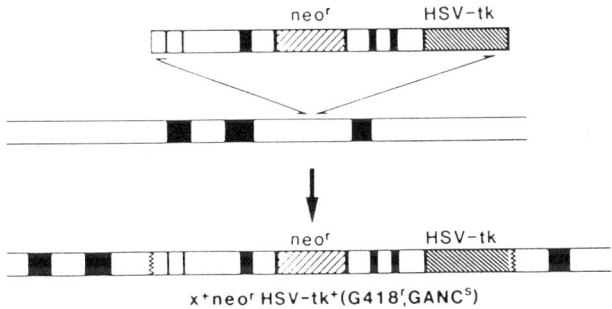

FIGURE 1-17. The PNS procedure used to enrich for ES cells containing a targeted disruption of gene *X*. a, A gene *X*-replacement vector, that contains an insertion of the *neo*^{r} gene in an exon of gene *X* and a linked HSV-*tk* gene, is shown pairing with a chromosomal copy of gene *X*. Homologous recombination between the targeting vector and genomic *X* DNA results in the disruption of one copy of gene *X* and the loss of HSV-*tk* sequences. Such cells will be *X*$^-$, *neo*^{r} and HSV-*tk*$^-$ and will be resistant to both G418 and GANC. b, Because nonhomologous insertion of exogenous DNA into the genome occurs through the ends of the linearized DNA, the HSV-tk gene remains linked to the *neo*^{r} gene. Such cells will be *X*$^+$, *neo*^{r}, and HSV-*th*$^+$ and therefore resistant to G418 but sensitive to GANC. Open boxes denote introns or flanking DNA sequences, closed boxes denote exons and cross-hatch boxes denote the *neo*^{r} or HSV-*tk* genes. (From Mansour et al., *Nature,* 336, 348—352, 1988. With permission.)

ing a functional neo-r (G418 resistance) and against those containing a functional HSV-tk (gancyclovir resistance, GANC-r) enrichment for cells containing the targeted event occurs.

The vector used for targeting the *int-2* gene, pINT-2-N/TK, was linearized and electroporated into ES cells. These cells were subjected to three growth conditions. Of the neo-r cells, 1 in 40,000 contained an *int-2* mutation. Of the 81 neo-5 GANC-r cell lines tested, 4 contained a mutant *int-2* gene. This represents a 2000-fold enrichment for ES cells containing the disrupted target gene. Southern blotting was used to confirm the disruptions of the target gene. Targeting did not appear to introduce gross rearrangements of the *int-2* gene.

This paper describes a combined positive and negative selection procedure that enriches for ES cells containing targeted disruptions. This procedure appears to be applicable to any gene. This has the potential to generate mice of any genotype desired.

Targeted Mutation of the *Hprt* Gene in Mouse Embryonic Stem Cells
T. Doetchman, N. Maeda, and O. Smithies
Proc. Natl. Acad. Sci. U.S.A., 85, 8583—8587, 1988   1-18

Gene-targeting was used to inactivate the hypoxanthine-quanine phosphoribosyltransferase (Hprt) in ES-D3 cells. The targeting construct contained a promoterless neo-r gene and had two short regions of homology with the endogenous Hprt gene. After electroporation of the targeting construct into the ES cells, G418 selection was followed by 6-thioguanodine selection.

For every 10^5 cells that survived electroporation, 1.3 to 5.3 had homologous recombination at the targeted Hprt locus. Of the six colonies obtained, four had the expected sequence, while two had small deletions at the border of one of the regions of homology.

This gene-targeting method is useful to decrease the background of nonspecific insertions and facilitate the detection of a specific homologous recombination event.

Germ Line Transmission and Expression of a Corrected HPRT Gene Produced by Gene Targeting in Embryonic Stem Cells
S. Thompson, A. R. Clarke, A. M. Pow, M. L. Hooper, and
D. W. Melton
Cell, 56, 313—321, 1989   1-19

Homologous recombination was used to correct a mutation in the HPRT gene of an ES cell line. The corrected cells were implanted into blastocysts to create chimeric mice that transmitted this trait to their progeny. Wild-type expression of the HPRT gene has been observed in these mice.

The ES cell line, E14TG2a, is deficient in HPRT expression. The sequence insertion vector, pDWM101, which contains the promoter and first 3 exons of HPRT was electroporated into E14TG2a cells. Selection was in HAT medium, as the targeted event results in HAT-R cells.

In the first experiment, 4.6×10^7 ES cells were electroporated and 8 HAT-R clones were selected. Of these eight clones, Southern hybridization analysis revealed that seven had the anticipated structure, while one clone had two vectors integrated in tandem. Cells from one of the corrected clones were used to generate chimeric mice, which could be detected by coat color.

One out of eight chimeric males tested demonstrated germ line transmission of the corrected gene. This male was bred to homozygous HPRT-deficient females. In the progeny, the corrected gene rescued the HPRT-phenotype and demonstrated wild-type level and spatial pattern of expression.

The targeting of genetic modifications by homologous recombination has been achieved in ES cells. The targeted cells can be used to create chimeric mice who transmit the modification in the germ line. The progene of these mice express the modified gene appropriately. This technique has proven successful with the HPRT gene and should soon be applicable to any gene of the mouse.

Production of Chimeric Mice Containing Embryonic Stem (ES) Cells Carrying a Homeobox *Hox 1.1* Allele Mutated by Homologous Recombination
A. Zimmer and P. Gruss
Nature, 338, 150—153, 1989   1-20

A general *in vivo* mutagenesis technique was developed and used to mutagenize the murine homeobox containing gene, Hox 1.1. A Hox 1.1 DNA fragment containing a 20-bp oligonucleotide in the coding region (Figure 1-20) was microinjected into ES D3 cells. These cells were screened for the homologous recombination event by PCR and Southern blotting. Single cells were picked by glass capillary tubes to obtain clonal lines.

The average frequency of homologous recombination following this procedure was 1 event per 150 microinjected cells. The ratio of homologous to illegitimate recombination was approximately 1 to 30. The cells containing the mutant allele appeared normal in culture. They were microinjected into the blastocysts of two mice litters. Four normal-appear-

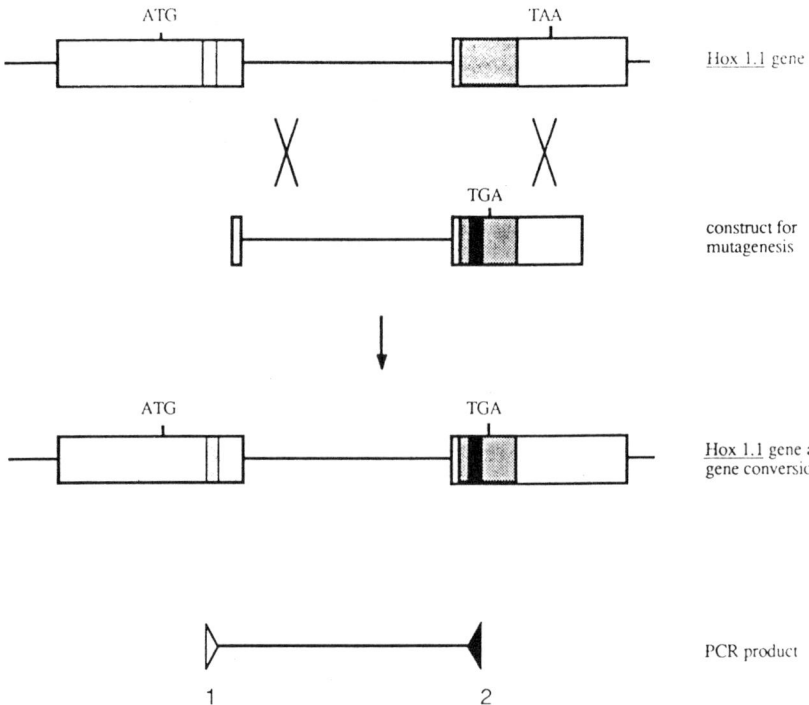

FIGURE 1-20. Strategy for the mutagenesis of the Hox 1.1 gene and its detection by PCR. An *Fsp*I fragment of the *Hox 1.1* gene (nucleotides 367-1937 relative to the translational start codon) was cloned into Bluescript (Stratagene) and a 20-bp oligonucleotide inserted into the *Eco*RI site of the homeobox. Homologous recombination of this fragment with the endogenous *Hox 1.1* gene in the recipient cells introduced this oligonucleotide into the endogenous gene. Consequently, transcripts from this allele had a frameshift mutation with a stop codon directly downstream of the oligonucleotide in-frame. To screen the recipient cells for homologous recombination events by PCR, we synthesized two primers of which the first (Primer 1; TTC CGC ATC TCA CCC TGG AT) was specific for the first exon of *Hox 1.1*, binding 5′ to the *Fsp*I site, and the second (Primer 2; AAT TGT GAG GTA CCG CTG AC) was identical to the inserted oligonucleotide. As the normal *Hox1.1* allele contained no binding site for primer 2 and the microinjected fragment contained no binding site for primer 1, only the mutant allele can contain both priming sites. The homeobox is indicated by a dark-shaded box. Primer 1 and its binding site are depicted as a light-shaded arrowhead and box, respectively. Primer 2 and the oligonucleotide are shown as a black arrowhead and box, respectively. (From Zimmer/Gruss, *Nature,* 338, 150—153, 1989. With permission.)

ing chimeric mice were obtained. Analysis of tail DNA confirmed that all four contained the mutant allele.

Using the method described above, homologous recombination can be used to create mutations of nonselectable genes in ES cells. These cells can then be used to create chimera that carry this mutation. The Hox 1.1 chimeric mice created in this experiment can be bred to create Hox 1.1-deficient animals to analyze the function of this gene. Mutational analysis of any mouse gene now appears possible with this method.

Production of a Mutation in Mouse *En-2* Gene by Homologous Recombination in Embryonic Stem Cells
A. L. Joyner, W. C. Skarnes, and J. Rossant
Nature, 338, 153—156, 1989

The putative mouse developmental gene *En-2* is homologous to the *Drosophila engrailed* gene, which is involved in the control of segmentation in the developing insect. *En-2* seems to be involved in establishing spatial domains in the brain of the early mouse embryo. To help study the function of *En-2,* mutations would be useful, but none are currently available. The work described here was designed to generate mutations in *En-2* in embryonic stem cells.

The strategy employed was based on the homologous integration of a vector containing a copy of *En-2* that had been disrupted by nucleotide sequences encoding resistance to neomycin. A vector containing the bacterial *neo* gene, driven by a fragment of the human β-actin promoter and flanked by fragments of *En-2,* was introduced into pluripotent embryonic stem cell lines. Identification of homologous integration events was based on detection of rearranged vector sequences using the polymerase chain reaction (PCR).

The efficiency of homologous recombination was about 1 per 3000 neo^{+r} colonies and 1 per 1.5×10^7 treated cells. Targeted insertion was confirmed by Southern blot analysis. Again using PCR, a rapid screening protocol was used that permitted the identification of individual targeted colonies within 2 weeks of the start of the experiment.

Preliminary data did not indicate clearly that mice with germ line mosaicism for *En-2* have resulted from the injection into blastocysts of cells harboring the mutation. Interpretation of this fact is currently unclear.

The general method presented here should be useful in screening homologous recombination events. Presumably, this will assist the development of mouse germ-line mutations.

♦One promising approach in defining the functional role of different genes expressed during mouse embryogenesis involves site-directed mutagenesis by homologous recombination of DNA sequences residing in the chromosome with newly introduced targeting sequences. The feasibility of such an approach depends on using a cell type that can successfully populate the germline when introduced into a host embryo. Embryonic stem (ES) cells are the cell type of choice in that these cells are derived from and are essentially equivalent to the inner cell mass cells of the mouse blastocyst. Furthermore, ES cells have been found to be capable of forming germline chimeras after injection into the blastocoel cavity of preimplantation host blastocysts, allowing one to derive a line

of mice directly from the injected cells. The experiments described in the above reports demonstrate that mutations can be introduced into the genome of ES cells by homologous recombination. Although the frequency of site-directed integration (relative to random integration) has been found to vary somewhat depending on the gene targeted and the method used for introducing the engineered construct into the ES cells (approximately 10^{-2} to 10^{-3} for direct injection and between 10^{-6} to 10^{-8} for electroporation), in all cases it has been sufficiently high enough to warrant continued use of this approach. It is important to note that no random integrations, in addition to the targeting event, have been found in any cells undergoing homologous recombination. These results indicate that it should be possible to make targeted alterations without the risk of deleterious consequences which might arise from random integrations. One problem that has been encountered is a consistent lack of germline contribution of the manipulated ES cells in the chimeras produced thus far. Only Thompson et al. (1989) have reported the existence of germline contribution from the introduced ES cells. The lack of functional germline contribution is likely to be due to karyotype instability induced by periods of prolonged culture coupled with stress associated with electroporation and selection schemes used for isolating the targeted cells. Nonetheless, with appropriate attention paid to karyotype, it should be possible to derive lines of mice carrying mutated genes of interest in the heterozygous state. When these mice are crossed to one another, embryos homozygous for the mutated genes will be produced, making it possible to begin to ascertain the functional significance of these expressed genes.
Terry Magnuson

Chlorambucil Effectively Induces Deletion Mutations in Mouse Germ Cells
L. B. Russell, P. R. Hunsicker, N. L. A. Cacheiro, J. W. Bangham, W. L. Russell, and M. D. Shelby
Proc. Natl. Acad. Sci. U.S.A., 86, 3704—3708, 1989   1-23

Chromosomal deletions are valuable for mapping genetic loci. In the mouse, *N*-ethyl-*N*-nitrosourea (EtNU), which induces mutations at high yields, results mainly in intragenic, and often in point, mutations. This paper provides evidence that chlorambucil, 4-{p[bis(2-chloroethyl)amino]phenyl}butyric acid, is more effective than any known chemical as an agent for inducing high yields of deletion mutations in mouse germ lines. This chemical has been previously reported as a mutagen, teratogen, and carcinogen in various test systems. In these experiments, single doses of chlorambucil were administered to male mice, and the effects on various germ-cell stages were investigated.

Dosages ranging from 10 to 25 mg/kg were administered; 60,750 offspring obtained from matings at various periods after administration were scored for mutations at 7 marker loci or for other visible mutations.

Mutations were induced in post-stem-cell stages, with maximum yield occurring after exposure of early spermatids. Matings involving these cells yielded about 1% mutations after low-dose exposure to chlorambucil. Spermatagonial stem cells did not seem subject to mutations induced by chlorambucil. Genetic, cytogenetic, and molecular criteria indicated that all but one mutation in post-stem-cell stages were deletions or structural changes. The stage-specific response to chlorambucil differed from all other chemicals tested.

Chlorambucil may be the agent of choice for the experimental induction of germ line deletion mutations in mammals.

♦ Radiation-induced multilocus deletions provide important genetic tools for characterization of large regions of the genome. This paper reports that a chemical mutagen, chlorambucil, is considerably more effective in inducing high yields of germ-line mutations than radiation or any other chemical investigated thus far. The maximum mutagenic effect appears to be in early spermatids. Homozygous lethality as well as cytogenetic and molecular data indicate that, in the case of some of the seven marker loci studied, multi-locus deletions or other structural changes have occurred following drug treatment. With the advent of long-range physical mapping coupled with walking or jumping libraries as well as yeast artifical chromosomal libraries, multi-locus deletion breakpoints become particularly valuable in molecular studies on the structure and function of regions of the mouse genome. Thus, effective procedures to induce deletion mutations in mouse germ cells will be extremely useful. *Terry Magnuson*

Amplification and Analysis of DNA Sequences in Single Human Sperm and Diploid Cells
H. Li, U. B. Gyllensten, X. Cui, R. K. Saiki, H. A. Erlich, and N. Arnheim
Nature, 335, 414—417, 1988   1-24

In humans, pedigree analysis is the only method that can be used to construct genetic linkage maps, and such analysis can measure genetic distances as small as about 1 cm. The work here was designed to permit the measurement of genetic recombination over shorter distances. To do this, the conditions required to analyze DNA sequences in single cells using the polymerase chain reaction were defined.

An analysis was performed of single cells from a co-cultivation of two different diploid tissue culture cell lines harboring two different β-globin alleles. Fifty cycles of DNA amplification were followed by dot blot detection of allele-specific sequences. Eighty-four percent of 37 cells analyzed reacted with a single allele-specific probe (19 with $β^A$ and 12 with $β^S$), while none of 12 control tubes, containing water instead of cells, were positive.

Single sperm cells from an individual heterozygous at both the low-density lipoprotein receptor and HLA DQA loci were next analyzed. When the first locus alone was amplified and detected, 55% of 80 individual sperm gave a hybridization signal, with 22 carrying 1 allele and 21 the other. One sample was positive with both allele-specific probes, while 16 control tubes were positive with neither probe. Simultaneous detection of DNA sequences at two loci on nonhomologous chromosomes in a single sperm was attempted next. Of 150 individual sperm tested, 61% hybridized at both loci, with near equal occurrence of the 4 possible types of gametes.

This method for the analysis of the genotype of single sperm at the DNA levels provides a powerful tool in human genetics. It permits the analysis of a large number of meiotic products from a single individual. It is possible that the present capability to type 500 sperm in a week can be enhanced. Analysis of DNA in single diploid cells by this method should be advantageous in studying developmental processes and could permit prenatal diagnosis of preimplantation embryos derived from *in vitro* fertilization.

Developmental Expression of PDGF, TGF-α, and TGF-β Genes in Preimplantation Mouse Embryos
D. A. Rappolee, C. A. Brenner, R. Schultz, D. Mark, and Z. Werb
Science, 241,1823—1825, 1988   1-25

While it seems likely that preimplantation mouse embryos synthesize growth factors, direct evidence supporting this has been impossible to obtain because growth factor transcripts present in low copy number are impossible to detect using current methods. Recent work from this laboratory has resulted in the development of a technique called single-cell RNA phenotyping, which permits the detection of low abundance mRNA from a single cell. This technique combines microtechniques for isolation of total RNA from 1 to 100 mouse embryos with reverse transcription and amplification of the transcribed cDNA using the polymerase chain reaction. Using this method, the experiments reported here analyzed the developmental expression of growth factor transcripts in mouse preimplantation embryos.

Expression of mRNA for transforming growth factor (TGF)-α, TGF-β1, and platelet-derived growth factor A chain (PDGF-A), but not of four other growth factors, was detected in single whole blastocysts. The translation of these transcripts was detected by immunocytochemical localization in permeabilized blastocysts, with the immunofluorescent signal for both TGF-α and PDGF antigens found in the perinuclear area of all blastocyst cells and the TGF-β1 signal found in 70 to 90% of cells. Both PDGF-A and TGF–α were found both in unfertilized oocytes and in blastocysts, while TGF-β1 transcripts appeared only in the zygote.

These experiments provide direct evidence for the production of growth factors by preimplantation embryos. These factors may function to induce early angiogenesis and decidualization of the uterus. Further study of growth factor receptors is required to determine whether functional intraembryonic targets for these factors exist.

Rapid Production of Full-Length cDNAs From Rare Transcripts: Amplification Using a Single Gene-Specific Oligonucleotide Primer

M. A. Frohman, M. K. Dush, and G. M. Martin
Proc. Natl. Acad. Sci. U.S.A., 85, 8998—9002, 1988   1-26

Despite the many recent advances in DNA cloning techniques, it is still difficult to obtain full-length cDNA copies of low-abundance mRNAs. The present work was designed to facilitate the cloning of such messengers. Based on the polymerase chain reaction (PCR), this technique, named "rapid amplification of cDNA ends" (RACE), was used to generate cDNAs by the amplification of a region between a single point in the transcript and either its 3′ or 5′ end.

To obtain 3′ ends, mixed RNAs were reverse transcribed to minus cDNA strands. The primer consisted of 17 dT residues plus an adapter sequence of 3 rare restriction endonuclease restriction sites (Figure 1-26). For the synthesis of the complementary strand, a gene-specific amplification primer was permitted to anneal with the minus strand; multiple rounds of PCR resulted in amplification of the double-stranded 3′ sequence. A similar strategy was used for amplifying the 5′ end (Figure 1-26). After amplification, the products were analyzed by Southern blot and were cloned. Full-length cDNA clones could be obtained from overlapping RACE products or by using primers derived from analysis of extreme 5′ and 3′ ends of RACE products and PCR.

The RACE technique was tested by generating cDNA copies of mRNA from the previously characterized mouse gene *int-2*, transcribed at low frequency. Twenty-nine independent *int-2* clones were isolated by screening of <0.05% of the cDNAs produced. Sequence analysis showed that the 3′ and 5′ ends of all four previously known *int-2* transcripts had been cloned using this technique.

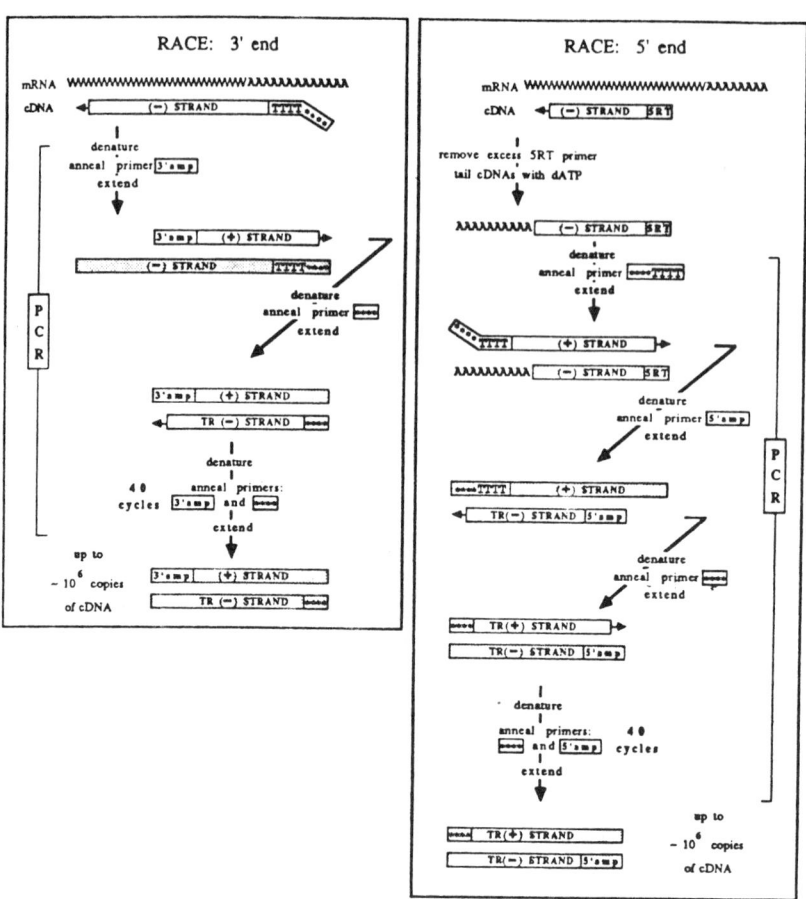

FIGURE 1-26. Schematic representation of the RACE protocol. Primers: ****TTTT (dT)$_{17}$-adaptor, 5'-GACTCGAGTCGACATCGATTTTTTTTTTTTTTTT-3'. This sequence contains the Xho I, Sal I, and Cla I recognition sites. ****, Adaptor, 5'-GACTCGAGTCGACATCG-3'. 3'amp (amp refers to amplification), specific to gene of interest, complementary to (–) strand. 5RT (RT refers ro reverse transcription) and 5¢amp, specific to gene of interest, complementary to (+) strand. Open rectangles represent DNA strands actively being synthesized; shaded rectangles represent DNA previously synthesized. At each step the diagram is simplified to illustrate only how the new product formed during the previous step is utilized. A (–) or (+) strand is designated as "truncated" (TR) when it is shorter than the original (–) or (+) strand, respectively. (From Frohman et al., Proc. Natl. Acad. Sci. U.S.A., 85, 8998—9002, 1988. With permission.)

Evidence here suggests that this 1-d protocol can result in the isolation of cDNAs from rare transcripts. This technique should facilitate the isolation of cDNAs representing variant mRNAs, such as those produced by alternative splicing or those transcribed from alternative promoters. Conceivably, this protocol might be used with primers based on amino acid sequence, or used to produce general cDNA libraries.

♦ The polymerase chain reaction (PCR) has dramatically changed experimental approaches to mammalian developmental genetics. For example, as discussed in the papers on gene targeting in embryonic stem cells, PCR is key to identifying those cells that have undergone the homologous recombination event. In addition to this application, the paper listed above by Li and colleagues demonstrates that construction of genetic maps in mammals will be easier and more accurate given that large numbers of single sperm can now be haplotyped using PCR. This will provide a much faster and more accurate method for measuring genetic recombination over shorter physical distances. In mice, this analysis can easily be extended to ovulated oocytes, thereby allowing for comparison of recombination frequencies between males and females. Another useful application of the PCR will be in analysis of gene expression. As evidenced in several papers published this year, there are many interesting genes that are being cloned in mouse, some examples of which include homeobox and paired-box genes as well as growth factor and proto-oncogenes. To begin to understand the developmental significance of these genes, it is important to be able to determine when they are expressed. However, to isolate quantities of RNA required to allow for detection of transcripts in early mouse embryos by Northern or RNAse protection assays, thousands of embryos are required. Although *in situ* hybridization has been valuable for examining tissuing-specific expression in single embryos, it is an extremely laborious approach for determining temporal patterns of expression. With the introduction of PCR, this latter task has become much easier as evidenced by the work of Rappolee and co-workers. Micro-techniques for isolation of total RNA from a single embryo coupled with reverse transcription and amplification of the transcribed cDNA were used by these investigators to detect selective expression of three growth factor genes. Although it is not possible to quantitate the level of expression easily (theoretically one transcript should result in detectable signal), the use of PCR will allow the investigator to determine what stages are worth studying in greater detail. Finaly, the paper by Frohman et al. demonstrates that rapid amplification of cDNA ends (RACE) can be used to clone full-length cDNAs from rare transcripts. A similar approach is also being used by this group to clone conserved genes in mice using degenerate primers. Although PCR has become an important tool for developmental biologists, some potential problems do exist in that non-specific background bands of the right size as well as different sizes have been observed in the amplified product. The appearance of these bands seems to be dependent on variables such as annealing temperature, primer concentration, number of amplification cycles, etc. In addition, amplification becomes less efficient with increasing fragment length, and a limited number of sequence errors (approximately 0.1%) can be introduced by the use of multiple rounds of amplification.

Nonetheless, despite these problems, PCR has made an important impact on developmental biology. *Terry Magnuson*

Mapping to Molecular Resolution in the *T* to *H-2* Region of the Mouse Genome with a Nested Set of Meiotic Recombinants
T. R. King, W. F. Dove, B. Herrmann, A. R. Moser, and A. Shedlovsky
Proc. Natl. Acad. Sci. U.S.A., 86, 222—226, 1989   1-27

This work was designed to facilitate the identification in mammals of developmental lesions with particular gene products. The experiments were based on a combination of male germ line point mutagenesis using *N*-ethyl-*N*-nitrosourea (EtNU), combined with fine-structure mapping analysis. These techniques were used to analyze the EtNU-induced quaking$^{lethal-1}$ mutation on chromosome 17.

Hybrid mice with multiple balanced homologs of chromosome 17 were intercrossed. Of 1346 total progeny recovered, 337 were recombinant in the region *T* to *H-2* (Figure 1-27). These recombinants permitted mapping resolution in the size range of single mammalian genes if recombinational hot spots were absent, as they seemed to be for this interval. Parents and progeny of this cross were sacrificed, with tissues taken for DNA isolation. Typing for chromosome 17 DNA markers was used to provide additional mapping data for these animals. Taken together, such crossing and analysis yielded a high resolution, nested set of recombinant chromosomes. The qk^{l-1} allele was found to be just proximal to the molecular marker *D17RP17*.

These experiments show the success of a meiotic fine-structure mapping strategy that can be used to closely define the location of developmental mutations in mice. The resolution achieved corresponds to physical distances that overlap the size of DNA fragments clonable as yeast artificial chromosomes. The identification of a tightly linked DNA marker to qk^{l-1} allele should facilitate the cloning of this mutation.

♦Although a number of mutations exist in the mouse genome, many of which result in recessive-lethal phenotypes, they are difficult to clone. To some extent, mutations caused by insertion of transgenes have circumvented this problem. However, specificity of tagging can be compromized if effects on expression extend beyond the site of integration. Furthermore, the low rate of insertional mutagenesis prohibits saturation of particular genetic regions with mutations. Although saturation mutagenesis can be achieved using chemicals such as N-ethyl-N-nitrosourea, the problem of gaining molecular access to loci defined only by point mutations still remains. The report listed above details a meiotic fine-structure mapping stategy aimed at overcoming this problem. The goal

Mapping cross:

Complete matrix of progeny:

		T+a	+1b	T1b	++a	T+b	+1a	T1a	++b
Parental types	T+a		SH *1*		SA *7*	SA *8*			SH *4*
	+1b	SH *1*			NH *13*	SB *10*			NB *18*
Recombinant in the T-1 interval	T1b				SH *2*				SB *11*
	++a	SA *7*	NH *13*	SH *2*	NA *17*	SH *3*	NA *16*	SA *9*	NH *14*
Recombinant in the 1-H-2 interval	T+b		SB *10*		SH *3*		SH *5(*)*		SB *12*
	+1a	SA *8*			NA *16*	SH *5(*)*			NH *15*
Double recombinants	T1a				SA *9*				SH *6(*)*
	++b	SH *4*	NB *18*	SB *11*	NH *14*	SB *12*	NH *15*	SH *6(*)*	NB *19*

FIGURE 1-27. Multiply heterozygous balanced lethal intercross and resulting progeny. Only chromosome 17 is drawn. Stippled bars, chromosome segments from the TWis line; solid bars, segments from BTBR. Centromeres are represented by knobs at the ends of the chromosome. T, T^{Wis}; l, qk^{l-1}; a, H-2^a; b, H-2^b. Gametic genotypes are shown at the left and top sides of the matrix; only T, qk, and H-2 loci are considered. Viable phenotypes are S, short tail; N, normal tail; H, H-$2^{a/b}$; A, H-$2^{a/a}$; B, H-$2^{b/b}$. Lethal gametic combinations are indicated by shading. Numbers in italics designate each particular genotype. (*) marks SH progeny that can be homozygous for *tf*. (From King et al., *Proc. Natl. Acad. Sci. U.S.A.*, 86, 222—226, 1989. With permission.)

was to be able to clone the quaking (*qk*) gene. Mice that are heterozygous at many loci (*T, qk, tf* and *H-2*) in the proximal region of chomosome 17 where *qk* is located were intercrossed. More than 300 recombinant progeny were recovered, thereby providing a mapping resolution on the order of single genes. Cloned DNA markers were analyzed retrospectively based on polymorphisms between the parental strains. One particular marker was found to be within 0.3 cM of *qk*. This distance is close enough to begin a chromosome walk with the intention of cloning the *qk* locus. The DNA samples saved from this large set of meiotic recombinants can now be used to map to high resolution any poly-

morphic marker on proximal chromosome 17. This mapping strategy depends on being able to detect polymorphisms between inbred mouse strains which is not always possible. Interspecific crosses may be more informative barring any differences in rearrangements or "not spots" of recombination between the two species being used. In either case, to make this technique applicable to different regions of the genome, molecular markers are needed, as well as easy progeny testing and easily scored flanking markers. Although molecular markers can be generated through chromosome microdissection, the latter two may not always be available.
Terry Magnuson

Antisense RNA Directed Against the 3′ Noncoding Region Prevents Dormant mRNA Activation in Mouse Oocytes
S. Strickland, J. Huarte, D. Belin, A. Vassalli, R. J. Rickles, and J.-D. Vassalli
Science, 241, 680—684, 1988   1-28

The mRNA for tissue plasminogen activator (t-PA) is present in primary mouse oocytes but is not translated until meiosis resumes just before ovulation. During the resumption of meiosis, the mRNA undergoes a 3′ polyadenylation and is concomitantly translated. It is later degraded and is undetectable in fertilized eggs. The present study employed antisense (as) t-PA RNA to study the mechanisms of these alterations in t-PA mRNA metabolism.

Capped, nonpolyadenylated transcripts complementary to the 5′-, the middle, and the 3′ portions of the t-PA mRNA were injected into maturing oocytes *in vitro*. All inhibited the appearance of t-PA activity in maturing oocytes, with the 3′-asRNA the most effective (97% inhibition). Overall protein synthesis was not affected by asRNA. Injection into primary oocytes of the 3′-as RNA, but not of the 5′- nor of the middle asRNA, resulted in a slow, gradual, decrease in size of t-PA mRNA. These shortened mRNAs were still present in secondary oocytes, although t-PA mRNAs were absent from control cells previously injected with 3′- or middle asRNA. Injection into primary oocytes of antisense RNA complementary to 103 nucleotides of the extreme 3′ untranslated region of t-PA mRNA prevented the polyadenylation, translation, and degradation of t-PA in secondary oocytes.

These findings provide evidence that the 3′ noncoding region of a maternal mRNA of mouse oocytes is critical for its translation and metabolism during meiosis. A ribonuclease activity in oocytes may cleave duplex (asRNA-mRNA) regions, resulting in a stable 5′ fragment and an unstable 3′ fragment, but only the 3′ portion of the mRNA may be accessible to hybridization in primary oocytes. The mechanism of posttranscriptional control in oocytes may be generally applicable to somatic cells.

♦ The growing mouse oocyte synthesizes large amounts of mRNA that are stored in the cytoplasm. Although a significant percentage is degraded after fertilization, some of it is recruited for translation at various times beginning with the resumption of meiosis and extending through to the 2-cell stage of embryonic development. Although no evidence exists for translation of maternally-derived RNA past the 2-cell stage, some has still been detected as late as the blastocyst stage. The mechanisms responsible for mRNA storage and recruitment are not understood, and this has been a difficult process to study until the recent identification of specific RNAs that are stored. One of these, which is the subject of the paper by Strickland et al. listed above, is the message that encodes tissue plasminogen activator (t-PA). This mRNA is present in germinal vesicle-containing mouse oocytes. However, translation does not occur until after resumption of meiosis which coincides with 3' addition of adenosine residues. Removal of the terminal 103 nucleotides prevents t-PA mRNA translation, polyadenylation and destabilization. Based on this evidence, the 3' noncoding region of t-PA mRNA is thought to be responsible for control of t-PA synthesis during oocyte maturation. The use of antisense RNA to begin to dissect the process of mRNA storage and recruitment was novel and will be applicable for the study of other such RNAs, one example of which includes the c-*mos* mRNA (also discussed in this Year Book). *Terry Magnuson*

The Dominant-White Spotting (*W*) Locus of the Mouse Encodes the *c-kit* Proto-Oncogene
E. N. Geissler, M. A. Ryan, and D. E. Housman
Cell, 55, 185—192, 1988

Mutations at the dominant white (*W*) locus of mice have pleiotropic effects on both embryonic development and hematopoiesis. The hematopoietic defect is known to be intrinsic to the pluripotent hematopoietic stem cell, but the mechanism of how the *W* gene exerts its various effects is largely unknown. Recently, a deletion mutant, W^{19H}, has been characterized, and has been used to map candidate DNA sequences for their proximity to the *W* locus. One such candidate gene is the *c-kit* proto-oncogene, which encodes a cell suface glycoprotein with kinase activity that may be a transmembrane receptor. The work described here was designed to investigate further the possibility that the c-kit protein is the gene product of the *W* locus.

Mouse sequences homologous to *c-kit* DNA were absent from the cell line W50A2T which carries the W^{19H} deletion. Two spontaneous *W* mutations showed evidence of genomic rearrangement when probed with sequences from the central segment of the *c-kit* gene encoding amino acids 165 to 791. Use of smaller probes showed that the rearrangement present in one of these mutants was in the region encoding amino acids 240 to 532 of the *c-kit* gene. The rearrangement in the other mutation was in the region encoding amino acids 342 to 791. In homozygotes for the W^{44} mutation, which corresponds to a very mild phenotype, *c-kit* mRNA levels were dramatically lower than in wild-type or W^{44} heterozygotes.

These results provide strong evidence that *c-kit* is contained within the *W* locus. This implies that the *W* locus is identical with the *c-kit* oncogene. This finding should inform future investigations of the *W* locus.

♦ Mice homozygous for the *W* mutation (located on chromosome 5) are sterile, show extensive white-spotting and have severe anemia that results in perinatal death. Based on the cell types affected (germ cells, melanocytes and hematopoietic lineage), it has been hypothesized that the *W* locus is involved in stem cell proliferation and survival during embryogenesis. In fact, it has been shown that the hematopoietic defect is first detected in embryonic yolk sac and persists into the fetal liver and bone marrow or spleen of post-natal mice. Because the anemia of *W* mutants can be completely cured by injection of wild-type blood-forming tissues, it has been suggested that the defect is intrinsic to stem cells of affected tissues and not to the surrounding environment. The report listed above by Geissler et al. presents convincing evidence that the protein encoded for by the c-*kit* proto-oncogene is the gene product of the *W* locus. The c-*kit* structural gene was shown to map within the W^{19H} deletion, and two other *W* alleles (W^{44} and W^x) shown clear evidence for c-*kit* gene rearrangement. What is needed now is a clear cut *in situ* analyses to determine the tissue-specific and temporal pattern of expression of this gene during embryogenesis. It is already known that c-*kit* is expressed in a broad range of cell types many of which are not obviously affected in *W* mutants. For example, the highest level of expression detected so far is in the brain. Consequently, an in-depth analysis of *W* mutants for more subtle defects is also needed in attempts to correlate function with expression.

It is known from other work that c-*kit* encodes a cell surface glycoprotein which possesses kinase activity. In addition, sequence analyses suggest conservation between colony-stimulating factor 1 and platelet-derived growth factor. These results suggest that c-*kit* may stimulate proliferation of cells in which it is expressed after binding of an appropriate ligand. Geissler et al. note that mutations at the Steel (*Sl*) locus resemble

W mutants except that the defect lies in the environment in which the melanoblasts, germ cells, and hematopoietic progenitors differentiate and proliferate. They then suggest that *Sl* mutations may reduce or eliminate the function of the ligand that interacts with the receptor encoded for by the *W* locus. this suggestion is interesting and definitely should be considered as the search for the c-*kit* ligand progresses. *Terry Magnuson*

---

undulated, a Mutation Affecting the Development of the Mouse Skeleton, Has a Point Mutation in the Paired Box of *Pax 1*
R. Balling, U. Deutsch, and P. Gruss
Cell, 55, 531—535, 1988

1-30

In *Drosophila* embryos, multiple segmentation genes govern metameric pattern formation. Genes homologous to many of these have been identified in vertebrates, and while many have expression patterns suggesting they play a role in development, an analysis of their function has been limited by a lack of appropriate mutants. The *Pax 1* gene of mice is homologous to the *Drosophila* paired box sequence, and has been shown to be expressed in ventral sclerotome cells and later in intervertebral disks along the entire vertebral column. The recessive mouse mutant, *undulated* (*un*), has vertebral malformations along the entire rostro-caudal axis. The work presented here tested the hypothesis that the *Pax 1* gene of *un* mice was mutated.

Analysis with 48 restriction enzymes and hybridization with a *Pax 1* probe showed that only a single difference was detectable between *un* homozygotes or heterozygotes and wild type mice. The altered Hae III fragment was amplified using the polymerase chain reaction and was sequenced. A single G to A base pair change was found at position 43 of the paired box of the *Pax 1* gene, resulting in the substitution of a serine for a glycine in the protein product. In all *Drosophila* and murine paired box-containing genes, this amino acid lies in a highly conserved region; the particular mutated glycine residue is 100% conserved in other paired box-containing genes.

These findings suggest that *Pax 1* is the prime candidate gene for *undulated*. This is based on the location of the gene, its expression pattern, and its mutation in *undulated* mice. The availability of this mutation should help address the mechanism of *Pax 1* function in murine development.

♦A mouse gene family containing a protein domain homologous to the *Drosophila* paired box sequence was previously identified by the Gruss lab and was discussed in the first Year Book on developmental biology.

The spatial and temporal pattern of expression of this family of genes, as well as the mouse homeobox genes, suggests a possible regulatory role during embryogenesis. As discussed in last years commentary regarding this work, the determination of function of genes known to be expressed in mammalian development is severly limited because of lack of appropriate mutants. This problem will hopefully be corrected via techniques of homologous recombination and the use of embryonic stem cells (see discussion on embryonic stem cells in this Year Book). The above listed report by Balling et al. have circumvented the problem of lack of mutations by demonstrating that the *undulated* (*un*) mutation located on mouse chromosome 2 has a point mutation in one member (*Pax-1*) of the paired box family. The reasoning that led to looking for a mutation in *Pax-1* of *un/un* mice was based on the fact that *Pax 1* maps to an area of chromosome 2 where *un* is located. Furthermore, *in situ* analyses indicated that *Pax 1* is expressed in tissues that are affected in homozygous *un* mice. Analysis of DNA from *un/un* mice then led to the identification of a point mutation that causes a Gly → Ser replacement in a highly conserved region of the paired box of *Pax 1*. Six other mouse strains carrying the wild-type gene do not show this change suggesting that it is not a common polymorphism found among mouse strains. The mutation lies outside the DNA binding domain of the paired box region, so it is not clear whether the mutation is affecting this property or another as yet unidentified functional domain of the *Pax-1* protein. Although the evidence strongly suggests that *undulated* and *Pax-1* are the same locus, it is not clear whether this particular allele results in a null phenotype. The fact that several alleles of the *undulated* mutation have been described in the literature, and that they result in a range of phenotypes, will be important in analysis of *Pax 1* protein function. The use of homologous recombination to produce a true null phenotype would also be something to consider in future work. *Terry Magnuson*

Legless, a Novel Mutation Found in PHT1-1 Transgenic Mice
J. D. McNeish, W. J. Scott, and S. S. Potter
Science, 241, 837—839, 1988   1-31

In the process of generating transgenic mice, insertional mutagenesis can occur. A transgenic mouse line is described in which homozygotes, but not heterozygotes, exhibit developmental abnormalities.

Transgenic mice were produced by injecting a DNA construct including a *Drosophila* hsp70 gene and a herpesvirus thymidine kinase gene. The F0 transgenic mice were normal. However, when the transgenic mice were mated, approximately 25% of the pups were abnormal (designated

PHT1-1). No mutant progeny were detected when PHT1-1 mice were mated to nontransgenic mice. DNA hybridization demonstrated that mutant mice were always homozygous for the transgene, while normal mice were heterozygous or nontransgenic.

The phenotype of the mutant mice included loss of most of the hindlimb, abnormalities of the forelimb, and craniofacial and brain defects. None of these mutants lived more than 24 h postnatally.

In the transgenic mouse line PHT1-1, a gene which plays an important role in murine development has been disrupted by insertion of the transgene. Identification, isolation, and molecular analysis of this interesting gene are now possible.

Role of Endogenous Retrovirus as Mutagens: The Hairless Mutation of Mice
J. P. Stoye, S. Fenner, G. E. Greenoak, C. Moran, and J. M. Coffin
Cell, 54, 383—391, 1988
1-32

One approach to identifying the genes involved in mouse development and physiology involves examining well-known mutations for the presence of specific transposable elements, such as retroviruses, that can serve as tags for gene isolation. To expedite this search, a set of oligonucleotide probes were developed that divided the nonectopic C-type proviruses into three families. These probes have been used to establish a causal relationship between a nonectopic proviral insertion (MX 40) and a previously well-known mutation, hairy (hr).

The proviral contents of HRS/J hr/hr, hr/+, and +/+ mice were compared by Southern analysis. The polytropic provirus probe JS5 detected a 3.1 kb reactive band in the hr/hr sample that was not present in the +/+ sample and was present at half copy level in the heterozygous sample.

In the HRA/Skh strain, a single hr revertant had been obtained. Mice homozygous for the reversion were compared to their heterozygous and to wild-type littermates. The revertant DNA contained a restriction fragment that was approximately 800 bp longer than the corresponding wild-type allele. This is the same size as the long terminal repeat (LTR) of MX 40. This fragment was cloned and sequenced and contained one copy of the MX 40 LTR, indicating that reversion had occurred by provirus excision.

These results demonstrate that the well-known mouse mutation hairy was caused by insertion of a retrovirus, MX 40, and that reversion in at least one case was due to excision of this provirus. This indicates the potential role of murine provirus in generating mutations and their usefulness in identification and study of the genes into which they insert.

A Transgene Containing *lacZ* Inserted into the *Dystonia* Locus Is Expressed in Neural Tube
R. Kothary, S. Clapoff, A. Brown, R. Campbell, A. Peterson, and
J. Rossant   1-33

Seven lines of transgenic mice were generated, carrying a mouse hsp68 promoter-lacZ hybrid gene. One of these lines, Tg4, expressed the lacZ gene in the absence of stress. A column of beta-gal activity on the dorsal midline was detected in these fetal transgenic mice. This expression was investigated.

The staining was detected along the medial ventral edge of the neural tube. It was initially detectable at day 9.5 and peaked on day 13.5, when staining was detected along the entire neural tube. By birth this staining was undetectable.

When the transgenic mice were interbred, 25% of the offspring had limb incoordination by day 10 to 12. Most of these affected progeny died by the time of weaning. Southern analysis demonstrated that affected mice were homozygous for the transgene, indicating recessive insertional mutation. Histological examination revealed a specific loss of dorsal spinal root sensory axons.

This syndrome appeared identical to a spontaneous mutation, dystonia musculoram (dt). Heterozygous dt mice were crossed with Tg4/-hemizygotes. This resulted in 25% affected progeny, indicating that dt and the Tg4 insertional mutation were allelic.

This insertional mutation of the dt locus will be useful in the cloning and molecular analysis of this interesting developmental gene. These results have also demonstrated that a mammalian transgene can respond to host regulatory elements. This can be exploited in the study of the regulatory regions of other mammalian genes.

Transgenic Mice Overexpressing the Mouse Homeobox-Containing Gene *Hox-1.4* Exhibit Abnormal Gut Development
D. J. Wolgemuth, R. R. Behringer, M. P. Mostoller, R. L. Brinster, and
R. D. Palmiter
Nature, 337, 464—467, 1989   1-34

The mouse homeobox-containing genes are thought to be important in the regulation of development, but there has been no direct demonstration that these genes can affect developmental processes. To study the consequences of overexpression of a homeogene, a Hox-1.4-containing DNA construct was introduced into the mouse germ line. To distinguish this gene from the endogeneous Hox-1.4 gene, 400 bases of SV40 DNA sequence were inserted into the 3′ untranslated mRNA region.

Two transgenic lines, 1974-3 and 1975-2, were characterized. In the 1974-3 line inheritance was via the Y chromosome, while it was autosomal in 1975-2. Northern blot analysis demonstrated expression of the transgene from both lines in the adult testes, as expected. No other expression was detected in adult mice.

In embryonic mice, the level of expression of the transgene was fivefold higher than the endogenous gene. *In situ* hybridization was used to analyze the excess expression. Expression of the transgene in the central nervous system and lung was elevated twofold higher than expression of the endogenous gene. The highest level of overexpression of the transgene was detected in the embryonic gut mesenchyme. The adult mice of both transgenic lines developed congenital megacolon.

Elevated expression of the Hox-1.4 transgene in the embryonic gut mesenchyme leads to the development of megacolon in adult mice. This is the first example of a mutant phenotype associated with the expression of a mammalian homeobox gene.

♦ A by-product of producing transgenic mice is the potential of generating mutant phenotypes. Such conditions can be caused by over-expression or mis-expression of a transgene. Alternatively, mutations can be caused by insertion of the foreign DNA into or surrounding an endogenous gene in such a manner that production of the wild-type gene product is disrupted. The power of generating mutations in this manner is that the foreign DNA can act as the molecular tag to allow for recovery and analysis of the preinsertion site which, in many cases, will constitute the wild-type allele of the mutated gene. The drawback to this approach is that it is somewhat haphazard in that there is no efficient way to screen for a desired mutation, and thus, one must study what happens to fall out. Nonetheless, a number of interesting transgenic-associated mutant phenotypes have been reported during the past year and some of these are discussed below.

McNeish et al. found that of the three transgenic lines produced by injection of a construct that carries a *Drosophila* heat shock gene and a herpesvirus thymidine kinase gene, one resulted in a homozygous mutant phenotype. The most striking defect was loss of most of the visible hindlimb structures. The forelimbs were also affected but the phenotype varied, ranging from absence of digit one to loss of the radius and all structures distal to the carpus. Craniofacial malformations such as cleft lip or palate were also observed. Brain abnormalities such as missing olfactory lobes and hemorrhagic protrusions from the cerebral cortex, hydrocephalus of lateral ventricles, thinning of the nasal septum, and abnormal brain shape were also detected. Although death is a postnatal phenomenon occurring within 24 h after birth, defects were detected as early as embryonic day 10 where abnormal mesenchymal cell death was observed. Although the insertion of the transgene appears to have disrupted a gene involved in limb and craniofacial development, the cause is still unknown. Furthermore it is difficult to visualize how a single gene product can affect such a diverse array of structures. Flanking sequences or the preinsertion site have yet to be cloned.

The average inbred mouse strain contains on the order of 0 to 6 ecotropic proviruses per genome whereas there are 40 to 60 nonecotropic (xenotropic, polytropic and modified polytropic) proviruses per genome. Stoye et al. have prepared a set of oligonucleotide probes that divide the latter group of proviruses into three different families, each with a small number of members that can be readily distinguished from one another by restriction mapping. In this report, evidene is presented showing a causal relationship between one particular proviral insertion and an autosomal recessive mutation known as *hairless* (located on chromosome 14). Mice homozygous for this mutation develop total alopecia and have a predisposition to thymic leukemia. These findings, together with sequence data, indicate that the mutation was caused by a naturally occurring provirus integrating into an intron or noncoding sequence. Preliminary analysis of other mouse mutations suggests that approximately 5% of the known recessive mutations in mice may be due to insertion of C-type proviruses. Many of these may be clonable since provirus integration does not tend to cause gross chromosomal rearrangements as is often seen with transgenes introduced into the nucleus by microinjection techniques.

Transgenic lines carrying the heat shock 68 promoter driving the *E. coli lacZ* gene were generated by Kothary et al. One line was found to express *lacZ* spontaneously, revealing a column of staining running along the medial ventral portion of the neural tube. Dorsal root ganglia adjacent to labeled levels in developing neural tube also showed infrequent spots of reaction product. Newborns showed no staining either in spinal cords or ganglia. Surprisingly, mice homozygous for the transgene survived to term but died within 4 weeks after birth showing signs of limb incoordination and severe loss of dorsal spinal root sensory axons. This phenotype is identical to that described for a spontaneous mutation of the mouse known as *dystonia musculorum*. Mice doubly heterozygous for the transgene and the dystonia mutation (*tg/dt*) showed the same phenotype indicating that the two mutations are allelic to one another. It is not clear whether neural tube specific expression of the *lacZ* reporter gene is activated by regulatory sequences of the *dystonia* locus or by sequences that are close to the *dystonia* locus but are unrelated. This would explain why expression of *lacZ* was found during midgestation but the obvious phenotypic abnormalities were not seen until after birth. It is also possible that subtle phenotypic effects may be occurring before birth.

Transgenic mice carrying a complete *Hox 1.4* gene together with sufficient upstream regulatory sequences and a 3' SV40 T-antigen coding region and poly(A) addition site were generated by Wolgemuth et al. Appropriate tissue-specific patterns of expression were observed with the transgene (highest in the testis and barely detectable levels in lung and kidney). The most striking of all expression data, however, was the intense labeling observed in gut mesenchyme of transgenic animals. This pattern of expression is significant in that a condition resembling congenital megacolon appears to be associated with the overexpression. It is not clear whether the condition is due to an adverse effect on the myenteric ganglion cells in the colon or to an effect on the various layers of the developing gut itself. In either case, this represents the first report of a mutant phenotype associated with the overexpression of a homeobox-containing transgene. This approach is likely to be valuable in attempts to determine mammalian gene function. *Terry Magnuson*

Lens-Specific Expression of Recombinant Ricin Induces Developmental Defects in the Eyes of Transgenic Mice
C. P. Landel, J. Zhao, D. Bok, and G. A. Evans
Genes Dev., 2, 1168—1178, 1988   1-36

The regulation of organogenesis at the cellular level is poorly understood in mammals. One approach to this problem is to introduce developmentally regulated production of a cytotoxic substance into the DNA of transgenic mice, resulting in specific cell ablation. In this series of experiments a modified ricin A toxic subunit was linked to the promoter and 5' flanking sequences of the alpha-A-crystallin gene as a probe of the development of the murine eye.

The transgenic mice expressed the ricin subunit in the lens and uniformly developed a severe microphthalmic phenotype, associated with lens disorganization and malformation, neurosensory retinal folding, and ectopic lens fiber cells in the eye. No defects were detected in other organ systems of the affected mice.

These results suggest that normal development of A-alpha-crystallin-producing lens fiber cells is necessary for eye morphogenesis and neuronsensory retina orientation and development. The engineered cell ablation approach employed in this study appears to be a useful tool in understanding the role of various cell types in organogenesis.

Targeting of an Inducible Toxic Phenotype in Animal Cells
E. Borrelli, R. Heyman, M. Hsi, and R. M. Evans
Proc. Natl. Acad. Sci U.S.A., , 85, 7572—7576, 1988   1-37

The absence of a genetic approach to the study of organogenesis in mammals has led to the construction of transgenic mice expressing toxic genes under the control of developmentally regulated promoters. A valuable additional technique would be the ability to induce the expression of the toxic gene at will. One possibility is the herpes simplex virus 1 thymidine kinase gene (HSV1-TK) system. This enzyme is, in itself, harmless to the cell, but can metabolize antiherpetic nucleotides such as ACV or FIAU to form substrates for cellular DNA polymerase, thereby blocking DNA replication within the transfected cell.

To test this technique, COS cells were transfected with a plasmid expressing HSV1-TK and then incubated with ACV or FIAU. ACV caused a 40% decrease in plasmid replication, while FIAU caused a 70% decrease in plasmid replication. Rat cell line 3B, which stably expresses a HSV1-TK gene, was incubated with ACV and FIAU. At 10 μM ACV, cell survival was 40%, at 0.1 μM FIAU cell survival was less than 40% and at 3.0 μM FIAU cell survival was less than 2%. Therefore, although HSV1-TK expression and incubation with nucleoside analogs are individually harmless to cells, expression of the HSV-TK in combination with the nucleoside analogs results in cell death.

To determine whether this system would also function in transgenic mice, the immunoglobulin heavy-chain enhancer and light-chain promoter were fused to the HSV1-TK gene and this construct was injected into fertilized mouse eggs. A transgenic founder established the 686 pedigree, which expresses correct TK transcripts in spleen and thymus. Transgenic mice and their nontransgenic littermates were subcutaneously implanted with a miniosmotic pump containing 50 mg/ml FAIU. One week later, all organs were normal except for the atrophied spleen and thymus in the transgenic animals. Cytometric analysis revealed 80-98% lymphocyte death in the thymus and 50 to 85% lymphocyte death in the spleen.

This report describes the creation of a suicide vector containing the HSV1-TK, whose toxic potential is activated in the presence of antiherpetic nucleoside analogs. The ability to control timing, dose, and delivery of the toxic dose allows lineage ablation studies at different stages as well as regeneration studies.

Targeted Ablation of Alpha-Crystrallin-Synthesizing Cells Produces Lens-Deficient Eyes in Transgenic Mice
S. Kaur, B. Key, J. Stock, J. D. McNeish, R. Akeson, and S. S. Potter
Development, 105, 613—619, 1989
1-38

Recombinant DNA and transgenic mouse technologies permit the ablation of specific cell types as probes of mammalian development and organogenesis. Murine eggs were microinjected with the cytotoxic diptheria gene fused to the alpha-A-crystallin regulatory region. Transgenic mice were generated which expressed this construct and were totally lacking all lens structure.

Of the 109 pups born, 8 carried the transgene and had eye malformations. Founders 4, 5, and 6 had eye size reduction, no lens, increased retinal cell density and whorling of retinal fiber layers. Founder 1 had an opaque lens. Founders 2, 3, and 8 were mosaics with a small highly vacuolated lens.

Ablation of lens cells during development results in a lens-deficient small eye with malformations of the iris and retina. Therefore, the lens plays an important role in normal eye development. These transgenic mice will be useful in the study of the role of the lens in early eye development.

Thymidine-Kinase Obliteration: Creation of Transgenic Mice with Controlled Immune Deficiency
R. A. Heyman, E. Borrelli, J. Lesley, D. Anderson, D. D. Richman, S. M. Baird, R. Hyman, and R. M. Evans
Proc. Natl. Acad. Sci. U.S.A., 86, 2698—2702, 1989   1-39

The study of stem cells and differentiation is difficult in mammals. However, cell-specific expression of HSV1-TK in transgenic mice allows conditional ablation of targeted cell types by treatment with the antiherpetic nucleoside ganciclovir. Therefore, HSV1-TK expression was targeted by an immunoglobulin promoter and enhancer to examine depletion and repopulation of the lymphoid system.

HSV1-TK expression was restricted to spleen, lymph nodes, bone marrow, and thymus in the KHTK transgenic mouse line. After ganciclovir treatment, greater than 99% of all lymphocytes were ablated. Non-expressing cells in the same mouse were not affected by this treatment. The wet weights of the spleen and thymus were reduced by 85% and lymph nodes by 80% as compared to control. The number of cells in the spleen were reduced by 85%, in the bone marrow by 85% and the thymus by 98%. The thymus was atophied, with little cortex and a hypocellular medulla. Flow cytometry indicated depletion of all lineages, with B-cells more sensitive to ablation than T-cells. Seven days after ganciclovir withdrawal, the total number of lymphoid cells recovered to almost 50% of the control group.

These results describe a transgenic mouse system that permits targeted conditional cell ablation. In the KHTK mouse line described in this paper, treatment with antiherpetic drugs causes ablation of B- and T-cell lineages. After drug withdrawal, sufficient progenitor cells remain to repopulate the lymphoid system. By controlling the time and degree of cell ablation with the HSV1-TK system, cell lineage, commitment, and organ plasticity can be studied.

Transgenic Mice with Inducible Dwarfism
E. Borrelli, R. A. Heyman, C. Arias, P. E. Sawchenko, and R. M. Evans
Nature, 339, 538—541, 1989
1-40

The pituitary gland is a major regulator of body physiology. The thymidine kinase obliteration system (TKO) was used to study the ontogeny of the growth hormone (GH) and prolactin- (Pr1) producing cells of the murine anterior pituitary. Transgenic mice were created, carrying HSV1-TK under control of either the rat GH or Pr1 promoter.

Transgenic mice expressing HSV1-TK in the GH-producing cells or somatotropes, were treated with the antiherpetic drug FIAU. These mice develop as dwarfs with atrophied anterior pituitaries lacking somatotropes and Pr1-producing cells or lactotropes. Transgenic mice expressing HSV1-TK in the lactotropes were also treated with FIAU. These mice were unaffected. Because toxicity depends on cell division, Pr1 expression and lactotrope differentiation appear to be post-mitotic events. After 6 weeks withdrawal of the dwarf animals from FIAU, the somatrotrope and lactotrope populations rebounded and the animals increased in size.

These results indicate that a stem-somatotrope population is a common precursor for the terminally differentiated somatotrope and lactotrope. This stem cell population has the potential to repopulate the anterior pituitary after depletion of somatotrope and lactotrope cells.

♦Mutant phenotypes produced by tissue-specific programmed cell death induced by a transgene which encodes a toxin was first introduced in 1987 by Palmiter and colleagues (*Cell,* 50, 435—443) and also by Breitman and co-workers (*Science,* 238, 1563—1565). In both reports, diptheria toxin (results in the NAD-dependent ADP-ribosylation of elongation factor-2, causing an inhibition of protein synthesis and cell death) was used to achieve tissue-specific genetic ablation. In the first case, production of the toxin controlled by elastase regulatory sequences resulted in an arrest of pancreatic development, and in the second case, the γ 2-crystallin promoter regulated the expression of the toxin gene thereby affecting lens development. The recent reports listed above by Kaur et al. and by Landel et al. extend Breitman's earlier results by using the αA-crystallin instead of the γ 2 promoter, and a range of eye-specific defects were observed. The variation in phenotype is presumably due to differences in expression of the two crystallin promoters, the consequence of ectopic expression of transgenes in some transgenic strains, or variations in the effects of different toxins. For example, Landel et al. used the castor bean, *Ricinus communis* (acts as a highly specific ribonuclease that catalyzes the inactivation of ribosomes by cleaving adenosine from the A4365 of the 28S ribosomal RNA), rather than diptheria toxin. In the report by Behringer et al. a similar approach was used to ablate growth hormone producing cells which resulted in dwarf mice. These reports are significant in that they demonstrate the potential for targeted cell ablation resulting from the ability of regulatory signals to confer specificity on toxin genes. One obvious difficulty is that ablation by the use of toxin genes results in a dominant-like phenotype. Consequently, if the expression is targeted for any early developmental event, it will be fatal and not possible to obtain transgenic lines for study. Thus, it is important that the manifestation of the toxic phenotype be inducible.

The papers from the Evans group (Borelli et al., 1988, 1989; Heyman et al., 1989) address this point by using the *Herpes simplex* virus 1 thymidine kinase (tk) gene as the toxin-producing gene (will metabolize nucleoside analogs toxic to dividing cells but are not used by the endogenous mammalian tk gene). Evans and co-workers were able to show that introduction of the *Herpes* tk gene under tissue-specific regulatory sequences, in conjunction with nucleoside analog treatment, leads to a drug concentration-dependent cytotoxicity and to cell-specific death (entitled by the authors as thymidine kinase obliteration or TKO). For example, when an immunoglobulin promoter was used, mice were created that expressed a massive depletion of B and T lymphocytes after treatment with the nucleoside analog. Similarly, when a growth hormone promoter was used in conjunction with nucleoside analogue treatment at appropriate times, mice developed as dwarfs lacking both somatotropes (growth hormone producing cells) and lactotropes (prolactin producing cells) in their pituitaries. In contrast, when a prolactin promoter was used the pituitary was unaffected, suggesting that prolactin expression and lactotrope differentiation occur post-mitotically.

The intriguing aspect concerning this approach is that precise timing of destruction can be achieved by choosing when to deliver the drug. Some drawbacks are that cells must be dividing to incorporate the toxic analogue into their DNA. In addition, the use of promoters with absolute tissue specificity and a low level of nonspecific expression is critical. Furthermore, the variable degree of penetrance and expressivity that have already been observed with other transgenes will make it difficult to have absolute control over the degree of ablation, thereby complicating interpretation of results. finally, regulatory elements for known genes must be used. This limits the approach to genes already identified by other means.
Terry Magnuson

Cloning Defined Regions of the Human Genome by Microdissection of Banded Chromosomes and Enzymatic Amplification
H.-J. Ludecke, G. Senger, U. Claussen, and B. Horsthemke
Nature, 338, 348—350, 1989

The molecular analysis of human genetic disease requires probes for defined regions of the human genome. Existing techniques are often inadequate. This report describes the dissection of the Langer-Giedion syndrome region of chromsome 8 from GTG-banded metaphase chromosomes and the isolation of clones for single-copy sequences from this region.

Normal chromosome spreads were G-banded with trypsin-Giemsa and dissected using an electronically controlled micromanipulator. The distal third of band 8q23 and the proximal two thirds of sub-band 8q24.1, approximately 10,000 kb, were dissected from 37 chromosomes and transferred to a collection drop in a moist chamber. The DNA, estimated at 0.5 pg, was digested with RsaI and ligated to a SmaI-digest of pUC. The inserts were amplified using the universal M13/pUC sequencing primers. After 26 cycles, the amplified inserts were released with EcoRI and cloned into pUC13. A total of 20,000 clones were obtained.

Of these clones, 50 were analyzed. The insert size ranged from 52 to 350 bp. Single copy sequences comprised 80% of the clones. The single-copy clones were used to probe Southern blots of DNA from two Langer-Giedion patients. Of the 40 clones, 11 were deleted in both patients and 9 in 1 patient only.

These results demonstrate that dissection of banded human chromosomes can produce thousands of region-specific probes within 10 d.

♦ One of the most promising approaches for isolation of genomic clones from specific regions of chromosomes is based on exploiting the positional information localized by genetic means as the basis for cloning strategies. The relatively novel techniques of chromosomal microdissection and microcloning were pioneered by Edström and colleagues on polytene salivary gland chromosomes from *Drosophila*. These techniques have been adapted to facilitate mammalian chromosomes and this methodology has proven invaluable for generating a series of molecular markers that are chromosome and region specific. Such markers have created a molecular entry point into several regions of the genome in both mouse and humans. The overall accuracy of generating physical markers closely linked to any region of interest depends on the resolution achieved by microdissection. This efficiency has always been compromised by the fact that banded metaphase spreads were not used for fear of decreasing cloning efficiency. Lüdecke and colleagues report in the paper listed above that metaphase chromosomes can in fact be used successfully for microdissection and microcloning. The use of the banded chromosome offers the distinct advantage that a discreet portion of the genome can be identified and isolated. The other novel aspect introduced by these investigators was the amplification of the genomic inserts by the polymerase chain reaction. The ability to amplify the dissected fragments means that fewer chromosomes have to be scraped in order to generate enough DNA for cloning purposes. However, an amplified library is likely to contain a number of clones that are repeated more than once. Therefore, it is not yet clear whether it may be more efficient in terms of number of independently derived inserts to isolate a greater number of fragments from banded chromosomes and clone these into a vector such as *lambda ZAP* for ease of screening. *Terry Magnuson*

Developmental Gene Expression 2

INTRODUCTION

The accurate expression of genetic information is essential for normal development. A gene product must appear at the correct time and in the correct place in order to ensure that development will proceed appropriately. The regulatory components that are necessary to direct such precise developmental control are slowly being identified.

Many levels of control are evident for the developmental regulation of gene expression. While most of the regulation seems to require some positive action, like the addiction of a regulative molecule, there is evidence that genetic rearrangements, genetic deletions, or the removal of a molecule-blocking expression, are also important. Most developmental gene expression is regulated at the transcriptional level. However, examples do exist for developmental regulation that occurs at the level of RNA splicing, transport, translation, or stability.

During the past year, much effort has been invested in further defining the molecular nature of the many factors that regulate developmental gene expression. Specific DNA sequences have been identified that are necessary for the appropriate developmental expression of certain genes. Specific factors have been identified that interact with DNA in a developmentally significant manner. One possible and exciting outcome of examining these many factors could be the realization that such factors are conserved throughout evolution and are fundamentally important for developmental regulation.

Interactions of Three Sequentially Expressed Genes Control Temporal and Spatial Specificity in *Aspergillus* Development
P. M. Mirabito, T. H. Adams, and W. E. Timberlake
Cell, 57, 859—868, 1989   2-1

In *Aspergillus nidulans,* three sporulation-specific genes, br1A, abaA, and wetA are important in the control of development. This paper reports on the primary sequence of the abaA gene, the inferred abaA polypeptide,

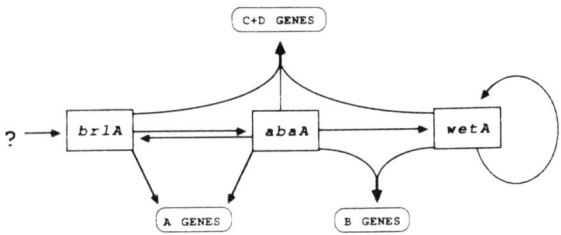

FIGURE 2-1. A model for genetic interactions controlling Aspergillus development. Single arrows indicate that the gene is sufficient to induce expression. Joined arrows indicated that the genes must function together to induce expression. Class C and D genes are differentiated by their patterns of expression in brlA, abaA, and wetA mutants during normal development. It should be noted that by our criteria, brlA and abaA are class A genes and wetA is a class B gene. (From Mirabito et al., Cell, 57, 859—868, 1989. ©Cell Press.)

and the effects of abaA induction under conditions that normally suppress conidiation.

The deduced ABAA polypeptide sequence contains 796 amino acids and is not homologous to other sequences in the GenBank and NBRF data bases. The abaA structural gene was fused to the alcA promoter and integrated into the argB locus to create the TPM1 strain, in which ABAA is nutritionally regulated. With threonine as a carbon source, ABAA is produced, triggering a block in hyphal elongation, vacuolization and abnormal septae. Induction of abaA activated brlA and wetA, as well as many other sporulation-specific mRNAs.

It appears that brlA, abaA, and wetA comprise the central pathway controlling development in *A. nidulans*. Expression of brlA in vegetative cells activates abaA which activates wetA and reinforces brlA expression. Sporulation-specific genes are activated by specific combination of these regulatory genes (Figure 2-1), setting up a regulatory cascade with feedback loops. Induction of abaA prior to brlA does not result in conidium formation, even though brlA is activated, indicating that the correct order of activation of these regulatory genes is important for proper differentiation.

A Large Cluster of Highly Expressed Genes is Dispensable for Growth and Development in *Aspergillus nidulans*
R. Aramayo, T. H. Adams, and W. E. Timberlake
Genetics, 122, 65—71, 1989   2-2

The 38-kb SpoC1 gene cluster of *Aspergillus nidulans* encodes at least 14 spore-specific transcripts that comprise more than 2% of conidial polyA+ RNA. In order to examine the function of these transcripts, a gene transplacement strategy was employed to delete the entire SpoC1 complex.

The SpoC1 complex was completely deleted from transformant TRA007, which contained no DNA sequences which hybridized to SpoC1 DNA probes and produced no SpoC1 RNA. However, there were no differences in morphology, growth, and spore production between TRA007 and wild-type *A. nidulans* strains. Spores from wild type and the TRA007 mutant strain were equally resistant to UV, high temperature, detergents, high salt, oragnics, and wet or dry long-term storage.

These results demonstrate that the SpoC1 gene cluster of *A. nidulans*, which represents 0.15% of a highly streamlined genome and codes for polyA+ RNAs that comprise more than 2% of conidial RNA, is dispensable for growth and development under laboratory conditions.

♦Although it has a relatively small genome, *Aspergillus nidulans* has a complex developmental program leading to spore formation. Genetic and molecular analysis have shown that many genes are involved in this developmental pathway. However, only three of these genes appear to be major regulatory elements in the central pathway. These genes are brlA, abaA, and wetA. They appear to make up a linear-dependent sequence resulting in conidiation. Their expression triggers the transcription of other conidiation-specific genes.

The report by Mirabito et al. examines the relationship of these genes by using transformation which allows their abnormal expression during hyphal growth. Based on this analysis, the authors are able to conclude that brlA and abaA positively control the expression of each other via a feedback loop, presumably to ensure the completion of this important developmental event. The wetA gene is a positive regulator of itself. Moreover, two additional important observations are described. The order of expression is crucial for the completion of conidia formation and the regulatory gene products must be partitioned during the final stages of spore differentiation. The results of these studies indicate an elegant complexity underlying this process.

In another report, Aramayo et al. examine the role of the spoC1 gene cluster which is developmentally regulated during spore formation and is highly expressed — making up greater than 2% of their conidial polyA+ mRNA. Interestingly, deletion of this 38-kb genetic region (which encodes at least 14 developmentally controlled genes) has no effect on growth, development, or the viability of the resulting spores. It should be noted that Mirabito et al. showed that the spoC1 genes were the only sporulation specific genes investigated that were not induced by forced hyphal expression of either brlA or abaA. Thus, although these genes are under strong developmental regulation, they may not be part of the main regulatory pathway. Nevertheless, the results do ask the significant questions of why these genes are expressed and what is their function. A similar situation has occurred a number of times in *Dictyostelium* where mutations in specific developmentally regulated genes have not blocked development. One suggestion is that these genes may be necessary for the organism's survival in nature but not under laboratory conditions. — *Stephen Alexander*

A Chemoattractant Receptor Controls Development in *Dictyostelium discoideum*
P. S. Klein, T. J. Sun, C. L. Saxe III, A. R. Kimmel, R. L. Johnson, and P. N. Devreotes
Science, 241, 1467—1472, 1988   2-3

Differentiation of *Dictyostelium* amoebae into a multicellular organism is triggered by starvation and coordinated by a cAMP signaling system. The effects of cAMP on development are mediated by a cell surface receptor (Figure 2-3). The cloning, sequencing, and functional analysis of the cAMP receptor are presented in this report.

Receptor specific antiserum was used to screen a lambda gt-11 library for the cAMP receptor cDNA. This cDNA can generate a 2-kb mRNA that is not detected in growing cells, is maximal at 3 to 4 h of development, and then decreases in abundance. Cell lines transformed with a vector that produces antisense receptor RNA do not express the cAMP receptor protein and do not aggregate nor develop following starvation. This suggests that the cAMP receptor is essential for development.

The deduced amino acid sequence of the cAMP receptor has 22% identity to bovine rhodopsin and is similar to other molecules believed to interact with G proteins. The protein is proposed to have 7 transmembrane domains, an extracellular amino terminus, and an intracellular serine-rich carboxy terminus, which is the site of ligand-induced phosphorylation. This is the first report of control of gene expression during development by a G protein coupled receptor.

◆ This paper describes the cloning and analysis of the gene encoding the cAMP receptor which is expressed during the aggregation phase of *Dictyostelium*. This central element of the chemotactic mechanism of this organism has also been implicated in regulating gene expression in both early and late development. Therefore, this molecular analysis represents an important step in the analysis of the development of this organism.

Sequencing of the cDNA for the cAMP receptor shows that it has a sequence and overall structural organization that is strikingly similar to other receptors such as bovine rhodopsin and B-adrenergic receptor with seven membrane-spanning regions and a C-terminal cytoplasmic tail. The tail has abundant serine residues that are the suspected targets of

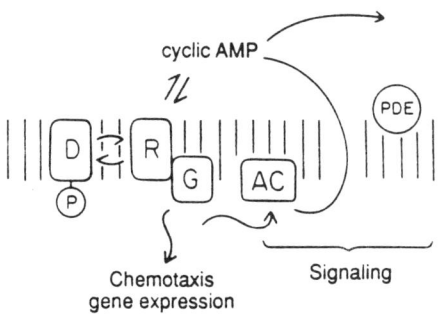

FIGURE 2-3. Diagram of *Dictyostelium* cyclic AMP signaling system. Interactions among receptor (R), modified receptor (D), guanine nucleotide binding protein or proteins (G), adenylate cyclase (AC), and phosphodiesterase (PDE). (From Klein et al., *Science*, 241, 1467—1472. With permission.)

phosphorylation during adaptation to cAMP. Antisense mutagenesis was used to block the expression of the endogenous gene and demonstrate that development is dependent on the expression of the cAMP receptor. A model is presented which suggests that the cAMP receptor is coupled to G-proteins as are other receptors that have similar structures. Indeed, this model is consistent with work of others showing that signal-transduction mutants of *Dictyostelium* that suggest that the effect of cAMP on chemotaxis and early gene expression is through a G protein-receptor complex (see review of Kesbeke et al., this volume). Further studies using homo-logous recombination with *in vitro* mutagenized cAMP receptor should allow the determination of the multiple roles this molecule plays in chemotaxis and gene expression as well as its interaction with *Dictyostelium* G-proteins. *Stephen Alexander*

Regulation and Function of G-Alpha Protein Subunits in Dictyostelium
A. Kumagai, M. Pupillo, R. Gundersen, R. Miake-Lye, P. N. Devreotes, and R. A. Firtel
Cell, 57, 265—275, 1989   2-4

The *Dictyostelium* cAMP receptor appears to transmit signals along two G protein pathways. Two G-alpha protein subunits, G-alpha-1 and G-alpha-2, have been cloned and their expression, regulation, and function examined.

The G-alpha-1 mRNAs are expressed in vegetative cells and increase in abundance by the loose aggregate stage. The G-alpha-2 mRNAs are barely expressed in vegetative cells and maximally expressed during aggregation. The expression of these two genes was examined in *Synag7* strains, which lack a protein necessary for receptor-mediated GTP-stimulated activation of adenylate cyclase.

In vegetative *Synag7* cells, G-alpha-2 was expressed at low levels. Upon starvation, there was only a slight increase in expression. Pulses of exogenous cAMP induced high levels of expression. Therefore, G-alpha-2 expression responded to cAMP pulses.

Expression of G-alpha peptides was also examined in frdA cells, which do not aggregate, display cAMP chemotaxis, express pulse-induced genes, and have defects in membrane-associated cAMP activation of GTP binding. Expression of G-alpha-1 was greater in frdA vegetative cells than in wild-type cells. G-alpha-2 expression was undetectable. Southern blot hybridization experiments indicated that the G-alpha-2 gene contains a 2.2-kb deletion in frdA mutants. This suggests that the fgdA gene encodes G-alpha-2.

The G-alpha-1 protein was overexpressed under the control of the actin promoter. Overexpressing cells were increased in size, were multinucleate, and did not aggregate. This suggests that G-alpha-1 functions in both growth and development.

The expression, regulation, and function of two *Dictyostelium* G-alpha proteins have been investigated. These proteins appear to play a central role both in signal trasnduction and in development.

♦Accumulated evidence in *Dictyostelium* indicates that cAMP exerts its effects on chemotaxis and gene expression via G proteins linked to the cAMP receptor (see reviews of the papers by Kesbeke et al. and Klein et al.). Now two Gα protein subunits have been cloned and this paper reports the patterns of expression of the genes in wild-type cells and mutants defective in signal transduction. They convincingly show that the phenotype of *frigid A* mutants is due to a lack of Gα2 subunits. Some of the mutants produce low levels of the Gα2 message while the one with the strongest phenotype, HC85, has a deletion of the Gα2 gene. The data indicate that it is the Gα2 subunit that is directly associated with the cAMP receptor and regulates the activation of the intracellular responses to cAMP. The Gα1 gene has been reintroduced and overexpressed in cells. These mutant cells have interesting phenotypes. Under certain conditions, the cells have trouble undergoing cytokinesis and become multinucleate. These strains produce fruiting bodies with altered ratios of spores and stalk cells. Thus, it is clear that this gene functions in both growth and development. The cloning of these genes should open the way, via gene disruptions and the expression of *in vitro* mutagenized proteins, to a detailed understanding of the G proteins role in *Dictyostelium* development and their relationship to the cAMP receptor.
Stephen Alexander

Developmental Appearance of Factors that Bind Specifically to Cis-Regulatory Sequences of a Gene Expressed in the Sea Urchin Embryo
F. J. Calzone, N. Theze, P. Thiebaud, R. L. Hill, R. J. Britten, and
E. H. Davidson
Genes Dev., 2, 1074—1088, 1988   2-5

The sea urchin CyIIIa cytoskeletal actin gene is activated at the late cleavage stage in the aboral extoderm cell lineages. The 2500-nucleotide 5' domain (Figure 2-5) of the CyIIIa gene is necessary and sufficient for correct spatial and temporal control of its expression during embryogenesis. Quantitative gel-shift assays were employed to examine the interactions between this regulatory domain and factors present at 24 h after fertilization, when the CyIIIa gene is active, at 7 h, when it is silent, and in the unfertilized egg.

A minimum of 14 DNA-protein interactions were detected in this region, including multiple protein-DNA complexes. The binding constant values, determined from competition studies, ranged from 2×10^4 to 2×10^6. The minimum factor prevalences were 2×10^5 to 2×10^6 molecules per embryo.

Four temporal patterns of factor binding were observed. The maternal pattern consisted of proteins detected in unfertilized eggs, which disappeared in early development. The early embryonic synthesis pattern consisted of proteins that were not detectable in the egg, but achieved maximal levels by 7 h. The late-cleavage pattern consisted of proteins not detected in the egg, detected at low levels at 7 h, and at high levels at 24 h postfertilization. In three cases, proteins detected at 7 h were no longer detected at 24 h.

Large increases in several factors occured after 7 h, which is approximately 2 h prior to transcriptional activation. Synthesis and accumulation of these factors may account for the expression of CyIIIa which occurs at this time.

♦ This paper presents the first detailed description of the regulatory apparatus of a gene expressed with strict tissue specificity in the sea urchin embryo. The CyIIIa actin gene is expressed only during larval development and only in a single cell type, the aboral ectoderm. Given such a "simple" pattern of spatial and temporal regulation, one might have expected this gene to be regulated by a single tissue specific, positive regulatory factor, and perhaps several general transcription factors. However, Calzone et al. have documented at least 14 separate sites of specific DNA protein interaction *in vitro*, including points of both positive and negative control (discussed further by Davidson, *Development,*105, 421—445, 1989). In part, this complexity may reflect the evolutionary

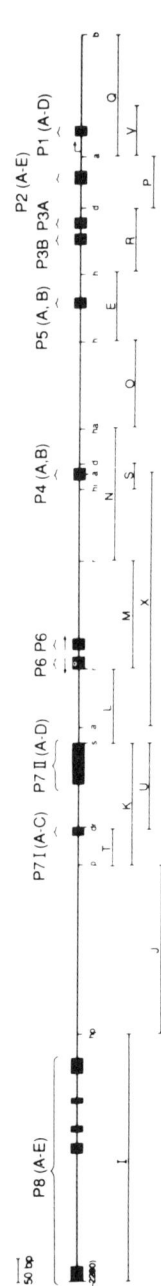

FIGURE 2-5. Map of CyIIIa cis-acting control elements. Black boxes indicate the location of the sequence elements in the 5'-flanking region of the CyIIIa gene that bind embryo nuclear or egg cytoplasmic proteins. The map summarizes results of extensive deletion and oligonucleotide competition experiments. In some cases, sequences required for binding did not completely overlap regions protected from DNase 1 in footprint studies. The complexes for each binding site are indicated above the map, as are the locations of transcription initiation and the exon 1/intron 1 boundary. There are two adjacent copies of the P6-binding site, located in inverse orientation. The probes are indicated below the gene map. Restriction sites used to construct the subcloned probes designated below the map are AvaII (a), BamHI (b), DdeI (d), DraI (dr), HindIII (h), HaeIII (ha), HinfI (hi), HpaII (hp), PstI (p), RsaI (r), and Sau3A (s). (From Calzone et al., Genes Dev., 2, 1074—1088, 1988. ©Cold Spring Harbor Laboratory.)

history of the family of five cytoplasmic actin genes, whose diverse patterns of expression are likely to have been achieved by accumulation and deletion of different regulatory elements. Among this array of putative regulatory elements are expected to be those that mediate the correct temporal and spatial expression observed *in vivo* for reporter genes controlled by it. Activation of the CyIIIa gene reflects specification of aboral ectoderm, and the establishment of the corresponding oral-aboral axis. Proteins that regulate this gene will provide starting points for elucidating the inductive interactions required for this specification.

An interesting observation in this work is that no maternal factors can be found in extracts of unfertilized eggs which form band-shift complexes which are the same as those in embryo nuclear extracts. This suggests that most regulatory proteins including "general" transcription factors are synthesized by zygotic genomes (or activated) during cleavage. Calculations from the *in vitro* estimates presented here for numbers of molecules per embryo indicate that even those proteins that are most abundant and accumulate most rapidly could be products of the typical "rare" message set, i.e., present at a few thousand transcripts per embryo.
Robert Angerer

Multiple Levels of Regulation of Tubulin Gene Expression During Sea Urchin Expression
Z. Gong and B. P. Brandhorst
Dev. Biol., 130, 144—153, 1988

Tubulin is an abundant maternal protein stored in the eggs of the sea urchin *Litechinus pictus* and synthesized during embryogenesis. Tubulin expression is regulated at several levels. The expression of tubulin genes during sea urchin embryogenesis was analyzed quantitatively.

Alpha- and beta-tubulin mRNA were quantitated by single strand tracer excess titrations and RNA gel blot hybridizations. Tubulin mRNA was constant at 1.3×10^5 transcripts per embryo during cleavage, increased during ciliogenesis, declined during midgastrula and then increased to approximately 6×10^5 per pluteus larva at 72 h postfertilization. During the cleavage stage, the rate of tubulin synthesis increased without a concomitant increase in its mRNA. During all the other stages tubulin synthesis paralleled tubulin mRNA synthesis.

Run-on assays in isolated nuclei were used to determine rates of tubulin gene transcription. During ciliogenesis at 12 h postfertilization, RNA synthesis contributed to the accumulation of cellular tubulin message. The accumulation of tubulin mRNA following gastrulation was due to increased message stability, as assayed by actinomycin D chase and RNA gel blot hybridization.

Tubulin gene expression during embryogenesis is regulated at several levels, with different regulatory mechanisms dominant at different developmental stages. During cleavage, translational regulation is paramount; during ciliogenesis, transcriptional regulation is paramount; during later stages, autogenous control of mRNA stability predominates.

♦ Much recent work on regulation of tubulin synthesis has focused on the autoregulation of tubulin mRNA stability by the soluble protein monomer pool which has been studied extensively in mammalian culture cells. This mechanism probably serves largely a homeostatic function and, although a similar mechanism may be important during urchin embryogenesis (see discussion by Gong and Brandhorst, *Mol. Cell. Biol.*, 8, 3518—3525, 1988), developmental and physiological regulation of tubulin synthesis suggests additional mechanisms are probably required in sea urchin embryos. In this paper, Gong and Brandhorst show that three different mechanisms of control — translational, transcriptional, and regulation of mRNA stability via the tubulin monomer pool — are predominant during three successive phases of normal development. At early stages, sea urchin eggs carry large pools of maternal soluble tubulin protein, but also maternal tubulin messenger RNA that is stable during cleavage, a combination which is incompatible with autogeneous regulation. Unlike culture cells, sea urchin embryos produce, in addition to cytoskeletal elements, motile cilia and apical sterocilia in the blastulagastrula period, and pluteus larvae differentiate a ciliated gut and elaborate bands and patches of cilia. In addition, deciliation of embryos, a process that leads to an increase in tubulin synthesis and regeneration of cilia, is controlled largely at the level of transcription (*Mol. Cell. Biol.*, 7, 4238—4246, 1987).

Superimposed on this multiplicity of regulatory mechanism is the additional complication that the sea urchin genome contains many copies of the α and β tubulin genes (Alexandraki and Ruderman, *Mol. Cell. Biol.*, 1, 1125—1137). Based on blot analyses of RNAs isolated from whole embryos at different developmental stages, Gong and Brandhorst suggest that the three regulatory phases described in this paper apply generally to the transcripts derived from different genes. However, some tubulin genes are differentially regulated both temporally (Alexandraki and Ruderman *Dev. Biol.*, 109, 439—451) and spatially (Harlow and Nemer, *Genes Dev.*, 1, 147—160, 1987), and in pluteus larvae tubulin mRNA accumulates to quite different levels in different tissues of the embryo (Eldon et al., *Genes Dev.*, 1, 1280—1292, 1987). The mechanisms underlying these tissue specific differences remain to be examined.
Robert Angerer

Transcription of the Spec 1-Like Gene of *Lytechinus* is Selectively Inhibited in Response to Disruption of the Extracellular Matrix

G. M. Wessel, W. Zhang, C. R. Tomlinson, W. J. Lennarz, and W. H. Klein

Development, 106, 355—365, 1989

2-7

Interactions between the cell and the extracellular matrix (ECM) are important for differentiation in some cell types. The influence of the ECM on gene expression during sea urchin development was investigated.

The ECM was disrupted with the lathrytic agent beta-amino-propionitrile (BAPN), which inhibits collagen deposition and arrests gastrulation. The levels of Spec 1, Spec 2, CyIIa actin, CyIIIa actin, and collagen were probed in *Stronglyocentrotus purpuratus*. The levels of metallothionine, ubiquitin, and LpS3 were probed in *Lytechnius variegatus* and *L. pictus*. Although gastrulation was blocked in all BAPN-treated embryos, no differences in the levels of these mRNAs were detected between treated and control embryos.

A differential cDNA screen was performed to compare poly A+ RNA from BAPN-treated and control *Lytechnius* embryos to discover transcripts which were sensitive to ECM disruption. The cDNA for a 2.1-kb mRNA was isolated that no longer accumulated following gastrulation arrest with BAPN. Removal of BAPN permitted gastrulation and accumulation of this transcript. The cDNA clone was homologous to LpS1, which is ancestral to the aboral Spec1-Spec 2 family of *S. purpuratus*. Isolated nuclei run-on assays demonstrated selective inhibition of LpS1 transcriptional activity by BAPN treatment.

Treatment with the ECM disrupting agent BAPN, which arrets gastrulation, did not cause major changes in transcription in three sea urchin species. However, one transcript, LpS1, was detected which was sensitive to BAPN treatment. This suggests that the transcriptional activity of at least one gene requires continued presence of an ECM-derived signal.

♦ Interactions of cells with the extracellular matrix (ECM) are critical for several events of morphogenesis of the sea urchin embryo, including migration of primary mesenchyme cells, skeletogenesis,1 and invagination during gastrulation (See also Solursh and Lane, this volume). Experimental treatments that interfere with the synthesis of individual components of the ECM generally lead to arrest at mesenchyme blastula stage. The molecular mechanisms and consequences of this arrest are unknown, but Wessel et al. have begun to explore them by identifying specific genes that fail to be activated when collagen metabolism is reversibly disrupted by BAPN. The first such mRNA they identified in embryos of *Lytechinus*

variegatus is the relatively abundant LvSf1 message, which is homologous to members of the extensively characterized family of Spec genes of *S. purpuratus*. This was unexpected because both early morphological differentiation of the ectoderm (flattening of these cells and consequent expansion of the blastocoel) and accumulation of other aboral ectoderm-specific mRNAs of *S. purpuratus* are not inhibited by BAPN. Specifically, BAPN inhibits transcription of LvS1, but does not affect accumulation of Spec1 and Spec2a counterparts in *S. purpuratus* embryos. (Accumulation of Spec1 and Spec2a messages is, however, not cell-autonomous and appears to require some kind of cell-cell or cell-substrate interactions; see Hurley et al., this volume.) These results are very interesting from the point of view of regulation and function of the Spec genes. First, the Spec gene family of *S. purpuratus* is the product of gene duplication and has been the site of unusual rearrangements of upstream sequences, at least some of which are required for normal regulation (W. Klein, personal communication). It seems likely that alterations in regulatory sequences in the genes of these two species, perhaps as a result of such rearrangements, may ultimately be related to differences in response to signalling from the ECM. Second, the single LpS1 gene of *L. pictus* encodes a tandemly duplicated protein whose structure differs greatly from that of the Spec genes, except for the calcium binding domains (Xiang et al. *J. Biol. Chem.*, 263, 17173—17179). The specific biochemical function of these proteins is not yet known. Clues to their function might emerge, however, if it were the case that expression of a small subset of genes encoding functionally related proteins were regulated by the ECM. Alternatively, regulation of LvS1 by the ECM may be representative of that of many aboral ectoderm-specific genes in this species. Finally, pursuit of the original expectation of these studies may lead to identification of genes expressed specifically in endoderm that are also regulated by the ECM and have important functions in morphogenesis during gastrulation.
Robert Angerer

The Expression of Embryonic Primary Mesenchyme Genes of the Sea Urchin, *Strongylocentrotus purpuratus*, in the Adult Skeletogenic Tissues of This and Other Species of Echinoderms
B. J. Drager, M. A. Harkey, M. Iwata, and A. H. Whiteley
Dev. Biol., 133, 14—23, 1989   2-8

Metamorphosis involves profound changes in morphology, tissue differentiation state, and behavior and therefore requires changes in gene expression. The echinoids metamorphose from free-swimming pluteus larva to adult benthic urchins. Available gene and antibody probes were used to investigate the persistence of expression of embryonic genes in the skeletogenic system of adult *S. purpuratus*.

Three embryonic primary mesenchyme-specific mRNAs were abundant in adult skeletal test and lantern tissues. These RNAs were not abundant in adult soft tissues. Homologous RNAs were also found in adult skeletal tissues of *S. droebachiensis, Evasterias troschellii,* and *Dendraster excentricus. In situ* hybridization was employed to localize these RNAs in regenerating spines of adult urchins. They were localized predominately to the calcoblasts that accumulated at the site of regeneration. In adult spines that were not undergoing regeneration, the most abundant RNA, SpLM18, was localized in a small population of noncalcoblast cells scattered through the shaft.

Despite the vast changes that occur during metamorphosis, expression of some echinoderm embryonic skeletal mRNAs continues in the adult. These same mRNAs are conserved across diverse echinoid species.

♦ Developmental biologists historically have been interested in primary mesenchyme cells because they differentiate autonomously from the micromeres which form at the vegetal pole of the 16-cell embryo and their fate is determined at this early stage. However, the cell biology of pimary mesenchyme cells is equally interesting. They are the unique architects of the calcite skeleton of the pluteus larva, and deposition of calcite spicules provides a simple morphological assay for the execution of genetic and biochemical events that are complex and only beginning to be elucidated. (For a review of the biochemistry of embryonic skeletogenesis, see Decker and Lennarz, *Development,*103, 231—247, 1988). Harkey et al. (*Dev. Biol.,* 125, 381—385) previously isolated recombinant DNA clones representing five abundant mRNAs which accumulate coordinately in differentiating primary mesenchyme cells. The accumulation of three of these same mRNAs in regenerating spines suggests that the same battery of genes is responsible for the biochemical events of formation of the adult calcareous skeleton. The cross hybridization at moderate stringency of these probes with RNAs from skeletogenic tissues of a member of a different order of echinoderms suggests that these proteins are remarkably highly conserved, and that proteins involved in biogenesis of calcite skeletons in members of other phyla are homologous to these. The authors point out that expression of these genes in embryonic primary mesenchyme cells is part of an autonomous developmental program, whereas it is under physiological regulation by external stimuli in the adult. However, the demonstration by Ettensohn and McClay that secondary mesenchyme cells can "convert" to a primary mesenchyme fate when the later cells are removed from the blastula suggest that activation of this program is also negatively regulated by cell-cell interactions in the embryo. Finally, although the authors remark that these gene products are involved in construction of a morphologically diverse set of embryonic and adult skeletal structures, it seems likely that they are involved only in the deposition of calcite per se, while modeling of the shape of these structures involves pattern formation mediated by specific cell-cell interactions utilizing other batteries of genes that remain to be identified. Robert Angerer

Developmental and Tissue-Specific Regulation of a Novel Transcription Factor of the Sea Urchin
A. Barberis, G. Superti-Furga, L. Vitelli, I. Kemler, and M. Busslinger
Genes Dev., 3, 663—675, 1989   2-9

Different histone gene families are expressed at different stages of sea urchin development. Two nonallelec pairs of late H2A and H2B genes have been isolated and characterized (Figure 2-9). Although these two gene pairs are regulated differently during embryogenesis, in adults both are expressed exclusively in the tube feet. A novel transcription factor was identified that interacts with the promoters of these tissue-specific late histone genes.

DNase footprint, gel mobility shift, and methylation interference assays were employed to detect the transcription factor in nuclear embryo extracts. In nuclear embryo extracts and in microinjected sea urchin embyos, the binding region of this factor was required for efficient transcription of the H2B-2.1 gene. All the recognition sites competed equally for factor binding and could substitute functionally for each other. However, the binding sites shared little homology. The factor increased in abundance throughout embryogenesis and was detected exclusively in the tube feet of the adult sea urchin, which is the specific site of expression of these late histone genes.

A tissue-specific sea urchin transcription factor, tissue-specific activator protein (TSAP), has been characterized. It may be involved in the regulation of other genes which have the same tissue-specific pattern of expression. Cloning of the TSAP gene is underway to test this hypothesis.

♦The promoters of all vertebrate and sea urchin histone H2B genes previously characterized contain the octamer sequence, ATTTGCAT, adjacent to the TATA box. This evolutionary conservation of promoter structure in the H2B genes suggests that the proximity of the octamer sequence to the TATA box is important for regulation of these genes. This suggestion is supported by data on the human H2B gene in which the octamer-TATA box juxtaposition confers cell-cycle specificity on transcription of the gene (LaBella et al., *Genes Dev.*, 2, 32—39, 1988). The

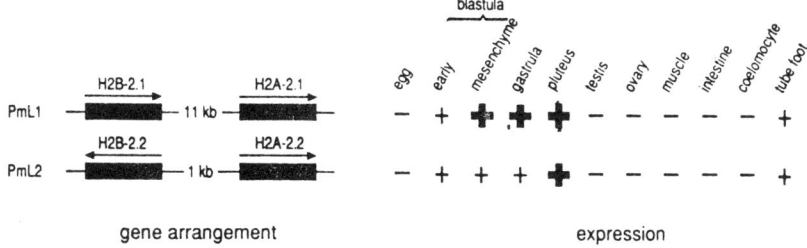

FIGURE 2-9. Genomic arrangement and expression pattern of the late H2A-2 and H2-B genes of the sea urchin *P. miliaris*. The H2A-2 and H2B-2 genes of cosmid PmL1 are coordinately regulated during embryogenesis, as are those linked on clone PmL2. The expression pattern of these two gene pairs is shown schematically. (+) Presence of the respective H2A-2 and H2B-2 mRNAs at different developmental stages and in adult tissues; (–) undetectable mRNA levels. (From Barberis et al., *Genes Dev.*, 3, 663—675, 1989. ©Cold Spring Harbor Laboratory.)

accompanying paper by Barberis et al. describes an exception to the structure of the H2B promoter in the late H2B-2 (and H2A-2) genes of the sea urchin. These authors have found a novel transcriptional element which is interposed between the TATA box and octamer element and which binds a tissue-specific activator protein (TSAP) found only in embryos and tube feet of adults. The authors present strong evidence that while both the octamer binding protein and TSAP are required for transcription, the octamer binding factor is ubiquitous, whereas the presence of TSAP correlates closely with the tissue-specific expression of the late H2B-2 and H2A-2 genes. They therefore suggest that TSAP may function as a "master regulator" for tube foot-specific genes in adults. In embryos, by contrast, H2A and H2B gene expression is temporally distinct, even though these genes bind TSAP at equal affinities. Thus, although TSAP appears to be necessary for the regulation of these genes during development, it is clearly not sufficient, and one must invoke an additional regulatory step to explain the developmental expression of these genes. Having identified TSAP, the first tissue-specific transcriptional factor reported in sea urchins, Barberis et al. will be able to explore such regulation in the ontogeny of a select tissue of the sea urchin. *Gary M. Wessel*

Three *Strongylocentrotus purpuratus* Actin Genes Show Correct Cell-Specific Expression in Hybrid Embryos of *S. purpuratus* and *Lytechinus pictus*
P. E. Nisson,, L. E. Dike, and W. R. Crain
Development, 105, 407—413, 1989   2-10

In hybrid *S. purpuratus/L. pictus* sea urchin embryos, the timing and accululatin of actin mRNA is the same as in normal *S. purpuratus* embryos. However, an *S. purpuratus* actin promoter linked to a bacterial CAT gene is not expressed in the correct cell types in *L. variegatus* embryos. To determine if *S. purpuratus/L. pictus* hybrid embryos, these were constructed in both directions and the expression of *S. purpuratus* CyIIIA, CyI, and M actin was examined.

In the hybrid embryos, CyIIIA, was detected only in the aboral ectoderm lineage, CyI in the gut, oral ectoderm and some mecenchyme cells of larva, and M only in two small cell clusters near the esophagus of the plutei. Therefore, all three of these genes were spatially regulated correctly in the hybrid embryos.

Therefore, the transacting regulatory factors which determine temporal and spatial expression patterns of these genes are present in the hybrid embryos. These factors must derive from transcription of the zygotic genome.

♦ The differential expression of members of the actin gene family has been examined extensively in the sea urchin *Strongylocentrotus purpuratus* and has been used as a model system to test for transcriptional regulatory elements in transgenic embryos. In particular, Eric Davidson and colleagues find that when the 5' flanking region (linked to a reporter gene) of the CyIIIa actin gene is microinjected into eggs of *S. purpuratus*, transcription is faithfully regulated both spatially and temporally. Yet when this same DNA sequence is microinjected into eggs of a distantly related species of *Lytechinus*, only the temporal expression is correct; spatial expression is aberrant (see Franks et al., 1989 Year Book). William Crain's laboratory has explored actin gene family expression using the classic technique of sea urchin interspecies hybrids. In these experiments, the *S. purpuratus* CyIIIa actin gene (along with the rest of the paternal genome) is combined with a *Lytechinus* maternal genome by interspecies fertilization. By *in situ* hybridization with *S. purpuratus* specific probes, they demonstrate that CyIIIa and all the other actin genes examined are correctly expressed, both spatially as well as temporally. When compared to the gene transfer experiments of Franks et al., these results suggest that the transcriptional elements regulating the spatial and temporal expression of CyIIIa are separable; the function of temporal regulatory elements is conserved in *Lytechinus* but functional spatial regulatory elements must be derived from the *S. purpuratus* genome (either directly or indirectly) after the interspecies fertilization. It follows then that the trans-acting factor(s) responsible for correct *S. purpuratus* spatial regulation must be synthesized or activated within the first few cleavages in development (since the aboral cell lineages are established by the 5-6 cleavage) and is consistent with the data of Calzone et al. (this volume) showing that several factors which interact with the CyIIIa regulatory domain are not present in the egg but appear early in development. The interspecies hybrid approach described by Nisson et al. also will be important for examining the regulation of genes that encode these CyIIIa trans-acting factors. *Gary M. Wessel*

A Hierachy of Regulatory Genes Controls a Larva-to-Adult Developmental Switch in *C. elegans*
V. Ambros
Cell, 57, 49—57, 1989   2-11

In *C. elegans* the proper timing of many postembryonic developmental events requires the activity of four heterochromic genes: lin-4, lin-14, lin-28, and lin-29. The hierarchical regulatory relationship between these genes was explored through epistasis analysis.

Single and multiple (2, 3, and 4) mutants of all of these genes were created and the timing of the larva-to-adult (L/A) switch compared. Loss of either lin-14 or lin-28 gene fuction led to a precocious L/A switch. Therefore, these genes appear to inhibit the L/A switch. On the other hand, in animals lacking lin-4 or lin-29 function, the L/A switch did not occur. Therefore, lin-4 and lin-29 appear to be required for the L/A switch to occur.

In the absence of all four heterochromatic genes, no switch occurred. Although strains containing only lin-4 activity did not switch, strains containing only lin-29 activity underwent the L/A switch. Therefore, lin-29 is necessary and sufficient for the L/A switch to occur in the absence of the other three heterochromatic genes.

Increased lin-14 and lin-28 expression blocked the switching activity of wild type lin-29. Therefore, lin-14 and lin-28 appear to prevent premature L/A switching by inhibiting lin-29. Elimination of lin-4 activity elevated lin-14 activity. Therefore, lin-4 appears to negatively regulate at least lin-14.

From these data, a model (Figure 2-11) of the hierarchical relationship of these four heterochromatic genes can be derived. At early stages, expression of lin-14 and lin-28 inhibit lin-29 activity, preventing the L/A switch. Later in development, lin-4 inhibits lin-14 and lin-28, allowing activation of lin-29, which triggers the switch.

♦This paper describes the genetic control of a larval-to-adult switch which involves several coordinate changes in the behavior of hypodermal cells during the fourth larval molt of *C. elegans*. The phenotypes of multiply mutant strains suggests a model wherein the switch is controlled by a stage-specific regulatory hierarchy of gene activity involving four previously identified heterochronic genes, *lin-4, lin-14, lin-28,* and *lin-29.* Previous genetic analysis of *lin-14* alleles led to the proposal that *lin-14* activity decreases during development to cause the expression of developmental programs in their proper sequence (Ambros and Horvitz, *Genes*

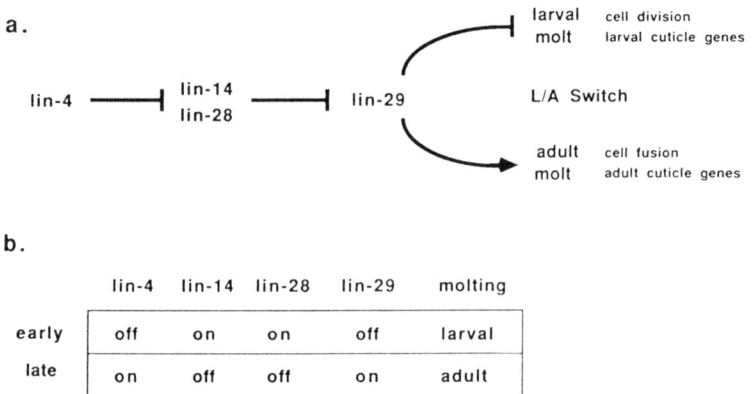

FIGURE 2-11. Model for the functional relationships among *lin*-4, *lin*-14, *lin*-28, and *lin*-29. The phenotypes of multiply mutant strains suggest a hierarchical regulatory relationship among genes controlling the timing of the L/A switch. (a) *lin*-4 is a negative regulator of *lin*-14 and *lin*-28; *lin*-14 and *lin*-28 both negatively regulate *lin*-29, a gene specifically required for execution of the switch. *lin*-29 wild-type function is proposed to positively regulate aspects of adult molting, including the expression of adult-specific cuticle genes and functions required for seam cell fusion. *lin*-29 is proposed to inhibit aspects of larval molting, including larval-specific cuticle genes and genes controlling seam cell division. These activities of *lin*-29 may be direct, as implied here, or via unidentified intermediate regulators. This model is a formal one, and makes no claim to specify the precise the precise molecular nature of the proposed interactions among genes and/or gene products. (b) The normal timing of the switch is porposed to result from the stage-specific activation of *lin*-4, leading to a corresponding temporal decrease in the combined activities of *lin*-14 and *lin*-28 and hence a stage-specific activation of *lin*-29. The model does not assume precise times during development that these gene activities would change, only that they would change between early and late stages of postembryonic development such that L4 cells would experience *lin*-29 activity and switch molting type. (From Ambros, *Cell*, 57, 49—57, 1989. ©Cell Press.)

Dev., 1, 398—414, 1987). This is an important paper because it begins to address one of the most central questions of developmental biology: what controls the timing of developmental events? *lin-14* is a key player in the developmental clock. It is negatively regulated by *lin-4* and it (and *lin-28*) negatively regulates *lin-29*. *lin-29* is required for the execution of the larva-to-adult switch. It will be interesting to see if similar heterochronic hierarchies are present in other organisms. *Joseph G. Culotti*

Transcriptional Regulation of a *Xenopus* Embryonic Epidermal Keratin Gene
E. A. Jonas, A. M. Snape, and T. D. Sargent
Development, 106, 399—405, 1989

The earliest example of *Xenopus* tissue differentiation is the activation of epidermal keratin genes, such as XK81A1, in the ectoderm of late blastula stage embryos. To localize cis-acting elements that control the tissue-specific expression of this gene, modified XK81A1 genes were injected into fertilized Xenopus eggs and expression monitored.

The XK81A1 promoter region, from −5900 to +26, was fused to a human beta globin gene and injected into fertilized eggs. Human beta globin was expressed and accumulated exclusively in the epidermis. Therefore, this region contains the elements necessary for correct expression of this gene. Injections of DNA containing 5′ deletions revealed that only DNA sequences 3′ to position −487 were necessary for tissue specific expression of this gene.

Therefore, epidermis-specific keratin expression is primarliy regulated at the level of transcription. The 5′ sequence elements between −487 and +26 are necessary and sufficient for this tissue specific expression.

Expression of Microinjected hsp70/CAT and hsp30/CAT Chimeric Genes in Developing *Xenopus laevis* Embryos
P. H. Krone and J. J. Heikkila
Development, 106, 271—281, 1989   2-13

Exposure of cells to environmental stress results in the synthesis of a small group of highly conserved proteins, the heat-shock proteins. Expression of these genes is often developmentally regulated during embryogenesis. In *Xenopus* heat-inducible expression of hsp70 is not detectable until the midblastula stage, while hsp30 expression is not inducible until the tailbud stage. In order to examine cis-acting sequences involved in the regulation of expression of the *Xenopus* heat chock genes , chimeric genes containing the *Drosophila* hsp70 and *Xenopus* hsp70 and hsp30 promoters linked to a CAT reporter gene were microinjected into *Xenopus* embryos.

Expression of either microinjected hsp70 promoter could be induced by heat after the midblastula stage of development, just as the endogenous genes become inducible at this stage. However, the microinjected hsp30 promoter also became active at this time, even though the endogenous hsp30 is not yet active. To examine the hsp30 gene for other regulatory regions, the intact *Xenopus* hsp30 gene was microinjected. However, these microinjected genes were also expressed after the midblastula stage.

These data suggest that endogenous hsp30 expression is inhibited in the embryo until the tailbud stage. This regulatory system was not effective in inhibiting induction of the microinjected gene.

♦The induction of heat shock promoters, either from *Drosophila* or from *Xenopus*, works when introduced into frog embryos by injection of cloned DNA into fertilized eggs. The developmental control of HSP30, however, fails; normally this gene is not inducible until the tailbud stage, but the injected form, either a promoter fusion or a complete HSP30 gene, is inducible immediately after the midblastula transition. The authors can only speculate as to the reason for this, but a clue may be adduced from their preliminary observation that HSP30 genes exist in a clustered arrangement in *Xenopus,* opening the possibility for long-range regulatory mechanisms that might not be adhered to by an introduced HSP30 gene present as a single copy on a plasmid. Another possibility suggested is differential HSP30 gene methylation, which would not be preserved in cloned DNA.

The keratin work, done in my lab, is the second example of correct tissue-specfic expression of a transgene in this system, the other being the α-actin gene (Mohun et al., 1989; Taylor et al., 1989). Keratin is a marker for epidermis, which differentiates autonomously but can be diverted to many other tissues, including muscle, by various inductions. The autonomous pathway leads to keratin expression while the inductive events repress these genes, so keratin regulation provides an opportunity to study several developmental control mechanisms at the molecular level. One interesting difference between the keratin and α-actin gene promoters is the apparent existence in the former of negative control elements whose removal leads to ectopic expression. The muscle actin gene, on the other hand, seems to be expressible only in muscle; removal of the essential control element (the proximal SRE) simply inactivates the gene in the embryo transformation assay. *Thomas D. Sargent*

Developmental Regulation of Two 5S Ribosomal RNA Genes
A. P. Wilffe and D. D. Brown
Science, 241, 1626—1632, 1988 2-14

Xenopus laevis contains two types of multigene families that encode 5S ribosomal RNA. Oocytes express both the large oocyte family of 5S genes and the smaller somatic 5S gene family. During embryogenesis, the oocyte genes are stably repressed, while the somatic genes continue to be expressed. There are only six nucleotide differences between the sequences of the genes in these two highly related families. This review discusses the molecular mechanisms involved in the establishment and maintenence of this differential pattern of expression during *Xenopus* development.

At least three factors, TFIIIA, TFIIIB, and TFIIIC, are required for transcription of the 5S genes. The binding of TFIIA alone is specific, but relatively weak. This affinity is increased by binding to TFIIIC. TFIIIB is required to form a competent transcription complex.

Developmental Gene Expression 73

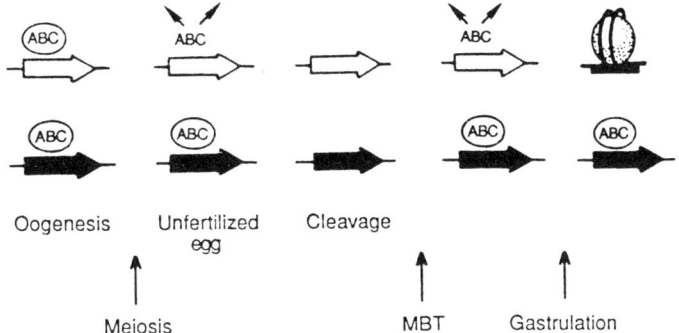

FIGURE 2-14. Model for the developmental control of 5S RNA gene transcription. This diagram summarizes the occupancy of oocyte (upper) and somatic (lower) 5S DNA with transcription complexes or chromatin during oogenesis and embryogenesis. A stable complex is represented by the factors A, B, and C encircled. The loss of the circle around a complex along with the arrows indicates unstable transcription complexes. These genes are still accessible to high levels of factors. The end result by late gastrulation is stable transcription complexes assembled on somatic 5s RNA genes and a repressed chromatin structure on oocyte 5S DNA. (From Wolffe/Brown, *Science*, 241, 1626—1632, 1988. With permission.)

Oocyte 5S RNA genes remain stably repressed in chromatin isolated from somatic cells, even in the presence of transcription factors and RNA polymerase III, unless histone H1 is removed from the chromatin. This implies that chromatin structure is involved in the maintenance of the stably repressed state. Replication through a transcription complex completlely removes that complex, indicating that at the time of replication the state of gene activation can be more easily altered.

Studies of *Xenopus* development have shown that during oogenesis transcription factors are abundant and both the oocyte and somatic 5S genes are expressed efficiently. When the oocyte matures, nuclear transcription is repressed. After fertilization, there are 12 rapid cell divisions without transcription. After the midblastula stage, transcriptional activation occurs. However, oocyte 5S RNA trancription is low and these genes are progressively inactivated into H1 containing chromatin. The end result is stable transcription complexes on the somatic 5S genes and stably repressed oocyte 5S genes (Figure 2-14).

The model proposed for the differential regulation of these two closely related gene families involves progressive limitation of transcription factors during development and a difference in stability oif the transcription factors on the two types of genes. Three of the six nucleotide differences between these two gene families are located in the 5' regions of the promoters (ICRs) and influence the stability of the transcription complexes. When a gene is not assembled into an active transcription complex it is available for assembly into inactive chromatin by histone H1.

♦Most research on the developmental regulation of gene expression focuses on protein-encoding genes that are transcribed by RNA polymerase II. This article, which is more of an overview than a specific research article, is included because it reports on the mechanistic detail responsible for the developmental regulation of 5S ribosomal DNA by RNA polymerase III in the frog, *Xenopus laevis*. The levels of control that are discussed include changes in the stability of specific protein-protein and protein-DNA interactions and how those changes affect chromatin structure. This model could just as easily explain how *cis*- and *trans*-acting factors interact to developmentally regulate the expression of RNA polymerase II-transcribed protein-encoding genes. In addition, it could have much broader relevance throughout phylogeny than just being limited to frog development. *Joel M. Schindler*

Allelic Exclusion in Transgenic Mice Carrying Mutant Human IgM Genes
Shaw A. C. Nussenzweig, E. Sinn, J. Campos, and P. Leder
J. Exp. Med., 167, 1969—1974, 1988   2-15

One of the two sets of parental Ig alleles is repressed through a poorly understood mechanism called allelic exclusion. To determine whether the functional Ig product of one allele suppresses a postrearrangement step necessary for the expression of the other allele, transgenic mice were created carrying a gene encoding exclusively membrane-bound human u heavy chains or exclusively secreted human IgM heavy chains. The expression of the endogenous and transgenic heavy chains were examined in the transgenic mice and in their hybrid progeny.

Expression of the membrane-bound human u chain in transgenic mice resulted in allelic exclusion of the endogenous mouse Ig heavy chain genes. However, expression of secreted human IgM heavy chains in transgenic mice did not prevent expression of endogenous mouse heavy chain genes. F1 hybrid animals that carried transgenes for both membrane-bound and secreted human u chains expressed both human heavy chains, but allelic exclusion of the endogenous mouse Ig heavy chain genes occurred.

Thus, the membrane-bound human u transgene can exclude endogenous mouse u production, but does not affect the expression of a productively rearranged secreted human u transgene. The simultaneous expression of the two rearranged transgenes in murine primary B cells suggests that allelic exclusion operates before the formation of a second functionally rearranged heavy chain gene.

Ig Lambda-Producing B Cells do not Show Feedback Inhibition of Gene Rearrangement
K. A. Gollahon, J. Hagman, R. L. Brinster, and U. Storb
J. Immunol., 141, 2771—2780, 1988   2-16

Plasma cells secrete either kappa or lambda light chains, but not both. This exclusivity is called kappa/lambda isotypic exclusion. A sequential model of initial rearrangement of kappa genes with lambda genes only available for rearrangement if both kappa genes are inactivated has been proposed to explain isotypic exclusion. To test this model, the expression of lambda genes in transgenic mice containing a rearranged kappa gene was studied.

Of the 15 transgenic lambda hybridomas investigated, 14 also produced kappa chains. Despite the presence of transgenic-kappa chains and endogenous heavy chains, most of these hybridomas also had rearranged endogenous kappa-genes. Flow cytometry and hybridoma analysis were used to analyze the spleen cells of normal mice. In nontransgenic mice, approximately 20% of the lambda-producing B cells produce both lambda- and kappa-chains.

These results demonstrate that lambda and kappa light chains can be expressed from the same cell in the presence of a heavy chain. This indicates that a functional kappa light chain associated with a heavy chain is not sufficient to turn off further light chain gene rearrangement in lambda+ cells. It is suggested that 2 lineages of B lymphocytes exist. One lineage, kappa, cannot rearrange lambda and has strong feedback control. The other lineage, kappa-lambda, can rearrange and express the lambda genes and does not possess strong feedback control.

♦ Following the successful rearrangement of one parental allele (found in both heavy and light chain loci), the opposing allele has been shown to be suppressed, a mechanism referred to as allelic exclusion. The advent of transgenic animals has advanced our understanding of this interesting biological process. This process starts by the random rearrangements on both heavy chain loci. This stops as soon as one of the two alleles generates a complete VDJ rearrangement. A new contribution, which utilized transgenic animals, demonstrated that the expression of only the membrane form of the mu heavy chain initiated the process which stopped any further rearrangements of the opposing heavy chain allele.

Following the successful rearrangement of the heavy chain loci, the light chain loci initiate a similar process. For quite some time it has been felt that this process was an orderly one in which the two kappa chain loci went first, and if unsuccessful, the lambda loci followed. As seen with the heavy chain loci, the successful rearrangement shut down the entire process, so that all mature B cells express only one light chain. Results from experiments utilizing transgene animals have suggested that this initial model may not be entirely correct. *Charles Snow*

Developmental Regulation of Embryonic Genes in Plants
C. Borkird, J. H. Choi, Z.-H. Jin, G. Franz P. Hatzopoulos, R. Chorneau, U. Bonas, F. Pelegri, and Z. R. Sung
Proc. Natl. Acad. Sci. U.S.A., 85, 6399—6403, 1988   2-17

Cultured carrot cells proliferate in clusters in medium supplemented with the auxin 2,4-dichlorophenoxyacetic acid. Without the auxin, carrot cell clusters form embrys that progress from globular to heart to torpedo stages. Two developmentally regulated genes, 8 and 59, were isolated and the molecular mechanisms underlying their expression were investigated in somatic and zygotic embryos.

As cells progressed from unorganized cell clusters through embryogenesis, the mRNA and protein products of genes 8 and 59 increased. These genes were expressed in plantlets, but not in seedlings or mature tissues such as leaf, petiole, or root. W001C cells grown in low density in the presence of auxin develop into globular stage embryos. The expression of genes 8 and 59 was very low in these cells. To determine whether auxin represses expression or whether progression to heart stage cells in necessary for expression, ABA 15 auxin, these cells proliferated as heart stage embryos and genes 8 and 59 were expressed. Therefore, auxin does not supress synthesis, suggesting that expression of these two genes is associated with heart stage embryos. Developing carrot seeds were also examined. Genes 8 and 59 were also abundantly expressed in zygotic embryos.

The temporal expression of two developmentally regulated genes was investigated during somatic and zygotic carrot embryogenesis. The parallel expression demonstrated in these studies indicated that somatic embryogenesis in the carrot is a useful model system for early plant development.

♦ It has been known for many years that plants can proceed through complete somatic embryogenesis beginning from a single somatic cell. Such somatic cells can be grown and maintained in culture. Plant hormones (auxins) play a crucial role in establishing whether cells will continue to grow or enter the developmental program of embryogenesis.

During early embryogenesis, the number of abundant gene products that undergo significant changes is limited, making the isolation of developmentally regulated genes difficult. Using the carrot as a model system, the authors present in this paper a novel approach to the isolation of differentially expressed genes and report on the temporal, stage-specific, and zygotic expression of two such isolates. This study is important because it indicates that carrot somatic embryogenesis can be a useful model system to investigate molecular aspects of plant embryogenesis. *Joel M. Schindler*

Regulated Genes in Transgenic Plants
P. N. Benfey and N.-H. Chua
Science, 244, 174—181, 1989   2-18

This review describes progress in the study of expression of genes in transgenic plants. The developmental and tissue-specific expression of 5-enolpyruvylshikimate-3-phosphate synthase (EPSP) is explored in detail and compared to that of the CaMV 35S enhancer.

Petunia EPSP promoter region fragments were joined to a CAT reporter gene and enzyme activity and RNA levels were analyzed in transgenic petunias. Sequences sufficient for tissue-specific regulation were contained in the 5′ promoter region, with at least one copy upstream of −800. CAT activity was low prior to flower opening and then increased rapidly. Only background synthesis was detected in transgenic tobacco plants.

The EPSP promoter region (−1800 to −800) was also linked to a GUS reporter gene and histochemical localization was performed on transgenic petunia and tobacco plants. Little staining was detected in the tobacco plant, while specific cell types stained heavily in the petunia at certain times in development. Sequences between −1800 and −800 of the petunia EPSP gene were sufficient for abundant developmentally regulated expression in certain tissues of the transgenic petunia plant. However, the expression from this promoter was also dependent on the transgenic host (petunia vs. tobacco).

The constitutive expression of the CaMV 35S enhancer joined to a GUS reporter gene was examined in transgenic plants as a control. GUS activity varied depending on the tissue type in both petunia and tobacco. Therefore, the 35S promoter does not appear to be constitutive in all plant cell types and can vary with the transgenic host species.

Transgenic plant studies were used to define the promoter sequences necessary and sufficient for cell-specific and developmentally regulated expression of the EPSP gene. Identification of these sequences necessary and sufficient for cell-specific and developmentally regulated expression of the EPSP gene. Identification of these sequences and the factors that interact with them is essential for understanding gene regulation in higher plants.

♦ Transgenic animals have proven to be valuable tools in dissecting some of the molecular components necessary for accurate temporal and spatial developmental expression of certain genes. Transgenic plants play an equally important role for understanding developmental gene expression in higher plants. In this paper, the authors report on the identification of sequences upstream of the promoter that are responsible for the developmentally regulated expression of the 5-enolpyruvylshikimate-3-phosphate synthase gene. These studies are important because they further substantiate the enormous progress made in the area of plant molecular biology and demonstrate how progress can be used to address questions central to developmental biology, including the identification of *cis*- and *trans*-acting elements that are responsible for regulating developmental gene expression. *Joel M. Schindler*

Developmental Expression of PDGF, TGF-Alpha, and TGF-Beta Genes in Preimplantation Mouse Embryos

D. A. Rappolee, C. A. Brenner, R. Schultz, D. Mark, and Z. Werb
Science, 241, 1823—1825, 1988

Preimplantation mouse embryos grow and differentiate in the absence of exogenous growth factors. The expression of endogenous growth factors was examined in single preimplantation embryos by reverse transcription (RT) followed by PCR amplification.

The system was validated through detection of beta-actin transcripts from a simgle mouse embryo. Three growth factor genes, TGF-alpha, TGF-beta-1, and PDFD-A, were expressed in uncultured mouse blastocysts, while EGF, bFGF, NGF-beta, and G-CSF were not. PDGF-A and TGF-alpha were detected in oocytes, disappeared during early cleavage, and reappeared in early cavitation blastocysts. TGF-beta-1 appeared after fertilization. The growth factors could also be localized by immunocytochemistry in permeabilized blastocytes.

The expression of some growth factors in mouse blastocysts suggests that they play a role in early mammalian embryogenesis.

♦ Studies on the *in vitro* growth and differentiation of early mouse embryos coupled with similar studies using embryonal carcinoma cells suggests that early mammalian embryogenesis can proceed in the absence of exogenous factors and that mammalian embryos may, in fact, produce their own endogenous factors to direct the process. The authors show in this article how the polymerase chain reaction (PCR) can be exploited to demonstrate the developmentally regulated expression of such endogenous factors. Using PCR, the authors detect levels of PDGF, TFG-α and -β in preimplantation mouse embryos, fail to detect transcripts for several other growth factors, and show that the transcript accumulation of the three positive growth factors is developmentally regulated, and not the same for all three. This article is important because it directly demonstrates the enormous power and value of PCR for the investigation of developmental events in mammalian embryos. *Joel M. Schindler*

Developmental Cell Biology 3

INTRODUCTION

Cells behave in developmentally meaningful ways in response to various stimuli. Such stimuli can induce responses or inhibit them, cause cells to move or to divide. The cellular response can be the result of altered gene expression but from a developmental standpoint, it is the cellular phenotype that is important.

The molecular description of several factors that effect the development behavior of individual cells or cell groups has had a profound impact on the entire field of developmental biology. Experimental evidence indicates that several growth factors, in particular, exhibit extensive molecular homology, if not identity, between species. These growth factors seem to represent large families of genes that serve multiple functions during different stages of development. Stimulating growth is one of them; inhibiting differentiation may be another. As additional members of these families are identified, their fundamental importance to development should emerge.

The definition of a stem cell is revisited as new factors that induce their differentiation are identified. How similar are such inducers? Do they share mechanisms or do they all function independently? What is the molecular relationship between stem cells for specific lineages, like hematopoietic stem cells, and entire organisms, like embryonic stem cells?

As the molecular definition of a cell continues to be refined with additional detail, so too will its behavior be more refined. For multicellular organisms, that behavior has important developmental consequences. The more we continue to learn about all aspects of cellular phenotypes and how they behave, the more we can hope to learn about the developmental consequences of that behavior.

Ammonia and Thermotaxis: Further Evidence for a Central Role of Ammonia in the Directed Cell Mass Movements of *Dictyostelium discoideum*
J. T. Bonner, D. Har, and H. B. Suthers
Proc. Natl. Acad. Sci. U.S.A., 86, 2733—2736, 1989   3-1

Ammonia (NH_3) is involved in orienting *Dictyostelium* cell masses and in positive photoaxis, by controlling cell speed. The affect of NH_3 on thermotaxis was investigated.

During the course of *Dictyostelium* development NH_3 production increased. When slugs were assayed with an NH_3 gradient, there was a peak ammonia concentration range which caused increased speed; below and above that range NH_3 caused decreased speed. As temperature increased, NH_3 production by cells also increased.

The greater the temperature, the greater the NH_3 production and the speed of the amoebae is NH_3-dependent. The result of this system is the thermotaxic response. This NH_3-dependent response system must be extremely sensitive, as differences of 4/10,000 to 5/10,000 of a degree across the tip of a slug are sufficient to cause an orientation response.

The Possible Role of Ammonia in Phototaxis of Migrating Slugs of *Dictyostelium discoideum*
J. T. Bonner, A. Chiang, and H. B. Suthers
Proc. Natl. Acad. Sci. U.S.A., 85, 3885—3887, 1988   3-2

The effect of NH_3 gas on phototaxis in *Dictyostelium* was investigated.

In the absence of exogenous NH_3 the mean slug speed was 1.31 ± 0.02 mm/h. After addition of NH_3 slugs migrated significantly faster, 1.45 ± 0.02 mm/h. In the presence of a system which enzymatically removed NH_3 as it was generated, all forward slug movement was halted. In the presence of a large amount of ammonia, slugs were unable to orient to light. If the slugs crawled over a surface inverted over an NH_3-generating solution, phototaxis was eliminated and gravity took over (Figure 3-2). The mean NH_3 evolved by slugs was 11 μM in the dark and significantly higher, 15 μM, in the light.

Therefore, the orientation of slime mold cell masses by NH_3 appears to be involved in phototaxis.

♦ The migrating slug of *Dictyostelium discoideum* is one of the alternative developmental pathways that the multicellular assembly can embark upon. It has been known for some time that ammonia is the morphogen responsible for this developmental decision. Much effort is being directed at determining the molecular and cellular mechanisms by which the slugs move. Both biochemical and genetic analysis of the glycoproteins in the slime sheath, or extracellular matrix, may be providing some answers (see the reviews in this volume on the papers by Alexander et al. and Smith et al.). Two papers by Bonner et al implicate ammonia directly in this process by acting to increase the speed of movement of cells in the

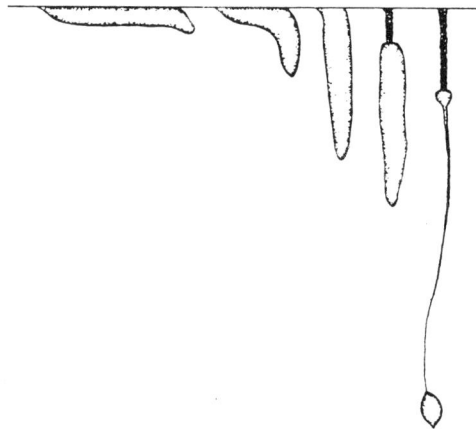

FIGURE 3-2. Semidiagrammatic drawing in a time sequence showing (from left to right) that if slugs are migrating on the ceiling of a dish containing NH_3, vapor, gravity appears to take over and the slugs point straight downward. Note the migration continues off the surface of the agar so that the fruiting body is suspended on a thread of slime sheath. (Also note the curious shape of the basal disc under these circumstances.) (From Bonner et al., Proc. Natl. Acad. Sci. U.S.A., 85, 3885—3887, 1988. With permission.)

processes of phototaxis and thermotaxis. The data presented show that ammonia increases the speed of migrating slugs and that removal of ammonia causes them to stop moving. Light and temperature both increase the production of ammonia. Increasing the concentration of ammonia to saturating levels results in the disorientation of the slugs to either unidirectional light or temperature gradients. These observations raise the question of how these external stimuli may control the production of ammonia and how the ammonia, in turn, may control the movement of cells. *Stephen Alexander*

Genetic Control of Cell Division Patterns in the Drosophila Embryo
B. A. Edgar and P. H. O'Farrell
Cell, 57, 177—187, 1989   3-3

In Drosophila, mitoses prior to interphase 14 are controlled by maternal products and are metasynchronous. After interphase 14, mitosis require zygotic transcription and occur in a highly ordered pattern, Mutations at the *stg* locus cause cell-cycle arrest in interphase 14. The *stg* gene was cloned and its phenotype and expression pattern were analyzed.

The *stg* phenotype was observed in live embryos under Nomarski optics and in fixed, stained embryos. Embryos homologous for *stg(7b69)* failed to initiate mitosis 14 and did not divide for the remainder of embryogenesis. Scanning microdensitometry was used to determine that the *stg* nuclei were 4C and had arrested in G2. When DNA was labeled, only polytene nuclei, which amplify their DNA without cell division, were labeled. Therefore, it appears that the DNA synthetic machinery functions in the mutant embryos. Despite cell-cycle arrest, the mutant undergo many aspects of normal development.

The *stg* gene was cloned by P element insertion. Genomic DNA jprobes from this region were used to isolate a 2308 bp cDNA with a long open reading frame. Comparison with the Dayhoff protein sequence bank indicated homology to the cdc25 gene of S.

In situ hybridization was used to analyze the *stg* expression pattern during development. Maternal *stg* RNA was present in the egg and was detected through mitosis 13, then it was degraded. Zygotically synthesized mRNA accumulated approximately 25 min prior to mitosis 14 only in cells that would divide. The order in which cells achieved abundant levels of *stg* mRNA predicted the sequence of mitosis initiation.

The *stg* locus encodes a conserved regulator of mitotic initiation. In the zygote, regulated *stg* expression governs the pattern of cell division during embryogenesis.

♦ The temporal and spatial patterns of cell division during Drosophila embryogenesis are fairly well understood. Initially a series of 13 very rapid synchronous nuclear division cyles occur within a syncytium. These divisions appear to require only maternal gene products. After cycle 13 cell division rates decrease and synchrony is lost, as cell division comes under zygotic control. At this time cell division becomes regionalized according to precisely defined spatial and temporal patterns. The location of these mitotic domains at each stage of embryogenesis have been mapped. Interestingly, at all stages control of cell division appears to be exerted at the G2/M boundary (the DNA content of resting nuclei is 4C), in contrast to yeast and many cultured cells where regulation appears to occur at the G1/S boundary.

The first suggestion that the *string gene* controls cell division derived from the observation that mutant embryos have fewer cells. Further analysis confirmed this hypothesis. The work reported in this paper lead to several interesting observations. One is that *string* has sequence similarity and is apparently the functional equivalent of the yeast cell cycle gene *cdc25*. Cloned *string* cDNA seem to be able to complement a *cdc25* temperature-sensitive mutant in yeast. Cell division is slower in partial loss-of-function string mutants, indicating that this may be a rate-limiting product. Null mutants divide at rates and patterns that are indistinguishable from wild type until the blastoderm stage (during this time divisions are under maternal control) and then cease dividing entirely. Measurement of nuclear DNA content confirmed that the cells arrest at the G2 stage of the cell cycle. However, perhaps unexpectedly, morphogenetic movements and differentiation into specialized tissues proceeds fairly normally in the mutant embryos, despite the reduced cell number. Therefore, unlike other animal systems, in Drosophila, cell division appears to have little influence on cell differentiation and determination. Perhaps the most significant finding of this study is the discovery that the spatial pattern of *string* gene expression in the postblastoderm embryo precisely matches the map of mitotic domains that had been previously established based on morphological observations. This regionalized *string* expression precedes actual cell division by about 30 min. These findings suggest a model whereby all factors required for cell division are synthesized constitutively in all cells and actual division depends only on the expression of *string*. If this is confirmed, the question will then shift to what controls *string* expression. Genes that influence dorsoventral polarity, gap and homeotic genes, are possible candidates. This important question will certainly be the focus of the next generation of experiments. *Marcelo Jacobs-Lorena*

The Xenopus *cdc2* Protein is a Component of MPF, a Cytoplasmic Regulator of Mitosis
W. G. Dunphy, L. Brizuela, D. Beach, and J. Newport
Cell, 54, 423—431, 1988

A cytoplasmic agent, M-phase promoting factor (MPF), induces entry into mitosis in Xenopus. Analysis of MPF was performed in a heterologous system combining the technical advantages of Xenopus and the genetic advantages of fungi.

A Xenopus cell-free system that recreated MPF-dependent motosis entry was used to assay the function of yeast *cdc* proteins. The product of the fission yeast suc1 gene, p13, inhibited MPF-induced nuclear disassembly. The dose of p13 required for inhibition was directly dependent on the dosage of MPF in the extract. Inhibition could be reversed by additional doses of MPF.

Antibodies to the yeast cdc2 kinase that interacts with p13 recognized 2 peptides of 34 and 33 kDa in Xenopus extracts. Columns containing yeast p13 specifically retained tightly bound Xenopus *cdc2.* Chromatography on p13-agarose also depleted MPF-activity.

These results demonstrate that the Xenopus *cdc2* protein is a component of MPF. The heterologous system employed in these experiments can be used to dissect the function and interaction of conserved elements that regulate cell cycle progression.

Purified Maturation-Promoting Factor Contains the Product of a Xenopus Homolog of the Fission Yeast Cell Cycle Control Gene cdc2⁺
J. Gautier, C. Norbury, M. Lohka, P. Nurse, and J. Maller
Cell, 54, 433—439, 1988   3-5

Entry into mitosis is controlled by the protein kinase cdc2-p34 in *S. pombe*. Entry into mitosis in Xenopus oocytes is controlled by MPF. Purified MPF consists of 32- and 45-kDa proteins and contains protein kinase activity. Immunochemistry was employed to determine the relationship between p34 and MPF.

Antibody to a highly conserved region of p34 was used on Xenopus immunoblots and ecognized a single protein of approximately 34 kDa. On blots of purified MPF, the same protein was recognized. After immunoprecipitation of purified MPF, a 34-kDa protein was bound.

A protein homologous to p34 of *S. pombe* is a component of the Xenojpus MPF. This demonstrates the conservation of mitotic control. This conservation will allow a combined approach, utilizing the biochemical and genetic approaches of several systems, to facilitate the study of regulation of mitosis.

Translation of Cyclin mRNA is Necessary for Extracts of Activated Xenopus Eggs to Enter Mitosis
J. Minshull, J. J. Blow, and T. Hunt
Cell, 56, 947—956, 1989   3-6

Invertebrate eggs contain cyclin proteins, which accumulate during interphase and are degraded at the end of mitosis. To search for related proteins in Xenopus oocytes, oocyte cDNA libraries were screened with a full-length sea urchin B-cyclin cDNA and a consensus cyclin oligonucleotide.

Several clones were detected that were positive with both probes. These clones were used to screen a full-length Xenopus ovary cDNA library. Clones containing the complete coding region of two highly homologous cyclins, Xlcyc1 and Xlcyc2, were identified and sequenced. Neither clone contained an upstream AUG. The predicted molecular weights were approximately 45 and 44 kDa. Cyclin mRNA was present in stage 3 oocytes and persisted to stage 11. RNase protection assays detected approximately 5×10^7 copies of each cyclin mRNA in unfertilized eggs.

The cyclins were subcloned for *in vitro* transcription followed by *in vitro* translation in reticulocyte lysates. These cyclins were compared to the products of translation of Xenopus egg cell-free systems. Cyclin polypetides comprised 8.7% of the methionine incorporated in the Xenopus cell-free system.

Antisense cyclin RNA was added to Xenopus cell-free mitotic systems. Addition of either antisense RNA separately, delayed, but did not block mitosis. Addition of both antisense RNAs, simultaneously, blocked chromosome condensation and nuclear envelope breakdown. Therefore, synthesis of at least one cyclin is necessary for entry into mitosis.

Xenopus oocytes contain two cyclins, which are homologous to each other, as well as to sea urchin and clam cyclins and *S. pombe* cdc13. This phylogenetic conservation of cyclin function suggests that cyclins occur as mitotic regulators in all eukaryotes.

♦Control of the mitotic cycle is important in the development of all multicellular organisms. Certain features of *Xenopus* eggs and oocytes have made this system especially useful in the study of such control. Maturation promoting factor (MPF) was discovered in 1971 as an activity that would trigger the resumption of meiosis in arrested frog oocytes. It has since been shown that something equivalent to MPF is probably present in all eukaryotic cells where it controls the transition from the G2 phase into mitosis. Dunphy et al. and Gautier et al. characterized *Xenopus* MPF, purified by different routes, and showed that it comprises two proteins, one of which (32 kd) is an amphibian version of $p34^{cdc2}$, a protein kinase encoded by the *S. pombe* cell cycle control gene cdc2. The function of the other MPF protein (45 kd) is not known, but one possibility, suggested by Gautier et al., is that it is a homolog of the protein encoded by the yeast gene nim1, which is about the right size and functions as an activator of $p34^{cdc2}$.

Cyclins are other proteins involved in controlling cell division. They were originally discovered in clam eggs, and have now been cloned from *Xenopus* by Minshull et al. Cyclin protein synthesis is required for the onset of an experimental version of mitosis carried out in egg extracts. The cyclin sequences show homology to the *S. pombe* gene cdc13, which interacts with cdc2 to control yeast cell division. Minshull et al. suggest that the function of cyclins is to activate MPF. However, the pathway probably involves additional components because there is a 2- to 3-hour lag between the end of the requirement for cyclin synthesis and the beginning of mitosis. These papers are a nice demonstration of how the individual advantages of two very different systems, in this case frogs and yeast, can be exploited to investigate a complex biological control pathway. *Thomas D. Sargent*

Mesoderm-Inducing Properties of INT-2 and KFGF: Two Oncogene-Encoded Growth Factors Related to FGF
G. D. Paterno, L. L. Gillespie, M. S. Dixon, J. M. W. Slack, and J. K. Heath
Development, 106, 79—83, 1989

It has been hypothesized that oncogenic transformation results from aberrations in the system that controls embryonic growth and differentiation. Many proto-oncogenes are differentially expressed during embryogenesis and appear to be embryonic regulatory molecules. The function of two transforming oncogenes, int-2 and hst/ks (kfgf), during embryogenesis was investigated in Xenopus.

The transforming factors were synthesized from cRNA in reticulocyte lysates containing canine pancreatic microsomes and serial lysate dilutions were applied to ectoderm explants. Mesoderm-inducing activity was assayed through visual inspection and RNase protection analysis of acardiac-specific actin. Bovine and kFGF were potent mesoderm-inducing factors, with specific activities of 1.3×10^7 U/mg and 3.2×10^6 U/mg, respectively, where one unit of inducing activity is the amount of 1 ml of medium that will induce vesicle formation in 50% of the explants. Int-2 was significantly less active, with a specific activity of 2.0×10^5 U/mg.

Bovine FGF synthesized *in vitro* was more active in mesoderm-induction than bFGF purified from bovine brain The titer was tenfold higher and a significant portion of the induced explants contained nototchord. Notochord was never observed in explants treated with int-2.

The mitogenic potential of these factors was assessed on C3H10T1/2 mouse fibroblast cells. Both bFGF and kFGF were potent mitogens. The activity of int-2 was tenfold lower.

These results demonstrate biological activity for int-2, which has previously been known for its involvement in viral oncogenesis. Its mesoderm-inducing activity in Xenopus suggests that it may be involved in mesoderm-induction during mammalian development. The *in vitro* synthesis followed by serial lysate dilution assay provides a simple, sensitive assay for mesoderm induction.

Presence of Basic Fibroblast Growth Factor in the Early Xenopus Embryo
J. M. W. Slack and H. V. Isaacs
Development, 105, 147—153, 1989   3-8

A small group of heparin-binding growth factors (HBGF) have been shown to have mesoderm-inducing activity in the Xenopus embryo. Another type of growth factor, TGF-beta-2, also has this activity. To determine which type of factor is responsible for mesoderm induction in the early embryo, soluble mesoderm-inducing activity was extracted from Xenopus ovary, egg, and embryo.

The recoverable mesoderm-inducing activity is bound to heparin. The content of this activity was 7 U/g wet weight in packed dejellied embryos or in unfertilized eggs. The content in ovary was significantly higher. No TGF-beta-like activity was detected.

Crude- and heparin-purified extracts were tested against a panel of antibodies to growth factors. The Xenopus activity reacted like bFGF and could be neutralized by anti-bFGF. Two anti-bFGF antibodies were used on Western blots of heparin-bound embryo material and recognized a band of approximately 19 K and another at 14 K.

The Xenopus activity was used to induce mesoderm from ectoderm explants. Low doses, 1 to 4 U/ml, had a ventral pattern, while higher doses, 8 to 256 U/ml, had an intermediate pattern. Occasionally, a dorsal pattern was seen.

These results are consistent with the involvement of bFGF as a morphogen in mesoderm induction in Xenopus embryos.

The Presence of Fibroblast Growth Factor in the Frog Egg: Its Role as a Natural Mesoderm Inducer
D. Kimelman, J. A. Abraham, T. Haaparanta, T. M. Palisi, and M. W. Kirschner
Science, 242, 1053—1056, 1988

3-9

Several growth factors are potent mesoderm inducers in Xenopus embryos. A Xenopus oocyte cDNA had been isolated that was homologous to a conserved region of mammalian bFGF. A probe corresponding to the carboxy-terminal bFGF-homologous region was used to detect bFGF-like Xenopus mRNAs on Northern blots.

Two transcripts were detected: a 2.1 kb transcript present from oocyte to stage 12 and a 4.2 kb transcript present in the oocyte and neurula. Xenopus oocyte and stage 17 DNA libraries were screened with the carboxy-terminal probe. One 4.3 kb cDNA, Lamda-40, was recovered with an open reading frame of 154 codons. The expected protein would be 84% identical to human bFGF. No secretory signal was detected.

This Xenopus bFGF was produced in *E. coli,* purified by heparin affinity and mesoderm induction of midblastula animal caps was assayed. Cardiac actin expression was enhanced in a dose-dependent manner by Xenopus bFGF. An antibody to a conserved region of bFGF was used for immunoblotting of heparin-purified egg or embryo extracts. A protein of approximately 15 kDa was detected from eggs through stage 17 embryos. There was approximately 100 pg (200 ng/ml) of FGF per egg.

A cDNA that encodes a Xenopus protein that is very similar to mammalian bFGF has been isolated. This protein is active in mesoderm induction. The presence of this factor in the oocyte and the lack of secretory signals may indicate that the regulation of mesoderm induction in Xenopus embryos is posttranslational.

Analysis of Competence: Receptors for Fibroblast Growth Factor in Early Xenopus Embryos
L. L. Gillespie, G. D. Paterno, and J. M. W. Slack
Development, 106, 203—208, 1989

Acidic and basic FGF are mesoderm inducers in Xenopus embryos. A putative receptor for FGF was characterized in Xenopus embryos.

Binding analysis of ectodermal explants with labeled aFGF indicated that specific sites were saturated at approximately 7 nM. There was a single class of binding sites with a dissociation constant of 1.4×10^{-10}. The number of binding sites was 3×10^8 per explant. Affinity labeling with a water-soluble crosslinker and labeled aFGF identified two bands of approximately 130,000 and 140,000 Mr. Competition experiments demonstrated that acidic and basic FGF shared the same receptor, but TGF-beta-2 did not.

The receptor density in animal hemisphere explants throughout the period of competence ws assessed. Receptor density increased from stage 7 to stage 8 (Figure 3-10) and then decreased to 10% of the maximum level. This expression pattern corresponds to the period of competence. All regions of the stage 8 embryo, corrected for surface area, were assessed for receptors. All regions contained receptors, with the highest density in the marginal zone, 70% of this level in the animal cap, and 40% in the vegetal cap. Dorsal and ventrical marginal zone regions contained equivalent numbers of receptors.

The FGF cell surface receptors of Xenopus embryos have been identified and characterized. This should facilitate studies of mesoderm induction in these embryos.

♦ Slack and Isaacs and Kimelman et al. demonstrate the existence in *Xenopus* embryos of functional FGF. The two labs disagree by a factor of 20 to 30 on the level of FGF present, but even the lower estimate should be sufficiently concentrated for significant inductive activity. Thus the case has grown quite strong for at least a partial role for FGF in mesoderm specification.

Gillespie et al. use ^{125}I ligand binding to quantify FGF receptors in *Xenopus* embryonic membranes at various times and positions. As shown in Figure 5 from their paper, there is a good temporal correlation between the competence of the embryo to respond to FGF and the abundance of its receptor. Another result in this paper is that the FGF receptor is present at roughly comparable levels (within a factor of about 2) in all three regions of the embryo they tested — animal, vegetal, and marginal zone. FGF does not appear to have an effect on vegetal cells, so it is not clear what role the receptor has in this region.

INT-2 and kFGF are oncogene products with homology to FGF. Paterno et al. used a rabbit reticulocyte *in vitro* translation system to

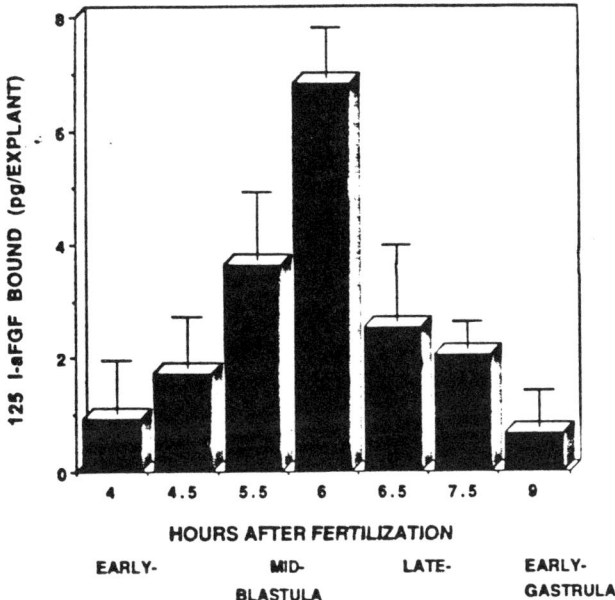

FIGURE 3-10. Stage-specific binding of 125I-aFGF to ectodermal explants. Embryos were maintained at 23°C and animal hemisphere explants were dissected at various times after fertilization. Explants were incubated with 7·2 nm-125I-aFGF ± 72 nm unlabeled aFGF for 45 min 4°C, washed extensively with NAM and counted. Specific binding was calculated by subtracting nonspecific binding from the totals. Each column represents the average value from 40 explants and five separate experiments. Standard deviations are indicated. (From Gillespie et al., *Development,* 106, 203—208, 1989. ©Company of Biologists.)

synthesize bioactive growth factor polypeptides from synthetic INT-2, kFGF, and bovine basic FGF mRNA. INT-2 homology to FGF is relatively weak, as is its inducing activity in the animal cap assay. kFGF is more similar to and works about as well as bovine bFGF. Unexpectedly, the *in vitro* synthesized bovine bFGF was found to induce α-actin about ten times better than the cognate protein purified from bovine brain. Translated bFGF also induces notochords fairly efficiently, something the brain protein does only rarely. This discrepancy complicates the interpretation of apparent qualitative and quantitative differences between inducers. *Thomas D. Sargent*

Expression of Epi 1, an Epidermis-Specific Marker in *Xenopus laevis* Embryos, is Specified Prior to Gastrulation
C. London, R. Akers, and C. Phillips
Dev. Biol., 129, 380—389, 1988   3-11

In Xenopus embryos the ectoderm opposite the site of sperm entry is destined to form neural ectoderm instead of epidermis. Therefore, ectodermal cells commit to a neural developmental pathway after fertilization. To determine when this commitment can initially be detected, a monoclonal antibody, Epi 1, that recognizes an antigen in epidermal, but not neural cells, was employed.

Blastomeres of eight-celled embryos were isolated and cultured. Most of the cultured animal cells expressed Epi 1, while the vegetal cells did not. Animal blastomeres from stages 6 through 11 were cultured until the midneurula stage and then fixed and immunohistochemically examined. Of the animal explants, 96% contained a discrete region that did not express Epi 1. When stage 10 embryos were examined, 7% of the ventral half had complete expression of Epi 1, while all dorsal-half epitheliums had well-defined regions without expression of this antigen.

A monoclonal antibody, Epi 1, that recognized an epidermal-specific antigen was used to examine early events in the selection of epidermal vs. neural fate. An event between fertilization and the third cleavage division alters the expression of the Epi 1 antigen. The events that control Epi 1 expression are not sufficient for neural commitment, but may establish a prepattern toward neural development.

Signals from the Dorsal Blastopore Lip Region During Gastrulation Bias the Ectoderm Toward a Nonepidermal Pathway of Differentiation in *Xenopus laevis*
R. Savage and C. R. Phillips
Dev. Biol., 133, 157—168, 1989   3-12

Xenopus embryo ectoderm requires interaction with chordamesoderm to develop along a neural pathway. However, the influence of earlier events on this process is not known. Expression of an epidermal specific marker, Epi 1, has been used to demonstrate that presumptive neural ectoderm is biased against expression of the Epi 1 antigen prior to contact with chordamesoderm. The source of this bias was investigated.

When ectodermal explants were removed prior to stage 10 and cultured to stage 14, a 10-cell-wide boundary region of graduated Epi 1 expression separated regions of nonexpression from regions of full expression. When explants were removed from stage 10.25 to 10.5 embryos and cultured, the boundary region was three cells wide. When explants were removed from stage 12 embryos and cultured, there was a sharp border with no region of graduation. Therefore, prior to interaction with chordamesoderm there is an increase in the size of the ectodermal region which does not express Epi 1 and a sharpening of the boundary between expressing and nonexpressing regions. This suggests that dorsal ectoderm may receive a signal through the plane of the epithelial layer at stage 10.5 which affects subsequent developmental steps.

Ventral ectoderm isolated prior to stage 10 expresses Epi 1 over the entire surface. Ventral ectoderm was used in a combination assay with blastopore lips, chordamesoderm, head mesoderm, and endoderm to determine which could inhibit Epi 1 expression. Dorsal mesoderm from stage 10.5 embryos inhibited Epi 1 express on more effectively than dorsal mesoderm from stage 11.5. Dorsal mesoderm had a small inhibitory effect without sharp pattern boundaries. Stage 12.5 anterior and middle mesoderm had little inhibitory effect, while posterior chordamesoderm was a very efficient inhibitor of Epi 1 expression, with a five-cell graduated boundary.

Stage 10 to 10.25 blastopore lip animal-side regions inhibited Epi 1 expression with sharp boundary regions. At stage 11 there was little inhibition, while stage 12 blastopore lips caused strong inhibition. This was not an artifact of lip cell migration.

Both invaginated chordamesoderm and blastopore lips inhibited Epi 1 expression on ventral ectoderm. The amount of inhibition varied with the position and developmental stage of the inhibiting tissue. This suggests that the blastopore lip region establishes a preneural bias in adjacent ectoderm pior to chordamesoderem interaction. This would indicate that dorsalization in Xenopus embryos proceeds by a successive series of events culminating in epidermal and neural plate formation.

Development of Neural Inducing Capacity in Dissociated *Xenopus* Embryos
S. M. Sato and T. D. Sargent
Dev. Biol., 134, 263—266, 1989   3-13

In the Xenopus embryo, dorsal mesoderm induces dorsal ectoderm to form neural tissue. To investigate the production of this neural-inducing signal, dorsal mesoderm differentiation was disrupted by embryo dispersion. The appearance of the neural-specific marker N-CAM was assayed.

Dispersion to single cells was accomplished by gentle agitation in calcium-free medium at stage 7 and continued until stage 10.5. Under these conditions alpha-actin, a muscle-specific marker, was not synthesized. However, N-CAM synthesis was not affected. Therefore, under conditions of complete dorsal mesoderm disruption the neural signal was still produced.

Neural induction occured in dispersed embryos that were reaggregated by early gastrulation. Therefore, loss of intercellular communication and disruption of dorsal mesoderm formation during stages 7 through 10.5 did not interfere with the formation of the neural inducer. This suggests that the ability to induce neural development could be acquired through the inheritance of specialized egg cytoplasm in the region of the embryo that will form dorsal mesoderm.

♦These papers deal with questions about the signals involved in the induction of the central nervous system. London et al. studied the expression in ectodermal explants of Epi 1, an extracellular glycoprotein that appears on nonciliated epidermal cells during gastrulation. By the 8-cell stage dorsal animal blastomeres have been programmed to express this marker poorly, compared to ventral animal cells. Thus the preneural ectoderm has been partially specified as "nonepidermal" long before gastrulation, when classical neural induction occurs. Note that this prepatterning does not include activation of neural-specific markers such as N-CAM or the suppression of epidermal keratin synthesis. They also mention that very small explants of dorsal ectoderm frequently express Epi 1 on most or all of the cells, so this prepatterning may depend on a kind of "community effect" as described for mesoderm specification by Gurdon.

Savage and Phillips show that ventral ectoderm explants, which normally express Epi 1 over their entire surface, can be induced by dorsal blastopore lip (DBL) tissue to form large clearly defined regions devoid of this antigen. As with mesoderm induction, suppression of epidermis, the "default" pathway for ectoderm, is an obligatory aspect of neural induction. London et al. show that this suppression is at least partially autonomous to the dorsal ectoderm, but the latter paper suggests that a signal might be emanating from the blastopore lip and moving laterally through the ectoderm to enlarge and sharpen the de-epidermalized zone. The chordamesoderm is usually considered the source of neural inducing signals, but this tissue works poorly in the Epi 1 suppression assay. The authors allude to unpublished results suggesting the DBL can "enhance and organize" N-CAM induction. If true, then the DBL may have a more direct role in neural induction.

Further questions about the role of mesoderm in neural induction were raised by work done in my laboratory by Sheryl Sato (Sato and Argent, 1989). She showed that although dissociation and dispersion of cleaving embryos through the early gastrula stage blocks the differentiation of muscle, this has no effect on the accumulation of the neural marker N-CAM. Symes et al. showed that notochord and other detectable mesodermal derivatives are also absent in such embryos, so an embryo deprived of all detectable mesoderm can still induce neural tissue. Either the neural tissue is induced by something other than mesoderm, or mesoderm exists but cannot be diagnosed by any of the terminal differentiation products characteristic of its derivatives (i.e., muscle, notochord, mesenchyme, etc.). I favor the latter alternative, and would speculate that a tissue is specified very early in cleavage, perhaps at the one-cell stage, that is able to manufacture neural inducer but cannot differentiate into traditionally recognizable mesoderm in the absence of cell communication prior to gastrulation. *Thomas D. Sargent*

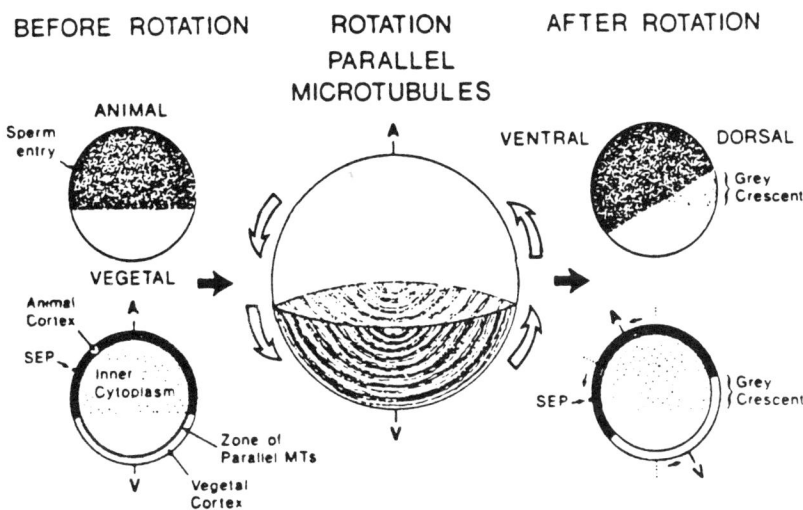

FIGURE 3-14. The cortical/cytoplasmic rotation. Dorsoventral polarity is specified by a 30° rotation of the cortex relative to the cytoplasm. Before rotation, the egg has a pigmented animal half and a nonpigmented vegetal half. A diagrammatic section shows the outer cortex and the inner cytoplasm, with the cortical thickness exaggerated for illustrative purposes. The cortex rotates so that the sperm entry point (SEP) moves vegetally. After rotation, the gray crescent is formed and represents the overlapping of pigmented animal cytoplasm by nonpigmented vegetal cortex. The gray crescent is a visible marker of the dorsal side in *Rana pipiens*, but is not seen in *X. laevis*, although the same cortical/cytoplasmic rotation occurs. During rotation, the zone between the cortex and the cytoplasm has parallel microtubules (MT), described in this paper. (From Elinson/Rowning, *Dev. Biol.*, 128, 185—197, 1988. With permission.)

A Transient Array of Parallel Microtubules in Frog Eggs: Potential Tracks for a Cytoplasmic Rotation that Specifies the Dorso-Ventral Axis
R. P. Elinson and B. Rowning
Dev. Biol., 128, 185—197, 1988   3-14

In frogs, a cytoplasmic rotation during the first cell cycle after fertilization specifies the embryo dorsoventral axis. The mechanism of this rotation is not known, but microtubules have been implicated. The involvement of microtubules in this process was investigated by immunocytochemistry and electron microscopy.

Fertilized eggs were analyzed for the presence of microtubules during the period of cytoplasmic rotation. A pattern of parallel microtubules was detected over the entire vegetal surface of the egg (Figure 3-14). These microtubules were close to the surface and in the same focal plane as the pigment granules. The microtubles could only be detected from 0.5 to 1.0

during the first cycle, which corresponds to the time of cortical rotation. The orientation of the microtubules was parallel to the direction of rotation in most eggs examined. The microtubules were localized to a shear zone approximately 1.4 ± 0.3 μm deep and 1.1 ± 0.4 μm thick.

Parallel microtubules were also detected in eggs activated by electric shock instead of fertilization. When fertilized eggs were treated with microtuble assembly inhibitors or with UV, no parallel cortical microtubules were detected and no cortical rotation occurred.

The rotation of the egg cortex relative to its cytoplasm specifies the embryo dorsoventral axis. This rotation is driven by an active motor. The transient array of parallel microtubules described in this paper may serve as tracks for the movement of the cortex relative to the cytoplasm.

Hyperdorsoanterior Embryos from *Xenopus* Eggs Treated with D_2O
S. R. Scharf, B. Rowning, M. Wu, and J. C. Gerhart
Dev. Biol., 134, 175—188, 1989   3-15

The normal dorsoventral and anteroposterior organizations of the amphibian embryo depend on the amount of cortical rotation after fertilization, Nieuwkoop center in the blastula and Spemann organizer in the marginal zone of the gastrula. Eggs were treated with D_2O, a microtubule stabilizer, to increase the size of their Nieuwkoop center. The effect of this treatment on embryo development was assessed.

Eggs were treated with 20 to 70% D_2O for a few minutes early in the first cell cycle after fertilization. The resultant embryos formed enlarged or twinned dorsal and anterior structures, with compensatory reduction of posterior and ventral structures. The embryos ranged from normal to cylindrically symmetrical with bands of cement glands and eye pigment. The affected embryos were termed hyperdorsoanterior.

Hyperdorsoanterior embryos occurred only when eggs or vegetal halves were treated with D_2O prior to 0.4 of the first cell cycle. Optimal conditions for hyperdorsoanteriorization were 50 to 70% D_2O for 2 to 6 min at 0.25 to 0.28 in the first cell cycle. In these eggs, there was increased rigidity, a precocious and random microtubule network, and reversal of the position of specification of the embryonic dorsal midline. Ventralizing treatments, such as UV and cold shock, antagonized the effect of D_2O.

The effects of D_2O appear to be mediated by egg rotation and response to rotation, such that abnormally large or multiple Nieuwkoop centers form. This results in abnormally wide or multiple Spemann organizers and finally in excessive dorsoanterior development at the expense of ventroposterior development. The results of this study demonstrate that all regions of the amphibian egg are initially capable of dorsal development.

♦About halfway through the first cleavage cycle, the thin cortex of amphibian eggs rotates relative to the underlying cytoplasm. The direction of this rotation determines the dorsalventral polarity of the embryo. This event has tremendous ramifications on development, so its mechanism is of great interest. The concerted rotation also presents an intereting problem in cell biology. Treatment of fertilized eggs with heavy water, D_2O, often (but not always) results in grossly exaggerated dorsal and anterior development. Scharf et al. show that this effect correlates with a precocious and semirandom polymerization of microtubules in the vegetal hemisphere. Normally, vegetal microtubules polymerize just before and during the rotation into a coherent radial pattern with two "hubs" located on the axis of rotation, as shown by Elinson and Rowning. The tubule mesh is located in a layer 1 to 3 μm deep. The shear zone of movement resides in this region, so the work involved in moving the large sphere of egg cytoplasm is generated near the microtubules. How this movement results in embryonic polarization is not known, but redistribution of cytoplasmic or cytoskeleton-associated determinant molecules is an attractive hypothesis. *Thomas D. Sargent*

Protein Kinase C Mediates Neural Induction in *Xenopus laevis*
A. P. Otte, C. H. Koster, G. T. Snoek, and A. J. Durston
Nature, 334, 618—620, 1988

Cell-surface-mediated signal transduction has been implicated in the induction processes of early embryogenesis. If growth factor receptor-type molecules are involved in embryonic induction, it should be associated with protein kinase C (PKC) translocation. Translocation of PKC was investigated during neural induction in *Xenopus*.

Whole embryos, uninduced ectoderm, induced neuroectoderm, and *in vitro*-induced neuroectoderm from stages 10 and 13 were assessed for PKC activity in both soluble and membrane fractions. The period of neural induction was characterized by a threefold increase in membrane-bound PKC activity and decrease in soluble PKC activity. This difference was also detected when uninduced and neuroectoderm were compared.

Uninduced stage 10 ectoderm was incubated with TPA, a potent PKC activator, and neural induction was examined. There was a dose-dependent neural induction response to TPA. Therefore, PKC translocation appears to induce neural differentiation.

Neural development-inducing mesoderm appears to induce translocation of PKC activity from the soluble to the membrane fraction. This indicates that growth factor-like receptors that cause PKC activation are involved in neural induction during Xenopus embryogenesis.

♦ It is well established that induction of the nervous system occurs under the influence of factors present in the chordomesoderm. However, little is known about the molecular nature of the inductive cues. Although the neural "inducer" is thought to be a diffusible molecule, its identification has been confounded by the ability of numerous artificial substances to induce nervous tissue formation.

This report tests the idea that receptor-mediated signal transduction may be an important response to neural induction. The authors measured protein kinase C activity in the ectoderm before and after induction. Protein kinase C translocation from the cytosol to the membrane is known to occur after stimulation of inositol phospholipid turnover by hormones and growth factors. The authors found that an increase in protein kinase C activity and the apparent translocation of the enzyme from the cytoplasm to the cell membrane followed neural induction. This correlation suggests that signal transduction may involve protein kinase C. Further evidence that PKC may play a role in the response to the neural inducer comes from experiments using an artificial activator of protein kinase C. When phorbol ester, a potent activator of PKC, was added to uninduced ectoderm, translocation of PKC from the cytosol to the membrane was observed. In addition, neural tissue was observed in the explants. These results demonstrate that stimulation of PKC activity can artificially trigger an induction-like response in the absence of the natural inducer. This suggests that ectodermal cells have an inherent response program that is turned on by induction involving stimulation of membrane-bound PKC.—*Marianne Bronner-Fraser*

Chemotropic Guidance of Developing Axons in the Mammalian Central Nervous System
M. Tessier-Lavigne, M. Placzek, A. G. S. Lumsden, J. Dodd, and T. M. Jessell
Nature, 336, 775—778, 1988   3-17

Axon guidance in the developing nervous system depends on the recognition of extracellular matrix and cell-surface cues, but may also depend on chemoattractant gradients released from intermediate or final targets. During spinal cord development, commisural axons are deflected toward the floor plate. To investigate the release of chemoattractant molecules, rat dorsal spinal cord explants were cultured with floor plate explants or conditioned media.

Dorsal explants cultured alone for up to 44 h had little or no outgrowth. Dorsal explants cultured 100 to 400 µm from day 11 floor plate explants had axon growth within 20 h. After approximately 39 h, these axons projected to the floor plate explant. In the presence of conditioned media, axon projection occurred, but the direction was random. The projecting axons were identified as commisural as they all expressed the TAG-1 glycoprotein.

These results demonstrate that the intermediate target, the floor plate, secretes diffusible factors that are selective chemoattractants for embryonic commisural axons. Therefore, chemotropism may be an important general mechanism of axon guidance in the central nervous system.

♦ The formation of connections in the nervous system depends on the proper outgrowth and projection of neuronal processes. Axons typically extend over long distance and follow characteristic trajectories to reach their targets. There are three possible mechanisms whereby an axon could be "guided" to its target: cell-cell interactions, cell-matrix interactions, or diffusible interactions. Various cell-cell and cell-matrix interactions have been found to operate during some aspects of neurite outgrowth. Because neuronal cell bodies are often distant from their targets, it is logical that some distant interactions also may be involved in axonal guidance. However, only a few examples of "chemotropism", all involving peripheral neurons, have been found.

This paper provides the first demonstration of a chemotropic agent in the central nervous system. In the spinal cord of the mouse, the authors have found that the floor plate, the ventral mid-portion of the neural tube, is a chemoattractant for commissural neurons in culture. Commissural neurons reside in the dorsal neural tube and project ventrally. After their processes traverse the floor plate, their surface properties are altered (from expressing TAG to expressing L1 glycoprotein) and they turn 90°. In this study, Tessier-Lavigne and colleagues have cultured the dorsal region of the neural tube, which contains commissural neurons, together with the ventral portion of the neural tube, which contains the floor plate. They find that the commisural axons will grow out of the explant and course toward the explanted floor plate in an oriented fashion. In contrast, other dorsal nural tube neurons are not attracted by the loor plate. This chemotropic efect was not mimicked by NGF or laminin.

This is an important finding because it represents the first example of chemotropism affecting a central population of neurons. Although it remains to be shown whether similar chemotropism functions *in vivo*, this *in vitro* system provides a useful assay for a diffusible attractant in the central nervous system. *Marianne Bronner-Fraser*

Purification and Characterization of Mouse Hematopoietic Stem Cells
G. J. Spangrude, S. Heimfeld, and I. L. Weissman
Science, 241, 58—62, 1988

To completely understand the developmental biology of the hematolymphoid system, the self-renewing pluripotent hematopoietic stem cells must be identified and isolated. Monoclonal antibodies that bind to surface differentiation antigens were used to remove differentiated cells from murine bone marrow. The remaining cells, termed Thy-1-lo Lin-, were enriched in clonal progenitors for spleen colonies, but remained heterogeneous. An antibody to an additional marker, Sca-1, was used to further subdivide this population.

Of the Lin- cells, 20 to 30% were Sca^+. From these Sca^+ cells, one splenic colony was observed per 10 i.v.-transferred cells. There was a tenfold enrichment in day 12 CFU-S vs. day 8 CFU-S. Limit dilution analysis indicated that at least one in five of these cells responded to the thymic microenvironment.

The Sca^+ cells were capable of multilineage reconstitution in lethally irradiated mice. Only 30 of these cells were necessary to rescue 50% of lethally irradiated mice. This represents a 1000-fold enrichment of progenitor cells as compared to bone marrow.

The Thy-1-lo Lin- Sca^+ cells identified and isolated in this study represent a pure population of mouse bone marrow hematopoietic multipotent stem cells.

♦ The search for the elusive hematopoietic stem cell has occupied scientists since the first demonstration that bone marrow transfers protected lethally irradiated animals. Intriguingly, it was shown that limiting numbers of bone marrow cells gave rise to colonies of myeloid-erythroid cells in irradiated individuals (Till and McCulloch, *Rad. Res.*, 14, 213) 1961) providing evidence for the existence of a pluripotential stem cell. Through the use of a variety of fractionation techniques, the putative pluripotential stem cell had been purified at least 200-fold as judged by formation of spleen colonies (Visser et al., *J. Exp. Med.*, 159, 1576, 1984). The main weakness of these earlier studies was that only splenic colony formation was analyzed and, therefore, whether or not a pluripotential stem cell was in fact purified could not be determined. A paper in the past year provides the best evidence to date, not only for the existence of such a pluripotential stem cell, but also that it may be possible to isolate the stem cell for use in clinical situations. This paper describes the use of monoclonal antibodies to purify a population of cells which posseses many of the characteristics one would expect for the pluripotential stem cell. Importantly, these purified cells were not only examined for splenic colony formation, but simultaneously for their ability to give rise to T and B cells. The authors estimate that they purified the stem cells at least 1000-fold. *Charles Snow*

Stimulation of B-Cell Progenitors by Cloned Murine Interleukin-7
A. E. Namen, S. Lupton, K. Hjerrild, J. Wignall, D. Y. Mochizuki, A. Schmierer, B. Mosley, C. J. March, D. Urdal, S. Gillis, D. Cosman, and R. G. Goodwin
Nature, 333, 571—573, 1988 3-19

A long-term lymphoid progenitor culture system was used to detect growth factors. A cDNA library was prepared from stromal cell line IXN/A6 and expressed in COS-7 cells. Biological activity was assessed from culture supernatants.

The positive pool was subcloned until a single positive clone, 1046, producing a growth-factor activity termed IL-7 was identified. Clone 1046 had an open reading frame of 462 bp. It contained a 25-amino acid leader sequence followed by 129 amino acids, with a predicted Mr = 14.9K. Two potential N-linked glycosylation sites existed at amino acid 69 and 90. There were six cysteine residues. Northern blots of polyA⁺ RNA from several mouse tissues were screened for 1046 expression. Both quantity and message size were tissue specific. The activity of this growth factor was 4×10^6 U/µg.

A novel growth factor, IL-7, which stimulates the proliferation of lymphoid precursors has been cloned and isolated.

Human Interleukin 7: Molecular Cloning and Growth Factor Activity on Human and Murine B-Lineage Cells
R. G. Goodwin, S. Lupton, A. Schmierer, K. J. Hjerrild, R. Jerzy, W. Clevenger, S. Gillis, D. Cosman, and A. E. Namen
Proc. Natl. Acad. Sci. U.S.A., 86, 302—306, 1989 3-20

Pro-B cells are committed to the B cell lineage, but have germ-line configuration of immunoglobulin genes. These cells differentiate into pre-B cells, which express the B220 antigen, and rearrange and express the heavy chain gene. Recombinant murine IL-7 supports extended growth of both pro- and pre-B cells. An IL-7 cDNA was used to isolate a homologous human clone from a human hepatoma cell line cDNA library.

The human cDNA was capable of encoding a protein of 177 amino acids, with a signal sequence of 25 amino acids. The predicted molecular mass was 17.4 kDa. There were three potential sites for N-linked glycosylation and six cysteines.

The human and murine sequences were 81% homologous in the coding region, 73% in the 5' noncoding region and 63% in the 3' noncoding region. Human IL-7 contained an insert of 19 amino acids, 96 to 114, not found in the murine sequence.

The human cDNA was expressed in COS-7 cells and was active in the murine pre-B cell proliferation assay. Human bone marrow cells were enriched for B-lineage cells on Percoll gradients. The recombinant human protein was active in this proliferation assay. Murine IL-7 was not able to induce proliferation in the human bone marrow assay.

A murine cDNA for IL-7 was used to isolate the corresponding human gene. The cloning and identification of the human gene will be useful for studies of hematopoiesis and regulation of the human immune system.

In Vitro Effects of Recombinant Interleukin 7 on Growth and Differentiation of Bone Marrow Pro-B- and Pro-T-Lymphocyte Clones and Fetal Thymocyte Clones
S. Takeda, S. Gillis, and R. Palacios
Proc. Natl. Acad. Sci U.S.A.,, 86, 1634—1638, 1989   3-21

To further examine the biological activity of IL-7, its effect on growth and differentiation of marrow pro-B clones (CB/Bm7, LyD9, LyB9), marrow pro-T clones (C4-77/3, C4-86/18, C4-95/16) and fetal thymocyte clones (FTH5, FTA2, FTD5) in the presence or absence of bone marrow stromal clone RP.0.10 was investigated.

Recombinant IL-7 alone stimulated DNA synthesis, but not growth of pro-B clones. The RP.0.10 cells did not support growth of pro-B cells. The combination of stromal cell monolayers and IL-7, stimulated growth in cultured pro-B cells. RP.0.10 cells in combination with LPS and IL-3 induced the pro-B cells to differentiate into IgM⁺ B cells. When IL-7 was added, there was a significant increase in IgM⁺ B cells, to 63%. In the absence of LPS, 20% of the cells differentiated. Cell-to-cell contact between the stromal cells and pro-B cells was necessary for rearrangement and expression of the immunoglobulin genes in the pro-B cells. Neither IL-7 nor RP.0.10 supported growth or proliferation of pro-T clones.

The interaction of stem cells with stromal cells and IL-7 may play an important role in the committment to the B-lymphocyte pathway.

Interleukin 7 (Murine Pre-B Cell Growth Factor/Lymphopoietin 1) Stimulates Thymocyte Growth: Regulation by Transforming Growth Factor Beta
D. Chantry, M. Turner, and M. Feldmann
Eur. J. Immunol., 19, 783—786, 1989   3-22

IL-7 promotes the growth of B cell precursors. The effect of this factor on the proliferation of murine thymocytes was compared to IL-1, a known thymocyte growth factor.

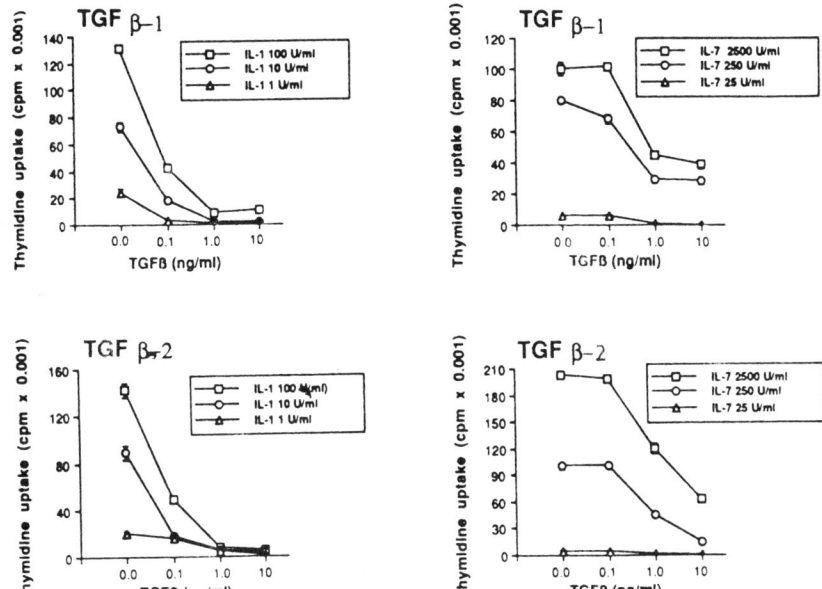

FIGURE 3-22. TGF-β inhibits thymocyte proliferation. Thymocytes were cultured with different concentrations of IL1β (100, 10, 1 U/ml) or IL7 (2500, 250, 25 U/ml) in the presence of increasing concentrations of TGF-β1 or TGF-β2 (10, 1.0, or 0.1 ng/ml). Proliferation was determined after 72 h. (From Chantry et al., *Eur. J. Immunol.*, 19, 783—786, 1989. ©VCH Verlagsgesellschaft mbH, D-6940 Weinheim.)

Thymocytes were cultured with IL-7 and there was a dose-dependent proliferative response, similar to that induced by IL-1. There was synergy between suboptimal doses of IL-1 and IL-7. The proliferative response to IL-1 and IL-7 was inhibited in a dose-dependent manner by TGF-beta (Figure 3-22). However, IL-7-induced proliferation was less sensitive to this inhibition, implying different activation pathways or action on different thymocyte subsets.

Murine IL-7 is a potent co-mitogen for murine thymocytes *in vitro*. It may play a role in interplay of stimulatory and inhibitory factors in the thymic microenvironment that leads to the committment to T cell differentiation.

Recombinant Interleukin 7, Pre-B Cell Growth Factor, Has Costimulatory Activity on Purified Mature B Cells

P. J. Morrissey, R. G. Goodwin, R. P. Nordan, D. Anderson, K. H. Grabstein, D. Cosman, J. Sims, S. Lupton, B. Acres, S. G. Reed, D. Mochizuki, J. Eisenman, P. J. Conlon, and A. E. Namen

The activation of resting T cells requires crosslinking of the antigen-specific receptor by its ligand and another signal which can be provided by IL-1 or IL-6. The pre-B cell growth factor, IL-7, was tested for this activity.

In the presence of ConA, IL-7 induced the proliferation of highly purified $CD4^+$ and $CD8^+$ T cells. This activity was accompanied by the induction of IL-2 receptors and IL-2 production. An anti-IL-2 MAb was added to the culture and inhibited the proliferative response to IL-7, indicating that this response is dependent on IL-2. Anti-IL-6 antibodies did not inhibit the prolifertive response to IL-7, indicating that IL-7 does not act through IL-6, but acts directly to stimulate T cell activation.

IL-7, initially defined as a potent growth stimuator for pre-B cells, is a potent costimulator of T cell activation.

♦ If the current concept of hematopoiesis is correct, a pluripotential stem cell releases daughter cells which become more differentiated stem cells giving rise to more restricted numbers of cells. Therefore, there must exist a stem cell which gives rise to only pre-B cells and pre-T cells, a lymphoid stem cell. A report in March of 1988 (Namen et al., *J. Exp. Med.,* 167, 988, 1988) described the purification, from SV 40-transformed bone marrow cells, of a factor (referred to at that time as lymphopoietin-1) which promoted the growth of pre-B cells in Whitlock-Witte cultures. In the past year, a complementary DNA clone for this factor has been isolated and this cDNA utilized to clone the human gene.

This factor has been shown to be distinct from all other isolated growth factors and has been labeled interleukin 7 (see Namen et al., *Nature,* 333, 571, 1988). What has become increasingly clear over the past several years is that lymphokines are pleiotropic in their activities and redundant in some of their functions with at least one other factor (see Paul, *Cell,* 57, 521, 1989). Over the past year, this characteristic of growth factors in general has been shown to apply to IL-7. Recombinant IL-7 was shown to support both the growth and differentiation of pro-B cells (pre-pre—B cells which have both the heavy and light chain genes still in germline configuration) provided that cloned bone marrow stroma cells were included within the cultures. This report found no evidence for IL-7 supporting either the growth or the differentiation of clones of pro-T cells or fetal thymocytes. However, recombinant IL-7 was shown to promote the proliferation of mitogenically stimulated murine thymocytes every bit as well as IL-1. In addition, IL-7 was shown to support the expansion of highly purified mature T cells to a mitogenic stimulus as well as IL-1 or IL-2. A more detailed understanding of all the activities of this new interleukin will most probably, be shortly available. *Charles Snow*

Clonal Deletion of B Lymphocytes in a Transgenic Mouse Bearing Anti-MHC Class I Antibody Genes
D. A. Nemazee and K. Burki
Nature, 337, 562—566, 1989   3-24

It is not known if B-cell tolerance is involved in natural tolerance. B-cell tolerance was examined in transgenic mice using genes for IgM anti-H-2-k MHC class I antibody. The founder transgenic mouse was homozygous for the H-2-d haplotype, had a high anti-H-2-k cytotoxic antibody titer in the serum, and produced the 3-83 idiotype.

Immunofluorescence studies demonstrated that the 3-83 idiotype was present only on B cells in the transgenic H-2-9/d mice. The transgenic mice were mated to heterozygous (H-2-d-/H-2-k) mice and three large litters were examined. The nontransgenic mice in these litters did not express 3-83. H-2-d/d transgenic mice had high serum concentrations of the 3-83 idiotype. The 3-83 idiotype was not detected in transgenic H-2-d/k littermates. Immunofluorescence of frozen sections indicated that the anti-self B cells could not be detected in the spleen and bone marrow of these mice.

These results demonstrate that the anti-self (anti-3-83 idiotype) producing B cells in the H-2-d/k mice were elminated or excluded from the spleen during development. Therefore, B cells appear to become tolerant in the presence of antigen even in the absence of antigen-reactive T cells.

♦This paper utilizes transgene animals to study the question of B cell tolerance. A considerable amount of controversy has existed concerning the possibility that clonal deletion participates in the process of B cell tolerance. Much of the more recent evidence has sided with the idea that many autoreactive clones are present throughout the lifespan of individuals and they are unable to respond due to the lack of appropriate helper T cells. The transgene experiments directly demonstrate that some self-reactive B cell clones are in fact deleted from the repertoire. *Charles Snow*

B-Cell Memory is Short-Lived in the Absence of Antigen
D. Gray and H. Skarvall
Nature, 336, 70—73, 1988   3-25

The interaction between antigen and specific B cells results in differentiation into antibody-producing plasma cells and memory cells. It is not known if maintenance of memory cell populations requires further antigen exposure. To investigate the *in vivo* half-life of memory B cells in the absence of antigen, thoracic duct lymphocytes from immunized K1b rats were transferred into sublethally irradiated congenic K1a rats. Reimmunization with antigen occured at 1, 3, 6, or 12 weeks and the donor (memory) response was assessed.

The memory cell response was rapid at 1 week, decreased by 3 weeks, and was minimal by 12 weeks after initial antigen exposure.

Memory B cell population maintenance requires continued antigen stimulation.

♦Although this may not be within the scope of our topic, a paper published this year provided the most interesting new information concerning memory B cells seen in a long time. The prevailing opinion on B cell memory maintains that certain daughter cells formed during an active immune response become long-lived cells which are responsible for B cell memory. This new report suggest that this may not be the case. Utilizing an adoptive transfer system, the authors demonstrate that "memory" B cells remain only as long as specific antigen is maintained within the system in a form which remains stimulatory to the B cells. In the absence of specific antigen the memory response is rapidly lost. These results may require a complete rethinking concerning our concepts of B cell memory. *Charles Snow*

Antibodies to CD3/T-Cell Receptor Complex Induce Death by Apoptosis in Immature T Cells in Thymic Cultures
C. A. Smith, G. T. Williams, R. Kingston, E. J. Jenkinson, and J. T. Owen
Nature, 337, 181—184, 1989   3-26

During development, T cells carrying anti-self receptors are eliminated through a clonal deletion process that is not understood. To determine whether antibodies to the invariant T cell receptor complex component CD3 could trigger cell death in immature T cells, organ cultures of fetal mouse thymus were incubated with anti-CD3 for 18 h.

Anti-CD3 treatment reduced cell yield by approximately 45%. Degradation of DNA occurred. Toluidine-blue stained sections and cytospins demonstrated apoptotic thymocytes. Electron microscopy indicated chromatin condensation and cell shrinkage after anti-CD3 treatment. This indicates that anti-CD3 treatment, which stimultes mature T cells, induces apoptosis in immature T cells.

Treatment with PMA and a calcium ionophore can mimic plasmamembrane stimulation in many cell types by increasing intracellular calcium levels. Treatment of immature thymocytes with PMA and ionophore, or with the ionophore alone, induced changes characteristic of apoptosis.

Treatment of immature thymocytes with antibody to CD3, the invariant component of the T cell receptor, induces apoptosis.

♦ A possible mechanism for the deletion of self-reacting T cell clones has recently been described. If self-reacting clones need to be selectively destroyed one would predict that the mechanism involved in the deletion of these cells should utilize the TCR. This recent paper provides evidence that signaling through the T cell receptor at a crucial time(s) early during the cell's development initiates an endogenous pathway of apoptosis. *Charles Snow*

Myeloid Leukemia Inhibitory Factor Maintains the Developmental Potential of Embryonic Stem Cells
R. L. Williams, D. J. Hilton, S. Pease, T. A. Wilson, C. L. Stewart, D. P. Gearing, E. F. Wagner, D. Metcalf, N. A. Nicola, and N. M. Gough
Nature, 336, 684—687, 1988   3-27

Maintenance of the totipotent phenotype in cultures of embryonic stem (ES) cells requires a soluble factor, differentiation inhibitory factor (DIA). Myeloid leukemia inhibitory factor (LIF), which induces differentiation in M1 myeloid leukemia cells, appears to be similar to LIF in size and glycosylation level. Purified recombinant LIF was substituted for DIA in the maintenance of totipotent ES cultures.

Labeled-LIF demonstrated specific high-affinity binding to cells from several ES lines, with a dissociation constant of approximately 90 pM. A concentration of 1000 to 5000 U/ml was sufficient to maintain the stem cell phenotype in ES cultures. After 22 generations of culture with DIF, ES cells were injected into blastocysts and contributed to many tissues of chimeric progeny mice.

Therefore, LIF, which induced differentiation of M1 cells, prevented differentiation of ES cells. Based on their similarities in composition and activity, it is possible LIF is the same molecule as DIA.

Inhibition of Pluripotential Embryonic Stem Cell Differentiation by Purified Polypeptides
A. G. Smith, J. K. Heath, D. D. Donaldson, G. G. Wong, J. Moreau, M. Stahl, and D. Rogers
Nature, 336, 688—690, 1988   3-28

Cultures of ES cells are maintained in a pluripotent state by 10 ng/ml

DIA, which suppresses their differentiation. This factor is a single chain glycoprotein of apparent Mr = 43,000, due to extensive glycosylation of a core Mr = 20,000 peptide. While investigating the activity of DIA in other systems, DIA was found to support growth of DA-1a MoMulv-induced leukemia cells. A cDNA clone for a factor with similar activity, human interleukin for DA cells or HILDA, was isolated and compared to DIA.

The sequence of the HILDA cDNA was identical to that of human LIF. Therefore, the HILDA cDNA was subcloned into an expression vector and transferred to COS cells. Medium conditioned by the transfected COS cells was capable of supporting nondifferentiated ES cells. The activity from the conditioned media was purified. Its apparent Mr = 45,000 and treatment with N-glycanase reduced it to Mr = 20,000. ES cells cultured in 10 ng/ml recombinant HILDA retained stem cell morpholgy. Competition studies indicated that DIA and HILDA/LIF interacted with the same receptor system.

DIA and HILDA/LIF are related multifunctional regulatory factors with biological activity in early embryogenesis and in hematopoietic stem cell systems.

Leukaemia Inhibitory Factor is Identical to the Myeloid Growth Factor Human Interleukin for DA Cells
J.-F. Moreau, D. D. Donaldson, F. Bennett, J. Witek-Giannotti,
S. C. Clark, and G. G. Wong
Nature, 336, 690—692, 1988   3-29

A human cDNA was isolated which encoded a factor, HILDA, which supported the proliferation of the murine leukemic cell line, DA-1a. This factor was found to be identical to leukemia inhibitory factor (LIF).

Lectin-stimulated C10-MJ2 T cells secrete a growth factor that supports the growth of DA-1a cells. Functional expression cloning in mammalian cells was used to isolate a cDNA, encoding this factor from a C10-MJ2 library. The gene was sequenced and it encoded a 202-amino acid protein that was identical to the predicted sequence of human LIF. Northern analysis identified a predominant 4 kb and a minor 1.8 kb mRNA species in C10-MJ2 cells. Recombinant HILDA was expressed in cos cells and displayed the size heterogeneity typical of glycoproteins.

These results demonstrate that the M1 differentiation factor LIF can serve as a growth factor for a myeloid leukemic line, DA-1a. Therefore, the same factor appears to be involved in regulating both proliferation and differentiation. The distinction between growth-promoting and differentiation-inducing activities appears to depend more on the target cell type than on the factor.

♦ *In vitro* differentiation of embryonic stem (ES) cells as well as embryonal carcinoma (EC) cells can be inhibited by culturing on embryonic fibroblast feeder layers. If cultured in the absence of feeders, differentiation occurs. Recently, it was found that Buffalo rat liver cells produce a factor, known as differentiation inhibitory activity (DIA), which suppresses spontaneous differentiation of ES or EC cells *in vitro*. The use of this conditioned medium removed the requirement for feeder-dependent growth, thus making it easier to manipulate the cells *in vitro*. In fact, ES cells have been propagated for over 30 generations in medium supplemented with purified DIA without overt differentiation. The reports listed above show that DIA and myeloid leukemia inhibitory factor (LIF) appear to be the same molecule. They both have similar biochemical and molecular characteristics, and both interact with the same cell-surface receptor system expressed by murine ES cells. In addition, like DIA, long-term maintenance of ES cell lines in LIF does not appear to alter their growth characteristics or their ability to contribute to all somatic tissues as well as the germ-line of blastocyst-injection chimeras. LIF is known to be a cytokine that induces macrophage differentiation of the M1 myeloid leukemia cell line, but in ES cells it has the opposite effect which is one of inhibiting differentiation. For the long term, the elucidation of the different intracellular signalling pathways which appear to interact with the same LIF receptor is an important goal. For the short term, the availability of purified, recombinant LIF will facilitate the generation and culture of ES cells which, with the advent of gene targeting experiments, is an important finding. *Terry Magnuson*

Expression of the FGF-Related Proto-Oncogene *int-2* during Gastrulation and Neurulation in the Mouse
D. G. Wilkinson, G. Peters, C. Dickson, and A. P. McMahon
EMBO J., 7, 691—695, 1988   3-30

The proto-oncogene, *int-2*, which is homologous to members of the fibroblast growth factor family, is expressed in early mouse embryos. Northern blot analysis and *in situ* hybridization (using the probes described in Figure 3-30) were used to investigate the developmental expression pattern of *int-2*.

Northern blots of mRNA from undifferentiated ES cells demonstrated low levels of *int-2* expression. Upon differentiation, expression of *int-2* inceased. *In situ* hybridization localized this expression to outer flattened cells surrounding inner clumps of apparently undifferentiated cells. The outer cells resembled parietal endoderm morphologically and expressed parietal endoderm markers. *int-2* probes were hybridized to mouse

FIGURE 3-30. Relationship between *int*-2 probes and the *int*-2 gene. The *int*-2 gene is iluustrated. Noncoding and coding exons are represented as open and hatched boxes, respectively. The jagged 5′ box is to indicate that alternative upstream sequences are utilized as exons in transcripts initiated at the promoter regions P1 and P2. Alternative poly(A) addition sites are indicated as A1 and A2. (From Wilkinson et al., *EMBO J.*, 7(3), 691—695, 1988. ©IRL Press, Oxford, England.)

embryo sections and expression was detected in parietal but not visceral endoderm. Therefore, *int-2* is expressed at low levels in the early postimplantation embryo. Following differentiation, *int-2* expression is increased in the parietal endoderm.

At 7.5 d, *int-2* transcripts were detected in cells migrating through the primitive streak. By 8 d, expression was detected in the extraembryonic mesoderm surrounding the exocoelomic cavity. After formation of a continuous mesodermal epithelium at 8.5 d, expression decreased. Therefore, *int-2* expression in extraembryonic mesoderm is correlated with migration.

At day 8.5 to 9.5, expression of *int-2* was detected in the neuroepithelium of the developing myelencephalon of the hindbrain, in cells adjacent to the otocytes. Expression was also detected in the lower surfaces of the endoderm of the pharyngeal pouches at 9.5 d.

There is a complex pattern of *int-2* expression in the developing mouse. *int-2* appears to be expressed at a low level in early development. As development progresses, *int-2* expression increases in specific cell types.

Multiple RNAs Expressed from the *int-2* Gene in Mouse Embryonal Carcinoma Cell Lines Encode a Protein with Homology to Fibroblast Growth Factors
R. Smith, G. Peters, and C. Dickson
EMBO J., 7, 1013—1022, 1988   3-31

The *int-2* proto-oncogene is a member of the fibroblast growth factor family, and is expressed during embryogenesis and in some mammary tumors. The most consistent expression pattern includes four sizes of mRNA: 2.9, 2.7, 1.8, and 1.6 kb. The structure and relationship of these RNAs was investigated in two ES cell lines, F9 and PCC4.

After induction, F9 cells express 2 *int-2* RNAs of 2.7 and 1.6 kb, while

PCC4 cells express all four size classes with 2.9 and 1.8 kb mRNAs predominant. RNase protection assays were used to dissect the structure of these RNAs. The four size classes of RNA initiate at heterogeneous cap sites within two distinct promoter regions. The downstream promoter is located in an intron in the transcripts that initiate at the upstream promoter. The second and third intron boundaries are the same in all transcripts. Two different polyadenylation signals, separated by 1100 bp, are used. However, all four classes of RNA encode the same protein.

Variation in cap site, polyadenylation sites, and splicing of the first intron are responsible for the structural complexity of the four message classes of int-2 mRNA. However, despite this complexity, all mRNAs encode the same product. The relevance of the different message types in regulation of expression is not known.

Four Classes of mRNA are Expressed from the Mouse *int-2* Gene, a Member of the FGF Gene Family
S. L. Mansour and G. R. Martin
EMBO J., 7, 2035—2041, 1988   3-32

Mouse embryos and endodermal cells derived from ES cells express four *int-2* mRNAs. To determine their structure, a cDNA library was prepared from PSA-1 endodermal cell mRNA and *int-2* cDNAs were isolated. The structure of the *int-2* mRNAs produced by these cells was analyzed by restriction mapping, Northern blot hybridization, and primer extension analysis.

As in the previous paper, two alternate start sites and two alternate polyadenylation sites were detected. However, alternate splicing was not detected in these cells, and a different start site for transcription was detected.

These results suggest that there are at least three different promoters that can control *int-2* expression and that all can function in the same cell type. All of the mRNAs analyzed so far can produce the same protein. The function of these alternative mRNAs remains unknown.

Expression Pattern of the FGF-Related Proto-Oncogene *int-2* Suggests Multiple Roles in Fetal Development
D. G. Wilkinson, S. Bhatt, and A. P. McMahon
Development, 105, 131—136, 1989   3-33

Normal expression of the proto-oncogene *int-2* is restricted to the mouse embryo. The expression of this gene during gastrulation and in the early-somite stage embryo has previously been characterized. The pattern of *int-2* expression during the rest of embryogenesis was examined through Northern blotting and *in situ* hybridization.

At 14.5 d, *int-2* transcripts were detected in the cerebellum, in cells adjacent to the ventricular zone in lateral sections. By 16.5 d, *int-2* was detected in cells distal to the ventricular zone in lateral regions, but in medial sections expression was adjacent to the lateral zone. This pattern suggested that *int-2* was being expressed in Purkinje cells. Immunocytochemistry confirmed that *int-2* was expressed in the Purkinje cells during this period. By 2 weeks postnatally, *int-2* expression could no longer be detected in the Purkinje cells.

At 14.5 d, *int-2* expression was also detected in the posterior retina and over the subsequent 2 d retinal expression increased. In the newborn mouse, *int-2* expression was only detected in the anterior periphery of the retina. In the mature retina, *int-2* was not detected.

At 14.5 d, *int-2* expression was also detected in the tooth mesenchyme. At 17.5 d, expression was detected in mesenchymal cells destined to form dentine and pulp. At 10.5 d, *int-2* expression was detected in the region of the otic vesicle that will form sensory tissue of the vestibular organ. At 17.5 d it was expressed only in the developing sensory regions of the inner ear.

The complex spatial and temporal pattern of *int-2* expression suggests multiple roles for the product of this gene during murine fetal development.

♦ The accumulation of *int*-2 transcripts in mammary tumors as a consequence of integration of the mouse mammary tumor virus suggests that the *int*-2 protein can induce mammary tumor formation if expressed. Sequence conservation (40 to 60% amino acid sequence similarity) suggests that *int*-2 belongs to a growing family of genes (at least five members have been identified) related to the fibroblast growth factors (FGFs). Although basic and acidic FGFs are known to have a variety of mitogenic functions, the normal role of *int*-2 in mouse is not clear. The above papers by Wilkinson and co-workers begin to address this question by examining tissue-specific expression using the techniques of *in situ* hybridization. These investigators show that *int*-2 transcripts accumulate within the parietal endoderm of extraembryonic structures, and in the developing embryo beginning with early gastrulation. The gene is active in newly formed migrating mesoderm but not in more organized mesodermally derived derivatives. In addition, *int*-2 transcripts have been found in early stages of Purkinje cell differentiation, in neuroepithelial cells of the hindbrain adjacent to the otocyst, and on the lower surface of endodermally derived pharyngeal pouches. Subsequently, the gene was found to be expressed in developing sensory regions of the inner ear (ampullae of semicircular canals, utricle, saccule, and cochlea), in the developing neuroblastic layer of the retina, and finally in tooth mesenchymal cells destined to form dentine and pulp of the incisor and first molar teeth. Based on these descriptive results, these authors suggest that *int*-2 may have multiple roles in the developing embryo. For example, the gene may be acting as an autocrine factor maintaining and possibly initiating the migratory state of mesoderm. Such a role, if it exists, must be selective since other migratory populations of cells, for example the neural crest cells, do not accumulate *int*-2 RNA. Moreover, expression of *int*-2 in the epithelial endoderm of the pharyngeal pouches and neural ectoderm of the hindbrain, as well as other areas, seems unrelated to areas of cell migration. Thus, it has been suggested that the gene may also function as an inducer of differentiation, much in the same way as has been suggested in *Xenopus* embryos where basic FGF can act as a morphogen, inducing differentiation of dorsal ectoderm to mesoderm.

The accompanying papers by Smith and co-workers and by Mansour and Martin show that teratocarcinoma endoderm cells express at least four different transcripts. However, it appears that the differences between the transcripts are localized to untranslated portions and that all four transcripts code for the same product, an FGF-related protein of 27,000 Da. The data from the two studies indicate that *int*-2 expression can be controlled by at least three different promoters and two different polyadenylation sites. The details of structures of *int*-2 transcripts in embryonic cells are not available, and it remains to be determined which of the three promoters are active in embryos. Although it is intriguing to suggest that different combinations of int-2 mRNA classes may be expressed in different cell types in the embryo (an issue that will almost certainly be resolved in the coming year by *in situ* hybridization studies of sections of mouse embryos employing probes that distinguish the different classes of *int*-2 mRNAs), this would not explain how the same protein is functioning in presumably different roles in different tissues of the developing embryo. An important question in this regard is the identification of the receptor(s) to which the protein is binding and an elucidation of the intracellular pathways that are activated by the binding of the growth factor. *Terry Magnuson*

Evidence for the Involvement of the Proto-Oncogene *c-mos* in Mammalian Meiotic Maturation and Possibly Very Early Embryogenesis
G. L. Mutter, G. S. Grills, and D. J. Wolgemuth
EMBO J., 7, 683—689, 1988   3-34

mRNA for *c-mos* accumulates and is stored in mature oocytes. The function of this message was investigated by Northern blots and *in situ* hybridization to determine its fate during subsequent development.

In oocytes undergoing meiotic maturation, *c-mos* transcript levels decreased rapidly. By the 2-cell stage, *c-mos* mRNA was not detectable by *in situ* hybridization. It remained undetectable through the blastocyst stage. Oocytes were cultured and c-mos mRNA decreased 18 to 43% after 7 to 17 h. This is the time of progression from metaphase I to metaphase II in the developing oocyte.

These results suggest a role for the product of the *c-mos* gene in the meiotic maturation of mammalian germ cells.

♦Mammalian oogenesis is a difficult process to study due to small amounts of readily obtainable sample material. Furthermore, identification of mutations that affect this process is complicated by the potential for loss of reproductive potential. The approach taken by the investigators listed above was a direct extension of earlier work on spermatogenesis where certain cellular oncogenes (c-*abl*, *int-1*, several members of the *ras* family, and c-*mos*) have been found to exhibit unique patterns of expression. These genes are of interest because of their proposed role in regulating cell growth, differentiation and proliferation. A 1.4-kb c-*mos* transcript was shown to accumulate in cytoplasm of growing oocytes, with levels beginning to drop after resumption of meiotic maturation. At the time of ovulation, a 1.65-kb transcript was observed to replace the smaller transcript detected in growing oocytes. The increase in size is likely to be due to polyadenylation of the smaller transcript. Transcript levels dropped below limits of sensitivity by the 1- to 2-cell stage of embryonic development, and the gene product was not seen again throughout the preimplantation stages. The drop in C-*mos* RNA coincides with a general pattern of oocyte mRNA turnover described by others. Although the c-*mos* protein is a putative serine kinase, nothing is known about its role in growing oocytes. It is possible that the c-*mos* gene product could play a role in the resumption of meiosis where both phosphorylation and dephosphorylation of several proteins occur. *Terry Magnuson*

Expression of Human Proteoglycan in Chinese Hamster Ovary Cells Inhibits Cell Proliferation
Y. Yamaguchi, E. Ruoslahti
Nature, 336, 244—246, 1988   3-35

Decorin is an extracellular matrix proteoglycan produced by fibroblasts. To assess the function of this protein, decorin was stably transfected into dihydrofolate reductase-deficient CHO cells in a system that allows amplification of transfected sequences upon exposure to increasing concentrations of methotrexate.

No decorin was expressed in nontransfected CHO cells. The highest level of expression in transfected and amplified cells was 25 pg decorin per cell per day. The amplified decorin-expressing cells spread out more on the substrate and formed a more even monolayer than CHO cells that did not express decorin. The expressing cells were threefold larger than control cells. This increase in area was proportional to the amount of decorin synthesized. The level of decorin expression was inversely correlated with the saturation density of the cells.

These results suggest a role for extracellular matrix components, such as decorin, in contact-mediated inhibition of cell proliferation.

♦ Decorin is a proteoglycan derived from fibroblasts, which has a core protein to which a single chondroitin/dermatan sulfate chain is attached. Little is known about the function of proteoglycans on the cell surface or in the extracellular matrix. This study has used a transfection paradigm to express decorin in cells which normally lack this molecule. The authors find that much of the decorin is secreted although some is retained on the cell surface. The ectopically expressed decorin causes profound changes in cell morphology which correlate with the level of decorin expression. CHO cells become progressively more contact-inhibited with higher levels of expression. These results are significant for several reasons. First, they show that transfected with a cDNA encoding a proteoglycan core protein can produce a functional proteoglycan. Second, the data suggest that decorin or other matrix molecules with which decorin interacts can affect cell morphology and behavior. It is possible that decorin provides an additional link between the cell surface and the extracellular matrix. Alternatively, decorin may promote expression of other matrix molecules which may in turn be responsible for contact inhibition. In support of the latter possibility, preliminary data suggest that decorin-transfected cells have increased fibronectin production. These experiments highlight the utility of using transfection to gain insight into the function of a molecule.
Marianne Bronner-Fraser

Maternal Controls, Cytoplasmic Determinants, and Imprinting 4

INTRODUCTION

It is self-evident that as an organism develops, the molecular components necessary for its biological success need to be provided by the organism itself. However, during embryogenesis, varying degrees of important developmental information is already contained within the egg, or genetically "preprogrammed". While in many cases these molecules have an early, transient effect on development, in other cases this developmental information can affect the organism during its entire lifetime.

A detailed and elegant picture is beginning to emerge regarding a class of maternal-effect genes in the fruit fly. The products encoded by some of these genes distribute in the early embryo in a manner that leads to the formation of gradients. These molecular gradients directly establish the overall body plan for the organism. Mutations in any of these genes lead to abnormal gradients and aberrant body plans in the adult organism. Thus, maternal controls can have a direct impact on adult morphology.

In the frog, certain maternal mRNAs remain localized within the egg. These molecules seem to be translocated to their appropriate locations rather than selectively stabilized there. The cytoskeleton has been implicated in this translocation process, although the exact mechanism is unknown.

Further studies continue to define the differences in the male and female genomes of the mammal. It is clear that each has different restrictions on expression early in development, although the mechanism that confers such restrictions is unknown.

Together, these types of studies directly indicate that the parental generation still retains a certain level of control over its offspring. The extent and mechanism of that control remains an open question.

A Gradient of *bicoid* Protein in Drosophila Embryos
W. Driever and C. Nüsslein-Volhard
Cell, 54, 83—93, 1988   4-1

Several lines of evidence indicate that the product of the Drosophila *bicoid* (bcd) gene is responsible for the anterior morphogenetic gradient in the embryo. The bcd gene has been cloned and its mRNA is located at the anterior tip of the oocyte. Therefore, this RNA cannot, itself, be responsible for the anterior gradient. However, the bcd protein may be the anterior morphogen gradient molecule. To investigate the distribution of the bcd protein, antibodies to bcd-fusion products were generated.

The antibodies recognized a 55- to 57-kDa doublet on Western blots of proteins from 0- to 4-h embryos. This protein was absent or reduced in most bcd mutants. The bcd protein could not be detected in oocytes and was in the nuclei of cleavage stage embryos.

Immunocytochemistry demonstrated that bcd protein was present in early embryos in a steep concentration gradient, with a maximum at the egg pole and a posterior exponential decay in protein concentration. The protein was detectable for approximately 30% of the length of the embryo.

These results suggest that the bcd protein gradient is generated by diffusion from mRNA translated at the anterior pole. An exponential decrease in bcd concentration over the embryo body would allow differential binding to promoter sites of varying affinity in different embryo regions. The properties of the bcd protein might provide a molecular basis for translating a continuous morphogen gradient into discrete gene activations.

The *bicoid* Protein Determines Position in the Drosophila Embryo in a Concentration-Dependent Manner
W. Driever and C. Nüsslein-Volhard
Cell, 54, 95—104, 1988   4-2

The bcd protein forms an exponential gradient along the anterioposterior axis of Drosophila embryos. To assess the correlation between bcd protein concentration and cell fate, bcd protein distribution was examined in embryos from females homozygous for mutations that affect anterior development and from females with up to four copies of the bcd+ gene.

Analysis of the bcd protein in various maternal mutants indicated that low levels of bcd, such as those normally found from 50 to 60% of egg length, were sufficient for *exu* and *swa* embryos (Table 4-2). Intermediate levels were required for head formation. The acron required the highest levels of bcd protein for normal development. The gradient was inter-

Table 4-2
Effect of Maternal Mutations on Pattern and bcd Protein Gradient

Class	Mutant Gene	Pattern	Gradient
Anterior	bicoid	Head and thorax absent, telson duplicated anteriorly	Absent
	exuperantia	Anterior reduced, gnathal and thoracic region enlarged	Shallow and low
	swallow	Anterior reduced, gnathal and thoracic region enlarged	Shallow and low
	Bicaudal-D weak	Anterior defects	Shallow and very low
	Bicaudal-D strong	Mirror-image duplication of posterior structures	Absent
	exu vasa	Duplication of gnathal and thoracic structures	Even distribution of low protein concentration
Posterior	oskar	Anterior normal, deletion of abdomen	Normal
	vasa	Anterior normal, deletion of abdomen	Normal
	valois	Anterior normal, deletion of abdomen	Normal
	tudor	Anterior normal, deletion of abdomen	Normal
	pumilio	Anterior normal, deletion of abdomen	Normal
	nanos	Anterior normal, deletion of abdomen	Normal
	staufen	Anterior truncated, deletion of abdomen	Reduced maximum
Terminal	torso	Anterior and posterior truncated	Normal
	trunk	Anterior and posterior truncated	Normal
	torsolike	Anterior and posterior truncated	Normal

From Driever/Nüsslein-Volhaid, *Cell*, 54, 95—104, 1988. ©Cell Press.)

preted autonomously. Medium or low protein levels were interpreted correctly regardless of what other levels prevailed in the rest of the embryo. There was a strong correlation between bcd protein concentration and the positions of anterior anlagen on the embryonic fate map.

These results indicate that the product of the bcd gene determines positon in the anterior of the embryo in a concentration-dependent manner. The mechanism of this determination is currently being investigated.

♦ It has been hypothesized for many decades that morphogen gradients are a way to generate diversity from an apparently uniform single-celled egg. A fair amount of circumstantial evidence accumulated in recent years to support this hypothesis, but a direct demonstration of such a morphogen gradient had not been available.

bicoid is a maternal gene required for the formation of the embryonic head and thorax. Genetic and cytoplasmic transplantation experiments suggested that the gene product forms a concentration gradient along the anterior-posterior axis of the embryo. The gene has been cloned and its RNA product was shown to be localized at the anterior end of the embryo (see also separate report in this volume on work by Macdonald and Struhl).

Using antibodies, the authors demonstrate that the *bicoid* protein is first synthesized in early embryos. Therefore, during oogenesis, either the maternal mRNA is not translated or the *bicoid* protein is unstable. An important finding was that the protein is distributed in the embryo according to a concentration gradient that exponentially decays toward the posterior pole. Since the RNA is detectable only along the anterior-most 20% of the embryo, while the protein can be detected up to 70% from the anterior end, the results suggest that the *bicoid* gradient forms by translation of an immobilized mRNA (the source) followed by diffusion of the protein toward the posterior of embryo. The balance of the rates of protein synthesis, diffusion, and degradation would determine the shape of the gradient. Although this model has not yet been directly tested, it provides a framework for further investigations.

Of prime importance was the finding that the positioning of the anterior embryonic structures (e.g., the cephalic furrow) was directly correlated with local *bicoid* protein concentrations. When the *bicoid* gene dosage was increased, the concentration of the protein was also increased and, as a consequence, the entire anterior embryonic pattern was shifted toward more posterior positions. On the other hand, when by genetic means the gradient was made more shallow, embryonic structures shifted anteriorly. This strongly suggests that cells "read" local concentrations of *bicoid* protein and differentiate accordingly, thus satisfying the concept that *bicoid* is a morphogen. In a set of papers that appeared later in the year, it was found that the *bicoid* protein directly affects the transcription of the embryonic gap class gene, *hunchback* (see separate report on work by Schröder et al., Driever and Nüsslein-Volhard, Struhl et al., and Hanes and Brent, elsewhere in this volume). Driever and Nüsslein-Volhard's findings also sharply contrast with the discovery that *nanos*, which was previously presumed to be a similar morphogen in the posterior part of the embryo, does in fact not play such a role but acts only indirectly by inhibiting activity of the maternal *hunchback* gene activity (see report on work by Hülskamp et al., Irish et al., and Struhl elsewhere in this volume). These papers represent very significant advances that are being made in the understanding at the molecular level of how the maternal information is translated into the establishment of the Drosophila embryonic body plan. *Marcelo Jacobs-Lorena*

Cis-Acting Sequences Responsible for Anterior Localization of *bicoid* mRNA in *Drosophila* Embryos
P. M. Macdonald and G. Struhl
Nature, 336, 595—598, 1988   4-3

Formation of an anterioposterior gradient of the *Drosophila* morphogen bcd requires the bcd mRNA to be restricted to the anterior pole of the embryo. To determine which portion of the mRNA was required for this localization, various portions of the bcd gene were replaced with analogous regions from other genes and the location of the transcripts in the embryo were monitored.

Sequences necessary and sufficient for the anterior pole localization of the bcd-fusion transcripts were contained in 625 bp of the 3' noncoding region of the bcd transcript. This region appeared capable of extensive secondary structure.

The existence of this disrete *cis*-acting signal supports the idea that bcd transcripts are selectively retained in the anterior of the oocyte after transfer to the anterior pole from the nurse cells. This would require fixed receptors capable of recognizing and binding the bcd transcript.

♦The *bicoid* gene product is believed to be a key morphogen for the determination of anterior structures of the embryo. Its mRNA is tightly localized in the anterior-most portion of the unfertilized egg (see separate reports in this volume). In embryos derived from mothers mutant in the *swallow* or *exuperantia* genes, *bicoid* mRNA localization is disrupted and development is abnormal, providing evidence for the importance of its localization. By what mechanism is the *bicoid* mRNA localized? The organization of the Drosophila egg chamber provides a possible clue. Most components of the oocyte are synthesized in the nurse cells and transported to the oocyte via cytoplasmic bridges. These communicate with the oocyte at its anterior end. Thus, it is possible that the *bicoid* mRNA that is synthesized in the nurse cells is trapped by component(s) of the oocyte cytoskeleton at the point of entry. This would require that these components be absent from the nurse cells, although they would not need to be localized within the oocyte. The *swallow* and *exuperantia* gene products may well represent such cytoskeletal components. the other component of the system is obviously the *bicoid* mRNA itself. The results of this study identify *bicoid* mRNA sequences that are necessary and sufficient for its localization. Thus, it represents a first step in the elucidation of this important mechanism of mRNA localization. Several pieces of the "puzzle" having been identified, it is likely that in the near future we may find out more of how they fit together.

The localization of the *bicoid* gene product contrasts with two other examples. In Xenopus, initially uniformly distributed *Vg1* mRNA becomes progressively more localized. In Drosophila the maternal *dorsal* mRNA is uniformly distributed at all times and it is the protein that becomes localized (see separate report by Steward et al. in this volume). Thus, a variety of mechanisms appear to operate, even within the same organism, to achieve localized expression of specific gene products. *Marcelo Jacobs-Lorena*

The *dorsal* Protein Is Distributed in a Gradient in Early Drosophila Embryos
R. Steward, S. B. Zusman, L. H. Huang, and P. Svhedl
Cell, 55, 487—495, 1988

To understand the influence of maternal information on early development, the establishment of dorsal-ventral polarity in Drosophila has been investigated. The product of the *dorsal* (dl) gene appears to be the last step in transmission of maternal dorsal-ventral polarity information. The temporal and spatial distribution of the products of the dl gene were examined during 0 to 3.5 h of Drosophila embryogenesis.

Labeled antisense riboprobes were hybridized to longitudinal embryo sections to assess the level and distribution of dl RNA. The level of dl mRNA was constant and its distribution uniform throughout developmental stages 1 through 9. The dl mRNA level began to decrease at stage 11 and was minimal by stage 14. The loss of this message also appeared to be uniform throughout the embryo.

To determine the distribution of the dl protein, antibodies to a dl-fusion protein were generated. In wild-type stage 14 embryo longitudinal sections, anti-dl stained the ventral portion of the embryo intensely, while dorsal staining was minimal. The dl protein was restricted to somatic nuclei. Cross-sectioning also revealed that the dl protein was distributed as a gradient along the dorsal-ventral axis.

Therefore, dl appears to function as a maternally derived morphogen in the establishment of the dorsal-ventral axis of the Drosophila embryo. The distribution of the dl protein along the dorsal-ventral axis suggests that position could be defined by the dl concentration gradient. The nuclear localization of this protein suggests that this information may be communicated through an effect on gene expression.

♦ Establishment of the embryonic body plan along the dorsal-ventral axis is mediated by products of 11 maternal "dorsal-group" genes and the "ventralizing" gene *cactus*. In embryos derived from mothers mutant in any of the dorsal-group genes, cells that would normally differentiate into ventral structures assume a dorsal fate. *cactus* mutants have the complementary phenotype; all cells assume the ventral fate. Similarity of phenotypes and genetic epistatic interactions suggest that the dorsal group genes belong to a regulatory cascade, with the *dorsal* gene placed at the end of this cascade. The *dorsal* gene had been previously cloned and shown to have sequence similarity to the avian *c-rel* protooncogene. *dorsal* RNA accumulates during oogenesis, persists through the blastula stage of embryogenesis, and is then rapidly degraded.

The most significant finding of this study was that the *dorsal* protein is asymmetrically distributed, forming a gradient along the dorsal-ventral axis of the embryo. This is the only maternally expressed gene for which asymmetric distribution along the dorsal-ventral axis has been demonstrated. Along the anterior-posterior axis, *bicoid* (see separate reports in this volume) and *caudal* had also been shown to form gradients. However, unlike these latter genes, *dorsal* mRNA is evenly distributed. As yet, no clue is available as to how the *dorsal* protein gradient forms from a uniformly distributed mRNA. Two main types of models can be postulated: either the uniformly distributed mRNA is differentially translated along the circumference of the embryo, or the mRNA is uniformly translated but the protein is differentially destablilized. The elucidation of this mechanism is one of the outstanding problems to be solved. The other question raised by the results is how that asymmetrical distribution is translated into differential cell determination. Taking into account that *dorsal* is at the end of the differentiation cascade and that the protein is nuclearly localized, it is tempting to establish a correlation of these properties with what is known about differentiation along the anterior-posterior axis. According to one model, *dorsal* would transcriptionally regulate downstream zygotically active genes that have a more specialized function in determination. Transcription of these genes would only occur if the *dorsal* protein is present above (in case of activation) or below (in case of inhibition) a threshold concentration. Candidates for such target genes are *zerknullt* and *decapentaplegic,* which are expressed only in dorsal regions, and *twist,* which is expressed only in ventral regions. These genes are active at about the time of maximal *dorsal* protein accumulation. Obviously, much work remains to be done to either prove or disprove this model. *Marcelo Jacobs-Lorena*

Posterior Segmentation of the Drosophila Embryo in the Absence of a Maternal Posterior Organizer Gene
M. Hulskamp, C. Schroder, C. Pfeifle, H. Jackle, and D. Tautz
Nature, 338, 629—632, 1989

Posterior development in the Drosophila embryo is controlled by the posterior group of maternal genes, of which the active component is the product of the *nanos* (nos) gene. In nos– embryos the product of the *hunchback* (hb) gene persists in the posterior portion of the embryo. To examine the role of nos in development, mutants in nos and hb were examined.

Flies were created that carried a hb-beta gal fusion product that persisted in the posterior region of embryos. In these mutants, the fushi tarazu pattern was partially fused and abdominal segments 3 to 7 were missing. This demonstrates that hb activity can suppress abdominal segmentation.

Therefore, flies were constructed carrying a chromosome that was mutant for both nos and hb (Table 4-5). The nos–/hb– fertile females were mated to heterozygous hb males. None of the nos–/hb– offspring developed a nos phenotype; instead, sementation was normal (Table 4-5). Therefore, loss of maternal hb function completely rescues the nos– phenotype. Therefore, the nos- phenotype is due to presence of hb protein in the posterior region of the embryo.

In the absence of maternally derived hb protein, nos, the active component of the posterior maternal genes, is unnecessary for a wild type phenotype. Therefore, nos does not function as a morphogen, but as a specific posttranscriptional repressor of hb. The spatial organization of the abdomen may not require a morphogen, but may be controlled at the zygotic level.

The Drosophila Posterior-Group Gene *nanos* Functions by Repressing *hunchback* Activity
V. Irish, R. Lehmann, and M. Akam
Nature, 338, 646—648, 1989   4-6

Of the seven maternal genes required for posterior development of the Drosophila embryo, the active component appears to be *nanos* (ns). ns mutations allow a uniform distribution of the product of the hb gene, rather than a gradient along the anterior to posterior axis. To study the relationship between hb and ns function, embryos lacking both maternal ns and hb function were created.

Progeny carrying maternal nos-hb– chromosomes were viable, indicating that that loss of hb function compensated for loss of ns function.

Therefore, the function of the ns gene product is to posttranscriptionally regulate the maternal hb product in the posterior of the Drosophila embryo.

Differing Strategies for Organizing Anterior and Posterior Body Pattern in Drosophila Embryos
G. Struhl
Nature, 338, 741—744, 1989   4-7

The interaction between the posterior maternal gene *nos* and the *hb* gene was examined by inducing ectopic expression of *hb* in *nos+* embryos and by removing hb expression from embryos lacking *nos* function.

The *hb* gene was placed under the control of the *hsp70* promoter and integrated into the germ-line by P-element-mediated transformation. A heat shock during the syncitial blastoderm stage generated uniform expression of *hb* throughout the embryo. The resulting larvae lacked abdominal segments and had a phenotype similar to that of *nos*– embryos.

Females homozygous for vasa mutations lay eggs in which the product of the *nos* gene is inactive. Irradiating larvae of the genotype *vas/vas hb/+* permitted single germ cells to become *vas/vas hb/hb* (Figure 4-7), lacking both *hb* and *nos* function. These progeny developed into viable flies.

Therefore, the role of the posterior determinant system in the Drosophila embryo is to prevent expression of the product of the *hb* gene from maternal transcripts deposited in the posterior half of the embryo.

♦ The establishment of the anterior-posterior embryonic body plan is dependent on two groups of maternal genes located at each pole of the egg. Detailed studies have identified the gene *bicoid* as a determinant of anterior structures; it is therefore referred to as an "anterior morphogen" (see separate reports in this volume). Less is known about the determination of posterior structures. A group of maternal "posterior" genes had been identified, including *vasa* and *nanos*, which are required for determination of posterior structures. Among them, *nanos* appears to be at the end of the hierarchy; the other genes of this group are believed to regulate *nanos* function. The zygotic "gap" gene *hunchback* is required for development of anterior structures. *hunchback* is also expressed in the ovary, but this maternal component is entirely dispensable for embryonic development. Normally, maternal *hunchback* activity present at the posterior part of the embryo disappears very early in development. However, in embryos derived from mutants for any of the posterior class genes, the maternal *hunchback* activity was shown to persist and abnormal development of posterior structures ensued. Hence, the possibility arose that the main function of *nanos* and other posterior group genes was to repress maternal *hunchback* activity. If this were true, then the posterior genes should be entirely dispensable in embryos derived from homozygous mutant *hunchback* mothers (zygotic *hunchback* activity is provided by a paternal gene introduced at fertilization). This hypothesis was proven to be correct by research independently undertaken by the thee laboratories.

The outcome of the experiments was very surprising. *nanos*, that had previously been assigned a posterior morphogen role, is now relegated to a secondary function of simply inhibiting maternal *hunchback* activity; when this activity is genetically suppressed, *nanos* (and other posterior group gene) function becomes entirely dispensable. The results of these studies suggest a model whereby abdominal differentiation corresponds to the "ground state" The normal function of the zygotic *hunchback* gene product would be to suppress posterior gene function in the anterior part of the embryo. Hence, persistence of maternal *hunchback* activity in the posterior part of the embryo would result in defective posterior development. *Marcelo Jacobs-Lorena*

Table 4-5
Phenotypic Rescue of Embryos Maternally Mutant for *nos* in the Absence of Maternal *hb* Expression

Number of host embryos	Adults (males/females)	Fertile females germ-line genotype (number)	Phenotype of embryos
418	162 (85/77)	Unknown (3) TM3/TM3^a (1) nos; hb/TM3^b (4) nos; hb/nos; hb^c (3)	ND wt^a wt and zyg. hb^b wt and mat. hb^c

Pole cell transplantation experiments show the phenotypic rescue of embryos maternally mutant for *nos* in the absence of maternal *hb* expression. For our experiments we used the nos^{L7} allele (R. Lehmann, unpublished data) and the hb^{9Q} allele[14]. Embryos maternally mutant for the nos^{L7} allele show complete lack of all abdominal segments[1,11]. To show that this phenotype is germ-line autonomous, we have tested this allele in a pole cell transplantation experiment of the type outlined below. We found that the phenotype is only dependent on the genotype of the germ-line, as shown in Fig. 2c. We noted, however, that the chromosome carrying the nos^{L7} allele appears to carry a second mutation, which leads to a much lower than expected frequency for the occurrence of homozygous mutant germ-lines in these experiments. The hb^{9Q} allele shows the lack of function phenotype of *hb* in homozygous mutant embryos[14]. But, it appears to have a weak neomorphic or antimorphic function, which is particularly evident in pole cell transplantation experiments. Embryos which are maternally mutant for this allele, but zygotically heterozygous, show a high-frequency lack of the second thoracic segment (M.H. and D.T., unpublished data). This feature served as a useful marker in our experiments. Both *nos* and *hb* are located on the third chromosome. A double-mutant chromosome was obtained by a recombination event between the two genes. This chromosome was brought over the *TM3* chromosome, which serves as a balancer to prevent further recombination. *TM3* is wild-type for both *nos* and *hb*. The pole cell transplantation experiment was essentially as described[14]. The donor pole cells came from embryos of a mating of *nos; hb/TM3* heterozygous flies. They were transplanted into embryos, which were maternally mutant for Ovo^{D1} (ref. 21). Host embryos (418) were injected, of which 39 per cent survived (columns 1 and 2). The females were backcrossed with heterozygous $hb^{9Q}/TM3$ males. Eleven fertile females were recovered, of which eight laid sufficient eggs to analyse between 30-150 embryos of each. The phenotypes were scored in cuticle preparations of first instar larvae before, or shortly after, hatching from the egg shell. a. The cuticle phenotypes of all embryos of this female were wild-type, about half of which survived to adults. The other half died as larvae, as is expected from animals homozygous for *TM3*. None showed defects in the second thoracic segment. b. The cuticle phenotypes of the embryos of this class of females were either wild-type (3/4), or zygotic *hb* (1/4). About 1/4 of the 'wild-types' did not survive, indicative of homozygousity for *TM3*. None showed defects in the second thoracic segments. c. The cuticle phenotypes of the embryos of this class of females were either wild-type, or showed the combined maternal and zygotic *hb* phenotype. None of them showed the *nos* phenotype. The 'wild-type' embryos showed lack of the second thoracic segment at a high frequency (~60%). A minor fraction of both classes of embryos showed weak abdominal defects (see Fig. 3). It is as yet unclear, whether this is due to an insufficient rescue, or to a weak residual activity of the hb^{9Q} allele. Several of the embryos of each female (20) were allowed to develop to adults. They were individually mated with heterozygous *nos/TM3* flies. The females of this progeny were then analysed for whether their embryos developed the *nos* phenotype. This was the case for a quarter of them, which is the expected frequency if their parents were all heterozygous for *nos/TM3*. This is the final proof that the mutant *nos* allele was originally homozygously present in the germ-line of the transplanted females.

ND, not determined; wt, wild-type.

From Hulskamp et al., *Nature*, 338, 629—632, 1989. With permission.

Maternal Controls, Cytoplasmic Determinants, and Imprinting 127

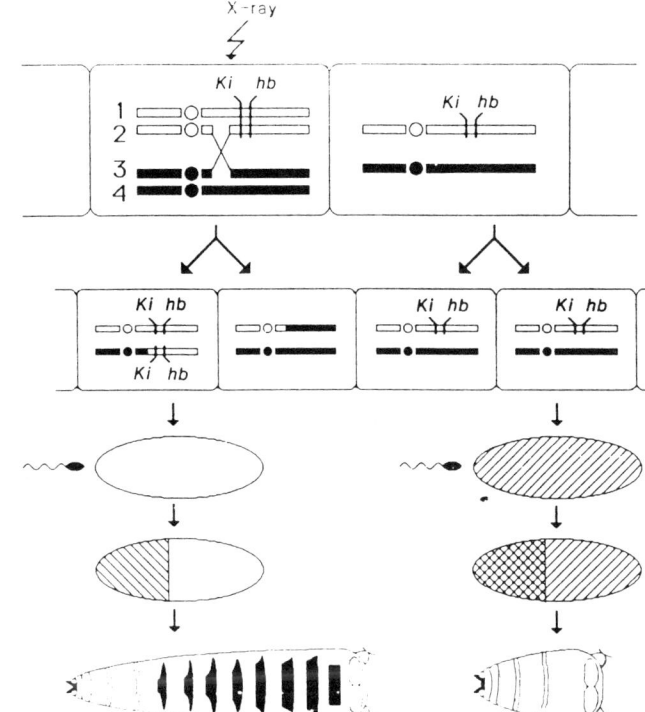

FIGURE 4-7. Induction of homozygous hb clones in the germ lines of females lacking the posterior determinant system. Female first and second instar larvae of the genotype vas^{PD}/vas^{PD}; $Ki\ hb^{14F}/+$ were irradiated with 1000 rad to induce mitotic recombination in single germ cells as shown at the top. The Kinked (Ki) mutation is a dominant bristle marker which is positioned just proximal to the hb mutation on the right arm of chromosome 3. Recombination between the Ki locus and the centromere will result in a homozygous vas^{PD}/vas^{PD}; $Ki\ hb^{14F}$ daughter cell, provided that chromatids 1 and 3 segregate together (the Ki and hb loci are tightly linked; hence, few if any recombination events are expected to occur between them). The descendents of this cell will form a germ-line clone giving rise to eggs lacking functional maternal transcripts of the hb gene. The other daughter cell will be homozygous for the wild-type Ki and hb alleles and, like the surrounding vas^{PD}/vas^{PD}; $Ki\ hb^{14F}/+$ cells, will give rise to eggs with ubiquitous, functional hb transcripts. Fertilization of vas eggs containing functional hb transcripts should generate phenotypically mutant embryos because the vas mutation prevents activity of the posterior determinant nos, thereby allowing the maternal hb transcripts (hatching) to persist and give rise to inappropriate posterior expression of the hb protein, as shown. Conversely, removing the wild-type allele of hb from the germ line via mitotic recombination should abolish the inappropriate posterior expression, whereas zygotic activation of the paternal hb gene (opposite hatching) under the control of the bcd gradient should suffice to generate the necessary anterior domain of hb expression. Note that the vas^{PD} mutation also prevents the formation of posterior polar granules associated with localized germ-cell determinants; consequently all of the progeny from vas^{PD}/vas^{PD}; $Ki\ hb^{14F}/+$ females are agametic, irrespective of the presence or absence of functional maternal hb transcripts. Note also that virtually all larvae derived from these females show a less extreme mutant phenotype that that observed in larvae derived from vas^{PD}/vas^{PD} mothers which carry two functional copies of the hb gene. In particular, they generally form a single band of abdominal denticles between the thorax and posterior terminalia in contrast to standard vas^{PD} mutant embryos which do so only occasionally. This slight degree of phenotypic rescue probably results directly from the lower maternal hb gene dosage in vas^{PD}/vas^{PD}; $Ki\ hb^{14F}/+$ females, as completely eliminating the maternal hb contribution, as described below, fully rescues abdominal segmentation. Approximately 800 female flies obtained from irradiated vas^{PD}/vas^{PD}; $Ki\ hb^{14F}/+$ larvae were outcrossed in small groups of 10 to 20 flies to wild-type males. Of the 70 such group matings which were performed, 24 gave rise to small numbers of wild-type larvae which developed in adult flies with normal abdominal patterns; all the rest produced only vasa mutant larvae (9 of these 24 matings gave rise to single surviving flies, 6 gave rise to 2 to 3 survivors, and 8 gave rise to 4 or more survivors). A total of 71 such adult flies were obtained: all were phenotypically Ki as was to be expected if they arose from mitotic recombination. In addition, all of the flies which survived long enough to be tested (55/71) were sterile, and in fact agametic, as was to be expected if they arose from vas mutant eggs. Finally, in two cases it was possible to identify the single female fly generatimng phenotypically wild-type larvae; in both cases, all but the few surviving progeny developed as vas mutant larvae. The phenotypic properties of the surviving flies therefore establish that these progeny arose from homozygous $Ki\ hb^{14F}/Ki\ hb^{14F}$ clones in the germ line of their homozygous vas^{PD}/vas^{PD} mothers. METHODS. Surviving flies were tested for fertility by outcrossing them singly to flies of the opposite sex for 5 to 7 d. Their ovaries or testes were then removed and examined under phase and Nomarski optics for the presence of germ cells. The frequency of clones obtained (~1 clone per 25 flies) is similar to that obtained in other studies; given that 24/70 group matings produced surviving adults, it is likely that more than one female contained a clone in some of these matings. (From Struhl, Nature, 338, 741—744, 1989. With permission.)

Differential Regulation of the Two Transcripts from the Drosophila Gap Segmentation Gene Hunchback
C. Schröder, D. Tautz, E. Seifert, and H. Jackle
EMBO J., 7, 2881—2887, 1988   4-8

The Drosophila gap gene hb is required for the anterior segment pattern of the embryo. The hb gene encodes 2 transcripts (3.2 and 2.9 kb) from different promoters. The temporal and spatial expression of these two transcripts during embryogenesis was examined by *in situ* hybridization of transcript-specific probes to embryo tissue sections.

The 3.2-kb RNA was maternally transcribed. Initially the distribution of this transcript was homogenous, then an anterior to posterior gradient was formed. Before the blastoderm stage, the 3.2-kb message became undetectable. At syncitial blastoderm the 3.2-kb transcript was detected in the anterior yolk nuclei and in one posterior and one anterior stripe.

The 2.9-kb RNA was the result of zygotic transcription. It was initially detected at stage 11 to 12. Fusions between this 2.9-kb transcript and lac Z were used to define its regulatory region. The regulatory region was approximately 300 bp upstream of the transcription initiation site of the 2.9-kb transcript. This region was sufficient to confer on the fusion transcript full regulation by bcd.

The zygotic 2.9-kb transcript is responsible for the gap gene function of kb and is regulated by bcd. The role of the 3.2-kb transcript is more complicated. Its maternal role is unclear. Zygotically it has a role in the formation of abdominal segments 7 and 8. The structure of these two messages suggest that they encode the same hb protein.

The Bicoid Protein is a Positive Regulator of *hunchback* Transcription in the *Drosophila* Embryo
W. Driever and C. Nüsslein-Volhard
Nature, 337, 138—143, 1989   4-9

The maternal gene *bicoid* (bcd) is required for proper development of the anterior of the *Drosophila* embryo. Messenger RNA of the bcd gene is localized at the anterior pole of the embryo, and the bcd protein is distributed in an anterior to posterior concentration gradient. To investigate the molecular interactions of the bcd maternal gene product which lead to activation of the zygotic gene *hunchback* (hb), binding studies between the bcd protein and the hb promoter were conducted.

Full length bcd protein was expressed in *E. coli,* combined with hb DNA segments, and immunoprecipitated with anti-hb antibodies. Two hb restriction fragments bound with high affinity; fragment A (–298 to –55) had a higher affinity than fragment B (–2172 to –2631).

DNase I footprint experiments demonstrated three regions within the

A fragment protected by the bcd protein: A1 (–280), A2 (–170), and A3 (–65). On the B fragment two protected regions, B1 (–2325) and B2, (–2280) were detected. An approximately 9 bp consensus protected sequence was derived, TCTAATCCC, which was best conserved in sites A2 and A3.

Promoter fragments from the hb gene were fused to a CAT reporter gene at position +107 and injected into the anterior of early cleavage embryos. CAT expression was measured at gastrulation. Expression was 50 times higher in embryos from wild-type females than in those from bcd- females, indicating bcd-dependent expression of CAT. Successive elimination of the A sites from the promoter fragments significantly reduced bcd-dependent hb-CAT expression. Similar results were obtained in Schneider cells.

These results demonstrate an interaction between a maternally derived regulatory protein, bcd, and a promoter from the zygotic gap gene, hb, which is involved in embryonic pattern formation. The bcd concentration gradient combined with promoters with varied affinities for bcd could restrict target gene expression to anterior embryo regions. This could establish the anterior to posterior axis of the embryo.

The Gradient Morphogen *bicoid* is a Concentration-Dependent Transcriptional Activator
G. Struhl, K. Struh, and P. M. Macdonald
Cell, 57, 1259—1273, 1989   4-10

The bcd protein is expressed in an anterior to posterior gradient and controls expression of the segmentation gene hb during Drosophila embryogenesis. The interaction between bcd and hb DNA was examined at the molecular level.

To determine which sequences were sufficient to generate the zygotic pattern of hb expression, short fragments of the hb promoter were inserted upstream of the TATA box of an hsp 70:lacZ (HSZ) reporter gene. These constructs were inserted into the Drosophila genome via P element mediated transformation and lacZ expression was monitored in early embryos. Anterior expression of the HSZ gene was dependent on a 123-bp core region normally located approximately 200 bp upstream of the proximal start site of hb transcription.

Transgenic flies carrying from zero to six copies of the 123-bp core fragment in front of the HSZ gene were generated. The posterior boundary of HSZ expression expanded with increasing numbers of 5′ 123-bp elements. Therefore, the activity of the hb promoter appears to be due to a series of at least partially redundant elements.

To determine whether the posterior boundary of expression of the hb and HB:HSZ gene was dependent on the bcd concentration gradient, the distributions of hb and HB:HSZ were examined in embryos from mothers carrying zero to six copies of the bcd gene. Embryos from mothers with no copies of the bcd gene did not activate expression of either gene. As the number of bcd genes increased, the domain of expression of the two genes expanded posteriorly. Therefore, the upstream hb regulatory elements respond to bcd in a concentration-dependent manner. The number of upstream elements also influenced the response to the bcd concentration gradient.

To determine if the interaction between the upstream elements and bcd protein was direct, bcd-dependent transcriptional activation was investigated in yeast. Fragments from the upstream regulatory region of the hb gene were fused upstream of the TATA box of the yeast his3 gene. The HB:HIS3 gene was integrated into the yeast chromosome at the his3 locus and these cells were then transformed with bcd coding sequences under the control of the yeast ded1 promoter. Activation of HB:HIS3 expression by bcd occurred when the 123-bp upstream region was present. Increased 5' sequences amplified this response. Therefore, bcd-dependent transcriptional activation occurs in yeast and requires the same upstream promoter elements.

To determine which domains of the bcd protein function in DNA binding and transcriptional activation, wild-type and mutant bcd coding regions were fused to the DNA binding domain of lexA alone or in conjunction with the transcriptional activation domain of GCN4. The hybrid proteins were tested for the ability to activate HB:HIS3 and a gene composed of a lexA binding site and the lacZ structural gene. In order to activate HB:HIS3, the hb regulatory sequences must be recognized and bound by the bcd derivative. In order to activate LEXA:LACZ, a transcriptional activation sequence must be present in the bcd derivative.

Intact bcd recognized, bound, and activated the target genes. Mutations in the bcd homeodomain region failed to interact with hb target sequences. bcd protein mutations, carboxy-terminal to the homeodomain, bound to hb sequences, but failed to activate.

The bcd protein appears to bind to the hb promoter in a concentration dependent manner. Therefore, the bcd anterior to posterior concentration gradient could limit the activation of the hb protein to anterior regions of the embryo. The bcd protein concentration gradient could control the activation of other genes in distinct spatial domains, depending on the affinity of each gene for the bcd protein.

DNA Specificity of the Bicoid Activator Protein is Determined by Homoedomain Recognition Helix Residue 9
S. D. Hanes and R. Brent
Cell, 57, 1275—1283, 1989   4-11

The bicoid morphogen contains a homeodomain, which may be directly involved in regulatory interactions with the hb promoter. To assess which bcd protein regions are required for DNA interactions, the bcd coding sequence was fused to either the lexA or GAL 4 activator DNA binding domains and transformed into yeast on 2-μm plasmids. Target plasmids containing a GAL1-lacZ reporter gene with upstream bcd binding sites were transformed into these yeast cells.

The bcd fusion proteins expressed in yeast stimulated expression of target genes containing upstream binding sites. When 10 copies of the bcd consensus binding site were upstream of the target gene, there is a 500-fold activation by LexA-Bcd.

The homeodomain recognition helix of the bcd protein is presumed to include residues 138 to 147 (position 1 to 10). To identify the residues that are important for specific DNA interactions, position 1 to 10 were each altered to alanines. When the lysine at position 9 of the recognition helix was changed to alanine, activation of targets containing bcd binding sites was abolished. When position 9 was changed to glycine, activation of Antp targets occurred.

These results demonstrate that the lysine at position 9 of the homeodomain recognition helix is crucial for the DNA specificity of bcd. When this lysine is converted to a glycine, the bcd protein recognizes Antp binding sites. This indicates that bcd and Antp class proteins make analogous contacts with the DNA and that residue 9 makes base-specific contacts. It is possible that all classes of homeodomain proteins make similar DNA contacts.

♦A crucial area of developmental biology relates to the question of how the information laid down during oogenesis is translated into patterns of embryonic development. The *bicoid (bcd)* and *hunchback (hb)* genes constitute one of the bridges between the maternal and zygotic programs of gene expression. *bcd* is a maternally expressed gene that encodes a protein containing at least three structural motifs: a "PRD-repeat" that is a histidine-proline polymer located at the amino terminus, a homeodomain in the middle of the protein, and an "opa-repeat" that is a polyglutamine repeat located near the carboxyterminus. Importantly, the *bcd* protein forms a sharp concentration gradient in the embryo with a peak at the anterior end and tapering to undetectable levels toward the middle of the embryo (see also two other reports on the *bcd* gene in this volume on work by Driever and Nüsslien-Volhard and by Macdonald and Struhl). Two transcripts coding for identical proteins are produced from the *hb* gene, one expressed both maternally and zygotically and the other only zygotically. The zygotic component is functionally the more important of the two, and its spatial pattern of expression entirely overlaps the regions of *bcd* expression. In fact, previous studies had shown that zygotic *hb* expression requires the *bcd* protein.

The paper by Schröder et al. demonstrates that a 300-bp element upstream of the *hb* transcription initiation site is sufficient to promote *bcd*-dependent transcription. The paper by Driever and Nüsslein-Volhard demonstrates further that the *bcd* protein interacts directly with a specific sequence of about 9 bp present in three copies within the 300-bp element. They also show that the 300-bp fragment can promote *bcd*-dependent transcription of a reporter gene in embryos and in cultured Drosophila cells. Deletion of one of the three 9-bp motifs abolishes the effect, suggesting that they are involved in the transcriptional activation. Struhl et al. studied in detail the DNA sequence requirements for *hb* promoter function and *bcd* protein sequence requirements for transcriptional activation of *hb*. Particularly revealing were studies where constructs containing *hb* promoter elements and *bcd* protein coding regions were cotransfected into yeast. *bcd*-dependent *hb* expression in yeast demonstrates that the transcriptional activation is mediated *directly* by the *bcd* protein. Moreover, the cooperative nature of the [*bcd* protein-*hb* promoter] interaction (the promoter is composed of multiple juxtaposed elements) supports a model whereby a smooth *bcd* gradient can be translated into sharp *hb* expression boundaries. The location of these boundaries would depend on the strength of the binding sites. Mutagenesis in the *bcd* protein coding region showed that the homeodomain is essential for binding to the DNA, while the sequence immediately downstream (toward the carboxy-terminus) is essential for gene activation. Interestingly, the PRD- and opa-repeats appear to play a lesser role.

The homeodomain codes for a helix-turn-helix motif. From studies of similar proteins in prokaryotes, the second helix was implicated in the DNA recognition function, with major contributions made by amino acids 1 and 2 of this helix. Hanes and Brent investigated in detail the interaction of the *bcd* homeodomain with the *hb* promoter. Surprisingly, it was the amino acid 9 (and not 1 and 2) that played the most important role in promoter recognition. Two major points were established by these authors. One was that changing a single amino acid in the homeodomain was sufficient to completely abolish [*bcd* protein-*hb* DNA] interaction. The second was that changing the amino acid in position 9 from lysine to glutamine was sufficient to switch the sequence recognition specificity of *bcd* to that of the *Antennapedia*-class homeoproteins, further confirming the critical nature of amino acids in this position of the homeodomain.—*Marcelo Jacobs-Lorena*

Identification of a Component of *Drosophila* Polar Granules
B. Hay, L. Ackerman, S. Barbel, L. Y. Jan, and Y. N. Jan
Development, 103, 625—640, 1988   4-12

Drosophila pole cells are precursors of the germ line. Maternal information necessary for pole cell formation is localized at the posterior egg pole. Polar granules are obvious candidates for determining factors in pole cell development. To investigate the role of polar granules in the formation of pole cells, an antipolar granule antibody, Mab46F11, was employed.

Staining with this antibody paralleled the localization of polar granules, nuclear bodies, and nuage. In the unfertilized egg, immunoreativity occurred exclusively at the posterior pole cortex. The nuclei that migrated to the pole cortex became immunoreactive. Once pole cells formed, antibody staining occurred exclusively in these cells. At the EM level, label was associated with polar granules and nuclear bodies in newly formed pole cells. At later stages, label was associated with nuage. This antibody did not recognize similar structures in other *Drosophila* species. During late embryo stages and throughout the larval stage, immunoreactivity was detected in nuclei. This general staining disappeared in adult somatic tissues.

To identify the 46F11 antigen, Western blotting of all tissues recognized by this antibody was carried out. In every case a 72,000- to 74,000-M_r protein was detected. The level of this protein paralleled that detected by immunocytochemistry.

Monoclonal antibody 46F11 is useful as a pole cell marker throughout development. It may be useful for the biochemical characterization of polar granules.

A Protein Component of Drosophila Polar Granules is Encoded by *vasa* and Has Extensive Sequence Similarity to ATP-Dependent Helicases

B. Hay, L. Y. Jan, and Y. N. Jan
Cell, 55, 577—587, 1988   4-13

Polar granules appear to be involved in the determination of the pole cells of Drosophila embryos. To understand the role of polar granules in pole cell formation, the components of polar granules must be defined. A monoclonal antibody, 46F11, which recognizes a polar granule antigen, has been used to clone and identify a component of the polar granules.

An ovarian cDNA expression library was screened with Mab46F11. The resulting cDNAs were used to isolate full-length cDNA clones from an ovarian lambda-gt11 library. The distribution of transcripts that hybridize to this probe was similar to the distribution of the Mab46F11 antigen. Northern analysis identified a major transcript of approximately 2.2 kb.

It was possible that one of the *grandchildless-knirps* class of maternal gap genes would encode this antigen, as mutant females have embryos which lack polar granules. Therefore, *tudor, valois, oskar, staufen,* and *vasa* flies were screened for the 46F11 antigen. This antigen was missing in *vasa* flies. The cDNA hybridized to the 35BC chromosomal region, which also contains *vasa*. Sequence comparisons indicated that the 46F11 antigen was *vasa*, with a predicted molecular mass of 71 kDa.

A large internal domain of the *vasa* protein has 27% amino acid identity with eIF-4A and 31% identity with human nuclear antigen p68. This domain includes sequences involved in binding of ATP. This suggests that *vasa* ia also an ATP-dependent nucleic acid binding protein. The carboxy-terminus of *vasa* contains multiple negatively charged residues, as found in some single-stranded nucleic acid binding proteins. Near the amino-terminus the sequence F(orS)RGGE(or Q)GG is repeated five times.

The gene for the polar granule component recognized by monoclonal antibody 46F11 has been cloned and identified as the product of the *vasa* gene. The failure of *vasa* embryos to form pole cells is consistent with the involvement of polar granules in the determination of pole cells.

The Product of the *Drosophila* Gene *vasa* is Very Similar to Eukaryotic Initiation Factor-4A
P. F. Lasko and M. Ashburner
Nature, 335, 611—617, 1988   4-14

The product of the *vasa* gene appears to be involved in the formation of posterior structures and pole cells in the *Drosophila* embryo. The cloning and characterization of this gene is reported.

Deletion analysis and chromosome walking in the 35BC chromosome region were used to isolate the *vasa* gene. An abundant 2.0-kb RNA was detected in mature ovaries. This transcript was missing from *vasa*-females.

In situ hybridization demonstrated abundant transcripts in stage 10 nurse cells. The transcript was exported to the oocyte. By cellular blastoderm stage this transcript was undetectable.

Embryo cDNA libraries were screened and a 2-kb cDNA with an open reading frame of 660 amino acids was detected. The predicted amino acid sequence had a relative molecular mass of 72,324. Comparison with the SWISS-PROT database indicated a 29% amino acid identity to murine eIF-4A, including the ATP binding site. An even greater similarity was detected between this protein and human nuclear antigen p68.

The sequence homology between *vasa* and eIF-4A suggests that vasa may be an ATP-dependent nucleic acid helicase active on a set of genes involved in the specification of posterior information.

♦The maternal genes that specify positional information along the anterior-posterior axis of the embryo can be divided into three classes: mutants in genes of the *anterior* class lack head and thoracic structures, mutants in genes of the *terminal* class lack extreme anterior and posterior structures, and mutants in genes of the *posterior* class lack abdominal structures. *vasa* is a member of the posterior class. As is the case for most (but not all) members of this class, *vasa* mutants also lack pole cells and polar granules. Pole cells are the precursors to the germ line. Polar granules are organelles that form during oogenesis, are localized at the posterior tip of the embryo where pole cells form, and are essential for the differentiation of functional pole cells.

Two laboratories have independently cloned the *vasa* gene, using two different approaches. Jan's group fortuitously isolated a monoclonal antibody that specifically stains polar granules. This antibody was used to isolate a cDNA clone. Ashburner's group used extant DNA probes and a deficiency, to "jump" from the alcohol dehydrogenase gene to the vicinity of the *vasa* gene.

Three main features enhance the importance of the reported experiments. (1) *vasa* is a member of a group of genes that plays an important role in the establishment of the embryonic body plan. (2) In early embryos the *vasa* protein is strictly localized to the polar granules, a structure intimately connected with germ cell differentiation. It is thus reasonable to assume that the *vasa* gene product is somehow involved in the process of germ cell determination or differentiation. (3) The predicted amino acid sequence of *vasa* shares extensive similarity with a family of proteins that bind to single-stranded nucleic and have ATP-dependent helicase activity. *vasa* also shares sequence similarity with eukaryotic initiation factor-4A. Since the polar granules contain RNA as well as protein, a number of possible roles for the *vasa* protein may play a role in localizing speific mRNAs to the germ plasm or of activating translation of specific mRNAs around the time of pole cell formation. One should look forward to the experiments that further define the role played by this key gene in early embryonic development. *Marcelo Jacobs-Lorena*

The Finger Motif Defines a Multigene Family Represented in the Maternal mRNA of *Xenopus laevis* Oocytes
M. Koster, T. Pieler, A. Poting, and W. Knochel
EMBO, J., 7, 1735—1741, 1988   4-15

A new class of nucleic acid binding proteins, characterized by multiple metal-coordinating loops or fingers, was initially identified in Xeonpus oocytes. Genomic and cDNA *Xenopus* libraries were screened for the presence of additional finger motif proteins with a synthetic oligonucleotide coding for the conserved H/C link found in several finger motif proteins.

A cDNA which encoded multiple copies of the finger motif was identified. This cDNA was used to identify many genomic and cDNA clones. Of these clones, 109 were characterized. These 109 clones derived from 14 separate genes which contained multiple finger motifs. Comparative sequence analysis indicated structural relatedness among these clones, with extended regions of sequence conservation. The transcription patterns of these genes varied during early development.

The abundance of finger motif proteins identified in the *Xenopus* oocyte in this study suggests that these proteins may play a role in the transmission of maternal information.

♦When a particular sequence motif is found to be characteristic of developmentally interesting proteins, it becomes possible, at least in theory, to clone functionally related proteins by using the conserved sequence as a probe. The zinc finger is just such a motif, and Koster et al. have generated a collection of finger-containing cDNA clones in this way. They made a degenerate oligonucleotide probe corresponding to the link region between fingers, screened a neurula cDNA library and isolated 14 clones with an average of 7 to 8 finger domains each. One clone, XLcNF1, which has 6 fingers, was used as a probe to screen an ovary cDNA library at relatively high stringency, yielding 50 positives representing a zinc finger subfamily. They also screened a partial *Xenopus* genomic library (100,000 plaques) with XLcNF1 and detected 35 strong positives. Library screening is not a very accurate way to measure gene copy number, but it is possible to roughly estimate from this hybridization frequency that the XLcNF1 subfamily of zinc finger genes has 20 to 200 members. Since the screening was done at relatively high stringency, only closely related genes were likely to have been detected, and the total number of zinc finger genes is likely to be substantially higher, perhaps on the order of several hundred to a few thousand genes.

This work illustrates the practicality of screening with a sequence motif probe. It also leads to a practical problem of how to assign functions to these genes and raises a theoretical question of how many regulatory genes are likely to exist in a vertebrate genome. *Thomas D. Sargent*

Developmental Expression of the Protein Product of Vg1, a Localized Maternal mRNA in the Frog *Xenopus laevis*
EMBO J., 8, 1057—1065, 1989   4-16

In *Xenopus* oocytes, Vg1 is a maternal RNA localized to the vegetal cortex. Vg1 cDNA clones have been isolated, and the predicted Vg1 protein is homologous to mammalian growth factor TGF-beta. Polyclonal sera was generated to a T7-Vg1 fusion protein.

Immunoprecipitation identified a 45-kDa protein from oocyte extracts, from embryo extracts a 43.5-kDa protein, and from *in vitro* translation of oocyte polyA+ RNA a 40-kDa protein. Synthesis of these proteins was specifically inhibited by Vg1 antisense ologonucleotides. Tunicamycin treatment revealed that the *in vivo* proteins were glycosylated versions of the *in vitro* 40-kDa protein. Both *in vivo* proteins were segregated in a membrane compartment and released by high pH treatment.

Synthesis of Vg1 proteins was first detected in stage IV oocytes and continued until late gastrula. The 45-kDa protein was more abundant in early embryogenesis, while the 43.5-kDa protein was more abundant in late embryogenesis. Vg1 was synthesized exclusively in the vegetal hemisphere, but diffused into the animal hemisphere of the oocyte.

The protein products of the maternal Vg1 RNA have been identified as 45- and 43.5-kDa glycoproteins, which are first synthesized in stage IV oocytes. The homology between Vg1 and TGF-beta suggests that this protein may be involved in mesoderm induction. Alternatively, as the protein is present in vegetal cells during the specification of endoderm, it may be involved in specification of endoderm with mesoderm arising from interactions between endoderm and ectoderm.

Localized Maternal mRNA Related to Transforming Growth Factor Beta mRNA is Concentrated in a Cytokeratin-Enriched Fraction from *Xenopus* Oocytes
M. D. Pondel and M. L. King
Proc. Natl. Acad. Sci. U.S.A., 85, 7612—7616, 1988   4-17

The maternal RNA Vg1 is localized to the cortical region of the vegetal pole of *Xenopus* oocytes. To determine if Vg1 RNA-cytoskeletal interactions may play a role in this localization, a fraction enriched in cytoskeleton components was examined for the presence of the Vg1 RNA.

A Triton X-100 insoluble oocyte fraction, composed primarily of the cytoskeletal proteins cytokeratins and vimentins, contained less than 2% of the total oocyte polyA+ RNA. However, Vg1 RNA was concentrated 35- to 50-fold in this fraction. When the salt concentration was increased during fractionation, the insoluble fraction was further enriched for cytokeratins and Vg1 RNA.

The localized maternal RNA Vg1 is concentrated in a detergent-insoluble oocyte fraction that is enriched in cytokeratins. Upon ovulation, cortical cytokeratin breaks down and Vg1 is released into the detergent-soluble fraction.

These results suggest that Vg1 may be immobilized at the vegetal cortex by a specific interaction with intermediate filaments, such as cytokeratins.

The Maternal mRNA Vg1 is Correctly Localized Following Injection into *Xenopus* Oocytes
J. K. Yisraeli and D. A. Melton
Nature, 336, 592—595, 1988

The maternal RNA Vg1 is homogeneously distributed in early stage *Xenopus* oocytes, but localized to the vegetal cortex in late stage oocytes. To study the localization process, an *in vitro* oocyte development system in vitellogenin medium was devised.

Stage III oocytes were cultured and endogenous Vg1 RNA was redistributed from homogeneity to the vegetal cortex in 6 d. *In vitro* synthesized Vg1 and globin mRNA were injected into stage III cultured oocytes. The distribution of the globin RNA remained homogeneous, while injected Vg1 was sequestered along with endogenous Vg1 RNA.

To determine whether localization requires translation, Vg1 RNA lacking the ribosome binding site, the start codon, and the signal sequence was injected into cultured stage III oocytes. Localization of this RNA occurred along with the endogenous Vg1 RNA.

Thus, it appears that localization of the Vg1 RNA within the *Xenopus* oocyte occurs through a process that specifically recognizes the Vg1 mRNA itself.

♦The presence of a mRNA highly localized in endoderm encoding a TGFβ-like polypeptide is a highly suggestive fact, given the mesoderm-inducing capabilities of endoderm and TGFβ2. Dale et al. have taken an important step towards understanding the role of Vg1 by showing that the predicted protein is actually synthesized in oocytes. Vg1 possesses a canonical proteolytic site that, by analogy to TGFβ, should be cleaved to release an active fragment in the 15-kDa range. However, their immunoprecipitation assays detected only an approximately full-sized presumed precursor protein. The failure to visualize the smaller band could be due to insufficient sensitivity, but if the smaller product exists at all, it is probably much less abundant than its precursor. Vg1 protein synthesized in the vegetal region of the oocyte diffuses to some extent into the animal hemisphere. If the protein is stable, then a significant amount might be accumulated in presumptive ectodermal cell cytoplasm by the end of oogenesis, something that would be difficult to reconcile with a role for Vg1 as a mesoderm inducer. A separate question concerns the mechanism whereby the Vg1 mRNA becomes localized. Yisraeli and Melton show that Vg1 RNA made *in vitro* and injected into oocytes gets localized if the oocytes are cultured under conditions supporting growth. The mechanism that moves Vg1 RNA to the vegetal pole is now known, but probably depends on active translocation rather than selective stabilization of the RNA in the vegetal pole. Pondel and King present evidence that intermediate filament cytoskeletal elements are involved in maintaining the Vg1 mRNA cortical vegetal localization. *Thomas D. Sargent*

Influence of Chromosomal Determinants on Development of Androgenetic and Parthenogenetic Cells
M. A. Surani, S. C. Barton, S. K. Howlett, and M. L. Norris
Development, 103, 171—178, 1988   4-19

Recent studies have demonstrated that in mammals the maternal chromosomes are required for the development of the embryo, while paternal chromosomes are required for the development of extraembryonic tissues. The influence of parental origin of chromosomes on development was analyzed in aggregation chimeras of fertilized embryos with either parthenogenetic or androgenetic cells.

In midgestation chimeras parthenogenetic cells were detected in the embryo and yolk sac, while androgenetic cells were detected in the trophoblast and yolk sac, but never in the embryo. Parthogenetic cells were detected in the yolk sac mesoderm, but not in the endoderm. Adult chimeras contained parthenogenetic cells, but never androgenetic cells.

Therefore, maternal chromosomes permit parthenogenetic cells to participate in the primitive ectoderm, but these cells do not participate in the primitive endoderm and trophectoderm lineages. Paternal chromosomes confer the opposite properties on androgenetic cells. The spatial distribution of cells with different parental chromosomes may occur because of differential gene expression or because of different responses to growth factors.

The Developmental Fate of Androgenetic, Parthenogenetic, and Gynogenetic Cells in Chimeric Gastrulating Mouse Embryos
J. A. Thomson and D. Solter
Genes Dev., 2, 1344—1351, 1988   4-20

Both the maternal and paternal genomes are necessary for complete embryonic development in the mouse. To analyze the contribution of each parental genome, chimeras were created between eight-cell embryos and single androgenetic, gynogenetic or parthenogenetic cells. The single-parent cells were marked with plasmid-carrying mouse beta-globin sequences. The tissue distribution of each cell type was analyzed at the late gastrulation stage by *in situ* hybridization to the marker sequences.

Androgenetic cells contributed to trophectoderm-derived tissue in the late gastrulation embryos. Parthenogenetic and gynogenetic cells contributed to the embryo and to extraembryonic mesoderm.

Therefore, important functional differences between the maternal and paternal genomes can be detected early in development.

Chimeras between Parthenogenetic or Androgenetic Blastomeres and Normal Embryos: Allocation to the Inner Cell Mass and Trophectoderm
J. A. Thomson and D. Solter
Dev. Biol., 131, 580—583, 1989

Complete development of the mouse embryo requires both maternal and paternal genome contributions. Androgenetic cells are capable of contributing to the trophoblast and yolk sac, and parthenogenetic cells are capable of contributing to the embryo and yolk sac. To determine how early this characteristic tissue distribution is established, chimeras were created between eight-cell embryos and single transgenic normal, parthenogenetic, or androgenetic blastomeres. The transgenic component was detected by *in situ* hybridization at the blastocyst stage.

There was no significant difference among the normal, parthenogenetic, and androgenetic chimeras in the number of transgenic cells that contributed to the inner cell mass or trophectoderm.

Since differences in tissue distribution between androgenetic and parthenogenetic cells have been detected at the late gastrula stage, but were not yet detectable at the blastocyst stage, both androgenetic and parthenogenetic cells must fail to participate normally in subsequent developmental steps. Tissue-specific selection must occur between the blastocyst stage and the late gastrula stage.

The Developmental Potential of Parthenogentically Derived Cells in Chimeric Mouse Embryos: Implications for Action of Imprinted Genes
H. J. Clarke, S. Varmuza, V. R. Prideaux, and J. Rossant
Development, 104, 175—182, 1988

To investigate the inability of parthenogenic embryos to develop to term, chimeras of transgenic parthenogenic embryos and normal embryos were created. The chimeric embryos were sectioned and examined by *in situ* hybridization up to 7.5 d of development.

At the blastocyst stage, parthenogenetic cells contributed to the trophectoderm and the inner cell mass. By 6.5 d of development, parthenogenetic cells were not detected in the extraembryonic trophoblast derived from the trophectoderm. In contrast, a 7.5 d, parthenogenetic cells contributed to all tissue descendents of the inner cell mass. Quantitative analysis indicated that parthenogenetic cells contributed as extensively as normal cells to all tissues derived form the inner cell mass at 7.5 d of development.

Therefore, the ability of parthenogenetic cells to contribute to the trophoblast is lost between the blastocyst stage and 6.5 d of development. This indicates that paternally derived genes are required prior to 6.5 d of development in the extraembryonic membranes, although not necessarily in the inner cell mass lineages. These results indicate that mouse development depends on the correct expression of imprinted genes.

Systematic Elimination of Parthenogenetic Cells in Mouse Chimeras
R. Fundele, M. L. Norris, S. C. Barton, W. Reik, and M. A. Surani
Development, 106, 29—35, 1989   4-23

The developmental potential of parthenogenetic cells was investigated in chimeras. The contribution of parthenogenetic cells to adult chimeras was assessed using GPI-1 allozymes as markers.

Parthenogenetic cells were detected in most organs and tissues in adult chimeras. However, parthenogenetic cells were generally detected in a limited number of tissues in each chimera. Chimeras with a significant contribution of parthenogenetic cells had reduced body weights. Parthenogenetic cells were detected consistently in brain, heart, kidney, and spleen and rarely in skeletal muscle, liver, and pancreas.

There appears to be a selection against parthenogenetic cells in chimeric mice. However, this selection is not uniform, and some tissues often contain parthenogenetic cells, while others consistently do not. Parthenogenetic cells appear to be eliminated sequentially from the chimera, beginning with trophoblast tissue, then other extraembryonic membranes, and finally the embryo. Paternally derived genes appear to be required for the development of extraembryonic membranes and for embryonic tissues derived from the primitive ectoderm lineage.

♦ Nuclear transplantation and genetic studies have shown that maternal/paternal nonequivalency exists in the mouse genome, and that this is due to imprinting of certain regions during gametogenesis. In general, embryos that contain only maternal chromosomes (parthenogenones and gynogenones) have poorly developed extraembryonic membranes, whereas embryos that contain only paternal chromosomes (androgenones) have well-developed membranes, but exhibit retarded embryonic development. Studies with chimeric embryos have shown that androgenetic cells contribute primarily to trophoblast derivatives as well as yolk-sac endoderm, whereas parthenogenetic or gynogenetic cells contribute mainly to the embryo proper and also the yolk-sac mesoderm. These differences in tissue distribution have been seen in gastrulating embryos, but not as early as the blastocyst stage, where the uniparental cells were found in both trophectoderm and inner cell mass. These experiments suggest that importnat functional differences between the maternal and paternal genomes exist early in development, and strong tissue-specific selection against the androgenetic or parthenogenetic components takes place in chimeras by the late gastrulating stage. The mechanisms responsible for the selection are not understood. Identification of precise cell types affected in mice carrying duplications and deficiencies of particular chromosomal regions will help to determine which imprinted chromosomal domains affect the development of specific embryonic tissues. *Terry Magnuson*

Cell Interactions 5

INTRODUCTION

Multicellularity, by definition, implies that cell interactions exist. If cell contacts disappear, the same cells are no more than a population of individual cells. Those single cells can interact in many different ways: as a sheet, as small clumps, as a cylinder, or as a large mass. Each type of interaction has functional implications. Throughout development, changes in how individual cells interact with their neighbors result in important morphogenetic alterations.

Cells interact with other cells for different reasons. Cell adhesion molecules have been extensively studied and are involved in the structural architecture that directly holds cells together. T-cell interactions in the thymus, on the other hand, are functional rather than structural in nature. The interaction is transient but critical. Neural crest cell migration and neuronal outgrowth involve cell interactions with both other cells and extracellular material. Both are structural as well as functional. Importantly, they demonstrate that during development, cell interactions must display some degree of plasticity. Cell movement, which is central to the development of almost all multicellular organisms, must be accommodated by plasticity in cell interactions. Not all cell interactions are adhesive or inductive; some are determinative. Cell fate can be established by the cells with which any individual cell interacts.

Some cell surface molecules have shown high degrees of homology between species, suggesting that they may play fundamental roles in developmental cell interactions. Others are included in growing multigene families. Functional analyses suggest that related functions may be performed by the same or similar gene products. Thus, similar proteins might perform different functions and different proteins might perform the same function.

Genetic and molecular biological investigation of cell interactions have elucidated some of the molecular components involved. In the cases of developmental immunology and developmental neurobiology, the detail has exposed the enormous complexity of cell interactions expressed in these developing systems.

Mapping of a Cell-Binding Domain in the Cell Adhesion Molecule gp80 of *Dictyostelium discoideum*
R. K. Kamboj, L. M. Wong, T. L. Lam, and C.-H. Siu
J. Cell. Biol., 107, 1835 — 1843, 1988   5-1

A *Dictyostelium* surface glycoprotein of M_r = 80,000 mediates EDTA-resistant cell to cell adhesion during aggregation. To characterize this molecule, full-length cDNA clones were isolated and sequenced.

The structure of the predicted gp80 protein contains a hydrophobic leader peptide, followed by a globular amino-terminal region, presumed to be involved in the cell to cell adhesion. This region is supported by a short stalk and followed by a carboxy-terminal membrane anchor.

To assess the function of these protein domains, three protein A/gp80 (PA80) fusion proteins were constructed, PA80I containing Val123 to Ile514 of gp80, PA80II containing Val123 to Ala258, and PA80III containing Ile174 to Ile514. PA80III did not bind to cells, while binding of the other 2 fusion proteins was concentration-dependent and could be blocked with cell cohesion-blocking antibody 80L5C4. PA80I and II significantly inhibited reassociation of cells. Labeled PA80II interacted with immobilized PA80I, PA80II, and gp80.

The cell binding domain of gp80 has been mapped to the amino-terminal globular region of this cell to cell adhesion protein of aggregating *Dictyostelium*.

The Contact Site A Glycoprotein of *Dictyostelium discoideum* Carries a Phospholipid Anchor of a Novel Type
J. Stadler, T. W. Keenan, G. Bauer, and G. Gerisch
EMBO J., 8, 371—377, 1989   5-2

The contact site A (csA) glycoprotein is an adhesion molecule expressed by aggregation stage *Dictyostelium*. It has a hydrophobic carboxy-terminus that is characteristic of membrane proteins with a phosphatidyl inositol (PI) anchor. The nature of the csA anchor was investigated.

The csA protein could be labeled with fatty acid, myo-inositol, phosphate, and ethanolamine *in vivo*. This indicated that this protein had a lipid membrane anchor. However, PI-specific phospholipase C did not cleave the csA lipid anchor. Therefore, it was cleaved with nitrous acid and was subjected to acetolysis. The lipids released were different from those released from the *Trypanosoma* variant cell surface glycoprotein, which contains a PI-glycan anchor. Therefore, the lipid anchor of csA did not appear to be PI. The lipid anchor demonstrated resistance to weak alkali and sensitivity to strong alkali, indicating that the fatty acid was in an amide bond. Sphinomyelinase released a lipid with the chromatographic behavior of ceramide.

These results suggest that the csA protein has a ceramide-based lipid glycan membrane anchor. It is possible that this represents one member of a new class of protein-linked lipid structures that are resistant to PI-specific phospholipases.

♦ Considerable evidence has accumulated implicating the cell surface glycoprotein gp80 in the EDTA resistant cell-cell adhesion of aggregation phase cells of *Dictyostelium discoideum*. Two new reports add valuable information to the picture we have of this molecule. The first paper identifies the cell-binding domain and establishes that protein-protein interactions are responsible for the homophilic binding. The second article shows that gp80 has an unusual phospholipid anchor linking it to the cell membrane.

The article by Kamboj et al. reports a complete cDNA sequence of gp80 and notes certain discrepancies with a previously published sequence — most notably in the 5' upstream sequence. gp80-fusion proteins were expressed and tested for their ability to bind to cells and block the binding of authentic gp80. They localized the cell binding region to the stretch of amino acids between valine 123 and leucine 173. Significantly, this is the same region recognized by a monoclonal antibody which is directed towards a protein epitope and which specifically blocks cell-cell adhesion. Furthermore, binding between soluble and immobilized fusion protein confirms that the cell binding activity is homophilic and protein-protein in nature. However, the cell-binding region has two sites of N-glycosylation; therefore, glycosylation could play a role in modulating cell-cell adhesion, as is the case with verebrate N-CAM.

In the second report, Stadler et al. have rigorously analyzed the lipid anchor that attaches the gp80 molecule to the cell membrane. Using a combination of radioactive labeling and both chemical and enzymatic degradation procedures, they conclude that the gp80 molecule is anchored by a ceramide based lipid molecule. This is in contrast to the previously characterized phosphatdyl inositol-glycan anchor that has been characterized on the surface of *Trypanosoma* variant cell surface glycoproteins. The data are consistent with the interpretation that the gp80 molecule is attached through an ethanolamine residue to a glycan-inositol-phosphate-ceramide. However, the complete structural characterization of the gp80 anchor remains to be determined. It is possible that this novel anchor molecule is the prototype of another type of anchor molecule attached to glycoproteins. For example, mammalian proteins that have lipid anchors that are resistant to phosphoinositol-specific phospholipase C may be of this type. *Stephen Alexander*

Echinonectin: A New Embryonic Substrate Adhesion Protein
M. C. Alliegro, C. A. Ettensohn, C. A. Burdsal, H. P. Erickson, and D. R. McClay
J. Cell. Biol., 107, 2319—2327, 1988

Gelatin agarose affinity chromatography was used in an attempt to isolate sea urchin fibronectin. Fibronectin was not isolated. This report describes the discovery of a new molecule, echinonectin (EN).

Embryo extracts were subjected to gelatin agarose affinity chromatography, and EN eluted in a sharp peak at 8 M urea. When this fraction was analyzed by SDS-PAGE in the presence of a reducing agent (DTT), a predominant protein band of 116 kDa was revealed. Under nonreducing conditions the protein migrated as a dimer of 230 kDa. Isoelectric focusing experiments indicated that the protein carried a net negative charge.

Protease mapping indicated that the protein differed structurally from fibronectin (FN). Polyclonal sera was raised to purified EN. There was little cross-reactivity between FN and EN.

Immunofluorescence demonstrated a punctate distribution pattern in the unfertilized egg. After fertilization, EN was located on the cell surface and was then concentrated on the apical surface of ectoderm cells during germ layer differentiation. Mesenchyme blastula-stage embryos were dissociated and bound at a higher level to EN than to FN. By EM, EN appeared as a bow tie-shaped protein composed of two U-shaped monomers attached by disulfide bonds.

Based on quantity, distribution and adhesive properties, the novel echinoderm protein, echinonectin, is proposed to serve as a substrate adhesion molecule during sea urchin development.

♦ The extraembryonic matrix of the sea urchin is assembled after fertilization and consists of the fertilization envelope (see Weidman and Shapiro, 1989 Year Book) and the hyaline layer. The hyaline layer is a complex mixture of extracellular matrix molecules which interact with blastomeres of the developing embryo. For instance, one of the major components of the hyaline layer is the glycoprotein hyalin, which is derived from the exocytosis of cortical granules at fertilization and which has an important cell adhesion role during development (McClay and Fink, *Dev. Biol.,* 92, 285, 1982); Adelson and Humphreys, *Development,* 104, 391, 1988). Alliegro et al. have isolated and characterized another component at the embryonic cell surface which they refer to as echinonectin. Echinonectin was found to have high affinity binding to embryonic cells, and because it was isolated by its affinity to denatured collatgen (gelatin), has the characteristics of a molecule which links cells to their collagenous extracellular matrix. Thus, echinonectin may be a member of the functional family including fibronectin, although it does not appear to be encoded by a gene from the same family. Another interest in echinonectin is that although the molecule is secreted into the extraembryonic matrix after fertilization, it is not derived from cortical granules as are other components of the hyaline layer. Instead, echinonectin is derived from cytoplasmic vesicles of the egg which achieve polarized secretion at the surface of the zygote after fertilization. The nature of the vesicles containing echinonectin is as yet unknown, but they must involve a complex biology of sorting and storing molecules during oogenesis for controlled release following fertilization. *Gary M. Wessel*

The Role of Secondary Mesenchyme Cells during Sea Urchin Gastrulation Studied by Laser Ablation
J. Hardin
Development, 103, 317—324, 1988   5-4

Sea urchin gastrulation includes primary invagination during which the archenteron forms and secondary invagination, during which the archenteron elongates and secondary mesenchyme cells (SMC) extend long filopodial protrusions toward the animal region. It has been suggested that these filopodia could exert a traction that elongates the archenteron. Laser ablation of gastrula SMC has been used to investigate the role of SMC filopodia in secondary invagination.

Ablation of the SMC at the onset of secondary invagination permitted archenteron elongation, although the tip of the archenteron was halted at the two thirds gastrula stage. Ablation of all SMC at the two thirds gastrula stage halted all further elongation. If a few SMC were retained at this stage, a slow elongation occurred.

Therefore, significant elongation of the archenteron can occur without SMC filopodia traction. Archenteron elongation appears to involve two processes, filopodia-indepdent elongation, which depends on active cell rearrangement and filopodia-dependent elongation, which depends on mechanical tension exerted by the SMC filopodia.

♦ Autonomous repacking of cells provides a motive force for the morphogenesis of epithelial tissues throughout phylogeny, e.g., imaginal disc evagination during insect morphogenesis, budding in *Hydra*, and archenteron elongation during amphibian gastrulation. During sea urchin gastrulation, archenteron elongation is accompanied by contractions of filopodia extended by secondary mesenchyme cells and also appear to provide force used for archenteron elongation. In this paper, Hardin uses laser microsurgical ablation of secondary mesenchyme filopdia to determine how these filopodia may participate in gastrulation. His results demonstrate that elongation of the first two thirds of the archenteron is independent of filopodial contractions from the secondary mesenchyme cells. However, filopodia are important for the elongation and perhaps directional placement of the archenteron to the oral primordium during the final one third of archenteron elongation. These filopodia may provide such force by a stabilized adhesion to the basal lamina (or exposed ectodermal cell surface) followed by cytoskeletal contractions. Alternatively, the filopodium may signal its cell and neighboring cells for directional elongation of the archenteron independent of filopodial contractions (see Solursh and Lane, this volume).

In many species of sea urchins, the archenteron does not extend straight to the animal cap as it does in *Lytechinus*, the species used in this study. Instead, the archenteron may turn as much as 90 degrees to meet the oral primordium at the equator of the embryo (e.g., in *Tripneustes esculentus*). If secondary mesenchyme filopodia universally participate in the extension and in the directionality of archenteron elongation, laser microsurgery experiments in such species would provide compelling support for the mechanism deduced by Hardin. *Gary M. Wessel*

Extracellular Matrix Triggers a Directed Cell Migratory Response in Sea Urchin Primary Mesenchyme Cells
M. Solursh and M. C. Lane
Dev. Biol., 130, 397—401, 1988   5-5

Sea urchin embryo primary mesenchyme cells form filopodia that attach to the lining of the blastocoel, which is the substrate for migration. These cells will migrate *in vitro* on a vertebrate fibronectin substratum. In this system, sea urchin blastocoel extracellular matrix (ECM) deposits trigger directed migration. The role of the ECM in this directed migration was explored.

Small latex beads were coated with ECM and placed near isolated primary mesenchyme cells. Upon filopodial contact, the primary mesenchyme cell moved toward the bead. Cells could be detected moving from bead to bead and collecting beads on their surface. As the beads were free-floating, they were not functioning as substratum. Scanning EM was used to examine the filopodia-bead interaction. Filopodia were closely associated with the fibronectin-coated substratum and passed underneath the beads, with surface features that reflected ECM-coated bead association.

Primary mesenchyme cell filopodial contact with ECM appeared to provide a stimulus *in vitro* that led to a cell migratory response. The role of the ECM in providing a localized directional stimuli for cell migration should be considered.

♦ Filopodia appear to play a major role in directing cell migration in embryogenesis during, for example, neural crest cell migration, neurite outgrowth, and amphibian mesodermal sheet extension. The mechanisms involved in filopodial-directed migration are unclear, but Solursh and Lane have devised a novel approach to examine this phenomenon. These investigators observed the migration of sea urchin primary mesenchyme cells in culture which were exposed to latex beads coated with embryonic extracellular matrix. They find that filopodia selectively contact beads that are coated with extracellular matrix molecules, and that these contacts lead to a directed migration of the cell to the beads. These results are important for several reasons. First, this directed cell movement is not mediated by an active filopodial contraction drawing the cell body toward the contact site (see also Hardin, this volume). Instead, there must be some signaling by the filopodial-bead contact to the migratory direction of the cell body. Second, the migratory behavior of the mesenchyme cells *in vivo* is similar to that seen by Solursh and Lane. Primary mesenchyme cells migrate with long filopodial processes to specific sites in the blastocoel indicative of directed cell migration. Even cells placed aberrantly in the blastocoel reach their correct destinations (see Ettensohn and McClay, 1989 Year Book). Solursh and Lane's approach should prove useful both in understanding the filopodial participation in directed cell migration and in elucidating the cell-extracellular matrix interactions that initiate this signaling. The later application should be particularly useful in analyzing the developmental roles of the increasing list of primary mesenchyme cell surface-specific and extracellular matrix molecules that recently have been identified and cloned. *Gary M. Wessel*

Altered Expression of Spatially Regulated Embryonic Genes in the Progeny of Separated Sea Urchin Blastomeres
D. L. Hurley, L. M. Angerer, and R. C. Angerer
Development, 106, 567—579, 1989   5-6

To examine the importance of cell to cell interactions in sea urchin development, 16 cell stage embryos were dissociated and incubated in calcium-free sea water. The levels of 11 different (Figure 5-6) mRNAs were compared in separated and control embryos at the mesenchyme blastula stage.

The levels of mRNAs that accumulate or decay early and uniformly in blastomeres were similar in separated and intact embryo cells. Messages that accumulate at the vegetal pole were present at normal or increased levels in separated embryo cells, as compared to intact embryos. The actin mRNA, CyIIa, which is normally restricted to mesenchyme cells, was expressed in four to five times as many cells in separated embryos (Figure 5-6A). The messages that are differently regulated in oral and

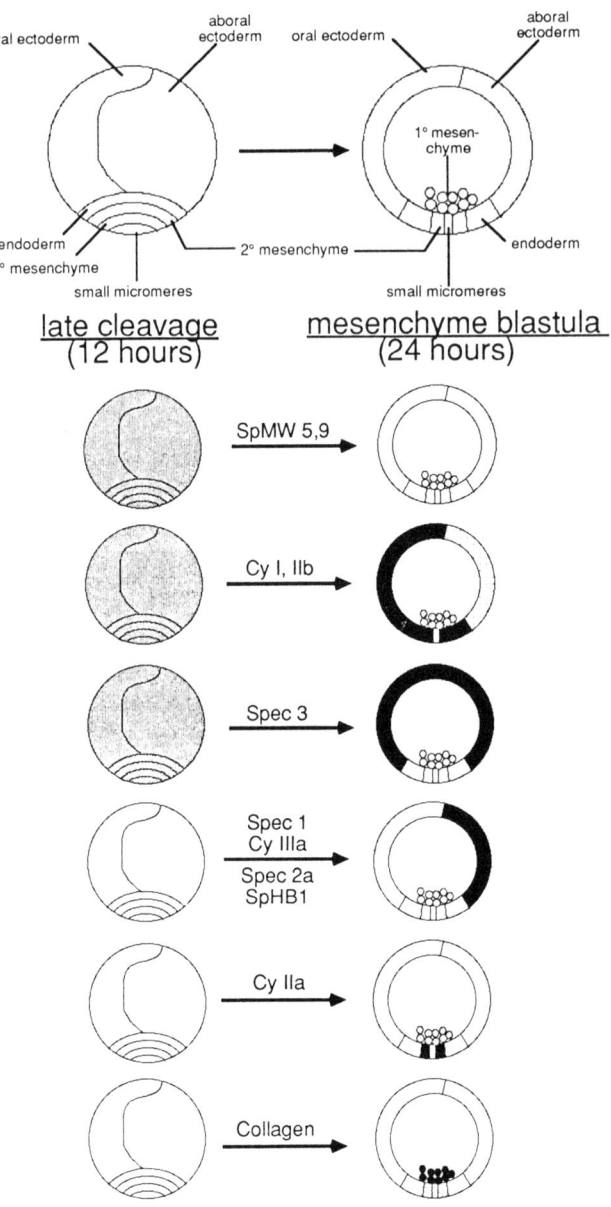

FIGURE 5-6. Distribution of individual mRNAs during cleavage and at mesenchyme blastula stage. The diagrams at top depict fates of regions of the 12-h (140 to 200 cell) embryo (left), and 24-h mesenchyme blastula (right). Below each diagram are shown the spatial distributions of the mRNAs determined at either 12 (left) or 24 (right) h of development. Unfilled regions indicate the absence of detectable mRNA and increasing abundance is denoted by progressively darker shading. (From Hurley et al., *Development,* 106, 567—579, 1989. ©The Company of Biologists Limited.)

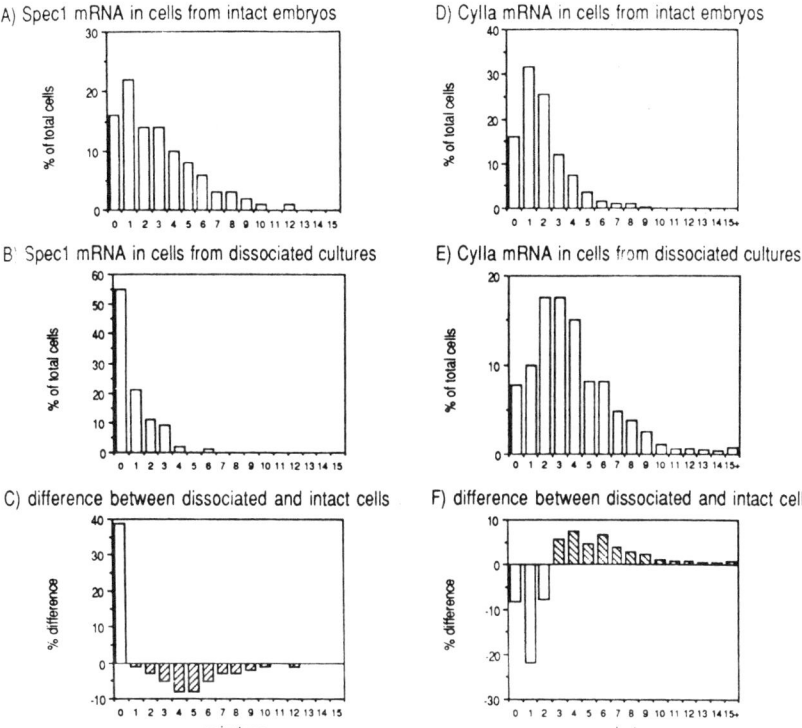

FIGURE 5-6A. Determination by *in situ* hybridization of the number of cells expressing Spec1 and actin CyIIa messages. Cells dissociated from inact embryos at 24 h (panels (A, D) or cells separated at 16-cell stage and cultured until 24 h (B, E) were hybridized with either Spec1 (A, B) or actin CyIIa (D, E) probes. To avoid bias in the histograms due to the presence of a few percent of large cells, values for grains/cell were converted to grains/arbitrary area using approximate cell areas determined with an ocular micrometer. The % of the total number of cells counted is shown on the y-axis for each of the grain density classes on the x axis. The total number of cells for each panel was (A) 370, (B) 501, (D) 522, (E) 754. Panels C and F are the difference histograms obtained by subtracting values for cells from intact embryos for those dissociated cells. This presentation illustrates the distribution of labeling intensities for dissociated cells (shaded bars) that replace a similar percentage of cells in samples from control embryos (open bars). Shaded distributions above the line represent the population of cells expressing an mRNA in separated cells but not in the intact embryo; distribtuions below the line represent the reciprocal case in which cells expressing an mRNA in the inact embryo fail to do so in separated cells. (From Hurley et al., *Development*, 106, 567—579, 1989. ©The Company of Biologists Limited.)

aboral ectoderm were neither expressed nor regulated properly in separated embryo cells.

Therefore, early interactions among sea urchin embryo cells are important for positive and negative regulation of the genes involved in specification of cell fate.

♦Blastomeres of the early sea urchin embryo exhibit plasticity in their potential, and cell-cell contact of the blastomeres is important in determining the fate of each blastomere (Horstadius, 1973; see also Livingston and Wilt in this volume). Hurley et al. have begun to examine how cell-cell contacts effect each blastomere at the molecular level, i.e., specific mRNA accumulation, which would signify developmental decisions well prior to the morphological manifestations described by Horstadius (for discussion of this approach, also see Gurdon, *Development,* 99, 286, and Wilt, *Development,* 100, 559). The data suggest that the earliest events of accumulation and decay of spatially uniform mRNAs during cleavage are cell autonomous, as is the initial accumulation of spatially restricted mRNAs expressed in cells at the vegetal pole or in whole ectoderm. However, accumulation of mRNAs that mark distinction between oral and aboral ectoderm is significantly reduced suggesting that some type of cell communication (either cell-cell or cell-extracellular matrix) is required to manifest the fate of these blastomeres. These observations may correspond to conclusions from experimental embryology that the animal-vegetal axis is maternally specified but the oral-aboral axis is established by inductive events and is labile. Another interesting observation presented here is the evidence for negative interactions among cells which in the intact embryo apparently repress the capacity of many cells to express the CIIa actin mRNA. Ettensohn and McClay (see this Year Book, 1989) have shown that secondary mesenchyme cells are repressed from adopting a primary mesenchyme fate *in vivo;* similarly, the observation here indicates that other cells of the embryo also have the capacity to express a message characteristic of secondary mesenchyme fate in the intact early embryo. Thus, although determination along the animal-vegetal axis and regulation of fate in experimentally manipulated embryos have traditionally been interpreted in terms of positive inductive influences, it is becoming clear that the interactions are complex and also include important negative influences. *Gary M. Wessel*

The *Caenorhabditis elegans lin-12* Gene Encodes a Transmembrane Protein with Overall Similarity to *Drosophila Notch*
J. Yochem, K. Weston, and I. Greenwald
Nature, 335, 547—550, 1988   5-7

The *lin-12* gene of *C. elegans* is involved in developmental decisions controlling cell fate. This gene has been cloned and the complete sequence is reported.

Genomic clones containing 13.5 kb and cDNA clones containing 4.5 kb of DNA from the *lin-12* region were sequenced. The predicted protein appeared to encode an integral membrane protein. A repeated motif found in epidermal growth factor was detected. Another region of this protein contained a repeated motif found in the yeast cell cycle control genes, cdc10 and SW16. There was extensive overall homology to the *Drosophila Notch* protein.

The similarity of *lin-12* and *Notch*, two genes involved in cell fate decisions, suggests that similar proteins may be involved in developmental decisions in many organisms.

♦ This paper describes the complete sequence of the *lin-12* gene of *C. elegans*, a gene that determines the fates of certain bipotential cells that require cell-cell interactions for the correct specification of their fates (Greenwald et al., *Cell,* 34, 435—444, 1983). *lin-12* exhibits overall sequence similarity to *Notch*, a *Drosophila* gene that is required for a decision between neural and epidermal cell fates (Wharton et al., Cell 43, 567—581, 1985; Kidd et al., *Mol. Cell Biol.,* 6, 3094—3108, 1986; Lehman et al., *R. Arch. Dev. Biol.,* 192, 62—74, 1983). Both predicted proteins have potential signal sequences at their N-terminus and potential membrane-spanning sequences internally. Both have EGF-like repeats extracellularly (*lin-12* has 13, *Notch* has 36). The *lin-12* EGF-like repeats are interrupted by 2 exons. Both predicted proteins include three tandem copies of a cysteine-rich extracellular motif called the *lin-12/Notch* repeat, and both have a putative intracellular domain which is primarily composed of a series of six copies of the *cdc10*/SW16 repeated motif implicated in cell cycle progression in yeast (Breeden and Nasmyth, *Nature,* 329, 651—654, 1987). The fact that *lin-12* and *Notch* appear to be transmembrane proteins suggests that they may serve as receptors for signals in cell-cell interactions that determine fate or possibly act as signals in these interactions. Mosaic analysis has shown that *lin-12* acts cell autonomously in cell-cell interations and is therefore probably a receptor (Seydoux and Greenwold, 1989). Mosaic analysis of *Notch* functions has given conflicting results (see Hoppe and Greenspan, *Cell,* 46, 773—783, 1986 vs. Technau and Campos-Ortega, *Proc. Natl. Acad. Sci. U.S.A.,* 84, 4500—4504, 1987), but its structural and functional (fate determination) similarity to *lin-12* suggests that it may also act as a receptor. The importance of this paper as stated in the last sentence is that "the remarkable similarity of *lin-12* and *Notch*, two genes involved in decisions of cell fate in distant phyla, suggests that proteins of this general structure may be ubiquitous and involved in other decisions of cell fate."
Joseph G. Culotti

---

glp-1 **and** *lin-12*, **Genes Implicated in District Cell-Cell Interactions in** *C. elegans*, **Encode Similar Transmembrane Proteins**

J. Yochem and I. Greenwald
Cell, 58, 553—563, 1989

Low stringency DNA hybridization was used to search for genes in the C. elegans genome that were related to lin-12, a gene that specifies cell fate during development.

Only one weakly hybridizing signal was detected in this screen. The next paper describes the identification of this gene as glp-1. The gene was sequenced. It was slightly smaller than lin-12 and there was approximately 50% overall homology between the exons of these two genes. The glp-1 gene, like lin-12, also appeared to be an integral membrane protein and to contain homology to epidermal growth factor, yeast cdc10/SW16, and Notch.

A gene that is very similar to lin-12 and to the Drosophila gene, Notch, (Figure 5-8) has been identified in the C. elegans genome. This gene appears to be the glp-1 gene, which is involved in control of gametogenesis. It is possible that a gradient of glp-1 product controlled by the distal tip cell regulates gametogenesis in each arm of the hermaphrodite gonad (Figure 5-8A).

♦ This paper shows that the *glp-1* gene in *C. elegans* is similar in structure to *lin-12* of *C. elegans* and *Notch* of *Drosophila*. *glp-1* like *lin-12* and *Notch* is known to act cell autonomously in intercellular signaling, probably as a receptor. While *lin-12* and *Notch* appear to be involved in fate determination, *glp-1* acts in the germline to receive a graded signal from the distal tip cell of the gonad which inhibits the germ cells from entering meiosis (Austin and Kimble, *Cell*, 51, 589—599, 1987). It also acts in the

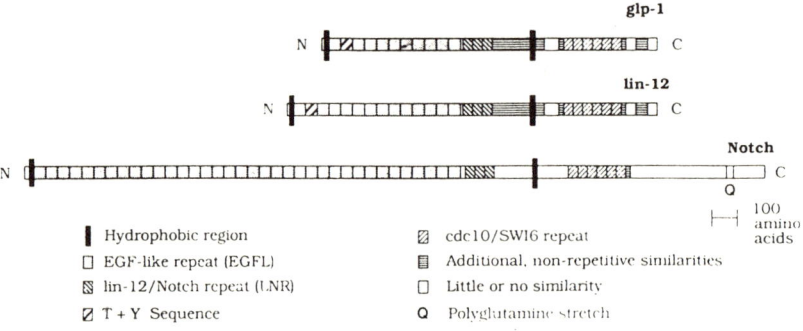

FIGURE 5-8. Structural similarities among the products of *glp-1*, *lin-12*, and *Notch*. The 1295 residue product of *glp-1* is compared in schematic form with the 1429 residue *lin-12* product and the 2703 residue Drosophila *Notch* product. The three products are aligned with respect to their presumed membrane-spanning domains. (From Yochem/Greenwald, *Cell*, 58, 553—563, 1989. ©Cell Press.)

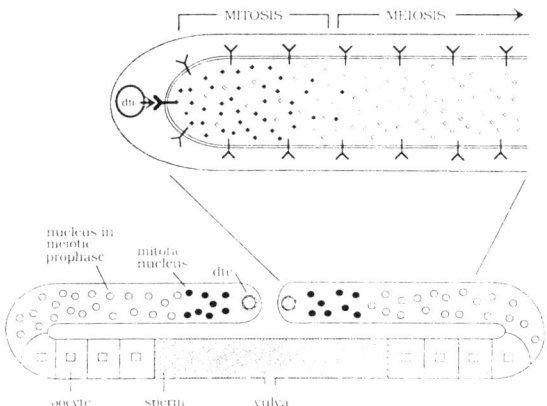

FIGURE 5-8A. A model for the function of *glp-1* in the germline. Part of the distal syncytium of one arm of the gonad of an adult hermaphrodite is shown enlarged above a schematic representation of the complete gonad, which is based on drawings by Hirsh et al. (1976), Kimble and White (1981), and Austin and Kimble (1987). *glp-1* gene products are represented by Y structures spanning the lipid bilayer that encloses the syncytium. The germ line nuclei of the syncytium of actually present in alcoves of the membrane (Hirsh et al., 1976). *glp-1* molecules at the distal end tip bind a singal molecule produced or organized by the somatic distal tip cell (dtc). Ligand-bound *glp-1* molecules activate second messenger molecules (closed diamonds) within the syncytium that inhibit meiosis or maintain mitosis of germ nuclei. The messenger molecules diffuse through the syncytium and decay to an inactive state (open diamonds) with time. Reactivation takes place only at the distal tip. (From Yochem/Greenwald, *Cell,* 58, 553—563, 1989. ©Cell Press.)

early embryronic pharyngeal precursor cell ABa (or one of its descendents) to receive a signal from a P1-derived cell enabling expression of anterior pharyngeal muscle cells (Priess et al., *Cell,* 51, 601—611, 1987). (The three cell embryo comprises ABa and ABp which are sisters, and P1, their aunt). *glp-1* action in ABa is strictly maternally inherited, that is, the progeny of mothers homozygous for the *glp-1* mutation are defective, regardless of their own genotype (Priess et al., *Cell,* 51, 601—611, 1987). This is an important paper because it shows that a number of cell surface molecules involved in intercellular signaling have similar amino acid sequences which include extracellular epidermal growth factor (EGF) domains (*gl-1* has 10, *lin-12* has 13, and *Notch* has 36), a hydrophobic putative transmembrane domain, and intracellular cdc-10/sw16 domains (*glp-1* has 6, *lin-12* has 6, *Notch* has 0). *lin-12* and *glp-1* have similar intron positions and are closely linked genetically and physically. These genes probably arose by gene duplication with *glp-1,* the likely ancestor to *lin-12,* since it is required during embryogenesis and during germline proliferation, whereas *lin-12* is a nonessential gene (Greenwald et al., *Cell,* 34, 435—444, 1983). *Joseph G. Culotti*

Transcript Analysis of *glp-1* and *lin-12*, Homologous Genes Required for Cell Interactions during Development of *C. elegans*

J. Austin and J. Kimble

Cell, 58, 565—571, 1989

This paper describes the cloning of *glp-1,* a gene involved in the development of the *C. elegans* germline.

The *glp-1* gene was cloned by a combination of proximity to the previously cloned *lin-12* and mutation analysis. The *glp-1* gene was found to be identical to a *lin-12* homolog detected in a screen of *C. elegans* (see previous paper). On Northern blots of adult hermaphrodite total RNA, a single 4.4 kb *glp-1* transcript was detected.

The levels of *glp-1* and *lin-12* transcripts were analyzed in staged wild-type hermaphrodites. The *glp-1* transcript was abundant in embryos, decreased during larval stages, and was abundant in adults. The *lin-12* transcript was abundant in embryos and was present at a low level in larvae and adults. RNAs were isolated from synchronized embryos and transcripts were quantitated. The *glp-1* transcript was abundant in early embryos, but not in middle and late stage embryos. The *lin-12* transcript was abundant in early embryos, less abundant in middle stage, and not abundant in late stage embryos.

To distinguish between germline and somatic expression, wild-type hermaphrodite RNA levels were compared to the levels in mutants with reduced germ cell number. The level of the *glp-1* transcript did not increase after the third larval stage in mutant animals, indicating that the increase in *glp-1* seen in wild-type animals is due to germline expression. However, the continued expression of *glp-1* in the mutant animals during this period indicates somatic expression. *lin-12* was expressed at similar levels during development in both wild type and mutant animals, indicating somatic expression. However, the level of *lin-12* expression was higher in wild-type adults, suggesting some *lin-12* expression in the adult germ line.

This paper reports the cloning of *glp-1* and its identity with a separately isolated *lin-12* homolog. The synthesis of *lin-12* in early embryos and of *glp-1* transcripts in somatic tissues, detected in these experiments, suggest that these genes have previously undetected roles in development. It is possible that these two similar genes have overlapping functions. Double mutation analysis is being carried out to investigate the orginal function of the common ancestor of these two genes.

♦This paper compares the developmental expression of *lin-12* and *glp-1* transcripts in wild type *C. elegans* and in mutants which fail to make germline cells. Both *glp-1* and *lin-12* are required for cell interactions that regulate cell fate (Austin and Kimble, *Cell,* 51, 589—599, 1987, Priess et al., *Cell,* 51, 601—611, 1987; Greenwald et al., *Cell,* 35, 435—444,

1983). In addition, both genes have been shown to function in the receiving cell rather than in the signalling cell by genetic mosaic analysis (Austin and Kimble,*Cell*, 51, 589—599, 1987; Seydoux and Greenwald, *Cell*, 57, 1237—245, 1989). *glp-1* acts in the germline of the developing gonad, to receive a signal from a somatic cell of the gonad which inhibits meiosis, whereas *lin-12* is needed for postembryonic interactions between certain somatic cells with equivalent development potentials for these cells to adopt different fates. One might, therefore, expect *glp-1* expression to be confined to the germline and *lin-12* expression to be confined to postembryonic somatic cells. While the expected expression patterns do occur, *glp-1* surprisingly is also expressed in the larval soma, and *lin-12* is expressed abundantly in early embryos (and may be expressed in adult germline). These results suggest that *lin-12* and *glp-1* may have overlapping functions, but the experiments do not provide the cellular resolution required to identify which cells produce each RNA. However, preliminary analysis of the *glp-1-lin-12* double mutant reveals defects not seen in either single mutant (J. Kimble, unpublished data), lending further support to the possibility that these genes may functionally substitute for each other in certain developing cells. *Joseph G. Culotti*

Cell-Cell Interaction in the Drosophila Retina: The *bride of sevenless* Gene is Required in Photoreceptor Cell R8 for R7 Cell Development
R. Reinke and S. L. Zipursky
Cell, 55, 321—330, 1988

5-10

The photoreceptor neurons (R-cells) of the Drosophila compound eye can be divided into three classes: R1 through R6, R7 and R8. The *sevenless (sev)* mutation results in loss of the R7 neuron. It has been proposed that the *sev* product is a receptor for a signal necessary for R7 development. This paper describes the identification and characterization of another mutation, *bride of sevenless (boss)*, also required for R7 development.

After screening approximately 10,000 X-ray mutagenized flies for a phenotype similar to *sev*, the *boss* mutation was detected. The phenotype of the *boss/sev* double mutant was identical to either mutation alone. In both *boss* and *sev*, R7 was missing while R1 through R6 and R8 were normal.

Mitotic recombination in marked flies was induced by X-rays to create mosaic eyes that contained both *boss–* and *boss+* cells (Figure 5-10). The mosaic eyes were sectioned and the phenotype and genotype determined. The results indicated that R7 development required *boss* expression in the R8 cell of the same ommatidia. The *boss* expression pattern in R1 through R6 was irrelevant for R7 development.

158   Year Book of Developmental Biology

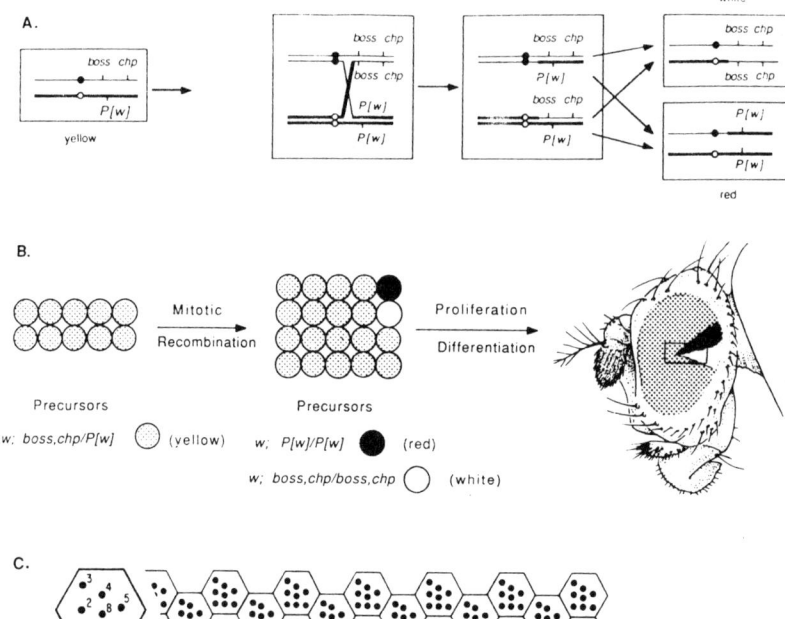

FIGURE 5-10. Genetic mosaic analysis. (A) Schematic diagram of a recombination event generating clones of different genotypes. As shown above, an X-ray-induced mitotic recombination event can result in progeny cells of different genotype. One daughter cell is homozygous for the recessive allele of interest, *boss*, and a closely linked mutation, *chp*, causing an altered cellular morphology. Thus, the *chp* phenotype indicates the boss genotype of each R-cell. This cell no longer carries a copy of the transposed *white* gene and will give rise to an unpigmented region of the compound eye. The other daughter cell is homozygous for the wild-type allele of *chp* and *boss*. These cells carry two copies of the transposed *white* gene, and thus form red clones in the adult eye. (B) Generation of precursors of different genotypes and the formation of a mosaic eye. The cells of the compound are derived from the eye imaginal disc. Dividing cells are subjected to X-irradiation during the first 48 h of development. Mitotic recombination in a heterozygous precursor cell produces daughter cells that proliferate during larval development producing clones of retinal precursors of different genotype. Pattern formation and differentiation commence during the third larval instar. Due to the lack of any strict cell lineage relationship between cells, ommatidia at the clone borders are often comprised of cells of different genotype (see C). Although the P[w]F4-3 permits identification of clones in the adult eye by color (due to expression in pigment cells), the low expression of this transposed white gene does not permit scoring of pigment granules in R-cells. The *boss* genotype of cells in inferred from the R-cell phenotype caused by the closely linked *chp* mutation. A second series of mosaics, analogous to that shown above, was generated by X-irradiating w;boss/P[w]D3. The expression of pigment from this transposed white gene is sufficient to permit scoring of pigment granules in R-cells as an indication of the boss genotype. The mosaic ommatidia shown in (C) is from the region outlined in the schematic diagram of the compound eye in (B). (C) Analysis of a mosaic ommatidia. Mosaic eyes were fixed and embedded in plastic. Thick sections (= 2 μm) were analyzed by light microscopy and the *chp* phenotype, indicating the *boss* genotype, of each R-cell was scored. Due to the variability of the *chp* marker, unambiguous assignments of the *chp* phenotype were determined by transmission electron microscopy of thin sections of the same eye. (●) Indicates a normal gene at the *chp* locus, whereas (○) indicates a mutant form. (From Reinke/Zipursky, *Cell*, 55, 321—330, 1988. ©Cell Press.)

This mosaic analysis demonstrated that cell to cell interaction between R8 and R7 is required for the development of R7. It is possible that the *sev* product is a receptor for *boss* or a *boss*-dependent signal produced by R8.

♦ The Drosophila eye is composed of about 800 simple eyes or ommatidia, each containing eight photoreceptor cells (R1 to R8). R7 is the primary UV light-sensitive photoreceptor. The authors used a powerful but simple screen to isolate mutants defective in R7 function, taking advantage of two unique properties of the eye system. One obvious but important feature is that photoreceptors are not essential for the fly's survival or reproduction. Another property is that R7 is the primary UV light-detecting photoreceptor cell. Moreover, when given a choice, wild-type flies are attracted to UV light ten times more effectively than to visible light. In flies defective in R7 function the opposite is true. A screen of 10,000 mutagenized lines yielded one mutant that lacked R7. Although the mutation has an identical phenotype to that of the previously characterized locus *sevenless*, the new mutation turned out to be in a different gene and this gene was named *bride of sevenless or boss*. Careful mosaic analysis has revealed that unlike *sevenless* (see separate report in this volume), *boss* is nonautonomous. That is, while expression of *sevenless* is required only in R7 itself for its own differentiation, *boss* expression is dispensable in R7 but absolutely required in R8. The *sevenless* gene had been shown to encode a putative transmembrane receptor protein with a cytoplasmic tyrosine kynase domain. The discovery that *boss* is required in R8 for R7 differentiation suggests a model whereby *boss* encodes an extracellular ligand (produced by R8) that is recognized by the *sevenless* protein (produced by R7). Moreover, *boss* must have a short range of action restricted to its own ommatidium since mosaic analysis showed that a mutant ommatidium completely surrounded by ommatidia containing *boss+* R8 cells still shows mutant phenotype.

The simple model suggested by the results require more work to further clarify the underlying mechanisms of cell differentiation. For instance, since *boss* has not yet been cloned, it is not known whether *boss* encodes the ligand itself or alternatively a transcription factor that activates other downstream genes coding for the ligand. Another issue that is not yet understood relates to the observation that *sevenless* protein is expressed not only in R7 but also in other photoreceptors that make contact with R8. Whey don't these *sevenless*-expressing cells also differentiate along the R7 pathway? What regulatory events take place in the prospective R7 cell after the *sevenless* protein receives the signal from R8? Answers to these and other important questions may allow us to make a quantal leap toward the understanding of the relationships between cell-cell communication and cell differentiation in the Drosophila eye. *Marcelo Jacobs-Lorena*

Mesoderm Induction in *Xenopus laevis*: Responding Cells Must be in Contact for Mesoderm Formation but Suppression of Epidermal Differentiation can Occur in Single Cells
K. Symes, M. Yaqoob, and J. C. Smith
Development, 104, 609—618, 1988

When *Xenopus* embryos are cultured in calcium- and magnesium-free medium, the blastomeres lose adhesion, but continue dividing. If the cells are allowed to retain contact, they will form mesoderm and epidermal structures when allowed to reaggregate. If the cells are dispersed, they will only form epidermis when allowed to reaggregate. This suggests that a mesoderm-inducing signal is transmitted when the cells are allowed to retain contact. To study the transmission of this signal, cells were completely dispersed and then exposed to XTC mesoderm-inducing factor (MIF).

Exposure of dissociated embryo cells to XTC-MIF did not induce mesoderm formation in these cells. However, MIF inhibited epidermal differentiation in reaggregated cells. This suggests that an early stage in mesoderm differentiation involves suppression of epidermal differentiation.

To determine whether MIFs can suppress epidermal differentiation in single cells, animal pole explants were completely dissociated and cultured with XTC-MIF or bovine FGF and then processed for immunofluorescence with an anti-keratin antibody. Both of these factors reduced the proportions of cells expressing keratin in a dose-dependent fashion. Therefore, both factors suppressed epidermal differentiation, although mesoderm formation was never detected. Later stage embryos required higher concentrations of MIF to suppress epidermal differentiation. By stage 12, epidermal differentiation could not be blocked (Figure 5-11).

Mesoderm induction appears to involve two steps. The first step can occur in single cells and suppresses epidermal differentiation. The second step requires cell to cell contact and results in the formation of mesoderm.

Loss of Competence in Amphibian Induction Can Take Place in Single Nondividing Cells
R. M. Grainger and J. B. Gurdon
Proc. Natl. Acad. Sci. U.S.A., 86, 1900—1904, 1989

Xenopus embryo ectodermal tissue loses its competence to form mesoderm during gastrula. The effect of intercellular interactions, cell division, and protein synthesis on this loss of competence were assessed in isolated animal cap ectoderm cells.

Ectodermal tissue was isolated from the stage of interest and either

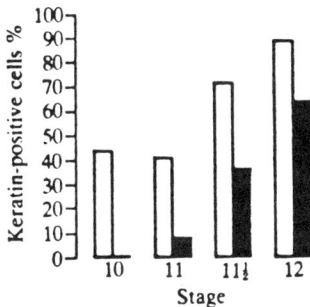

FIGURE 5-11. Almost all animal pole blastomeres are committed to epidermal differentiation by stage 12. Animal pole blastomeres were obtained from ebryos at the indicated stages and dispersed by incubation in CMFM. They were cultured in the absence (open bars) or presence (solid bars) of 2.6 µg ml^{-1} partially purified XTC-MIF (20 units ml^{-1}). At stage 10 XTC-MIF suppressed epidermal differentiation in all the cells, but at later stages increasing numbers of cells become refractory to the factor. (From Symes et al., *Development,* 104, 609—618, 1988. ©The Company of Biologists Limited.)

cultured or subjected to the experimental treatments. The ectoderm was then placed in contact with inductively active vegetal tissue to assess competence to form mesoderm. Muscle-specific actin RNA and protein synthesis were the mesoderm-specific markers examined.

When animal cap ectoderm was isolated from the rest of the embryo at stage 9 and cultured to stage 11, it was no longer competent to form mesoderm. If animal cap cells were dissociated, competence was lost at the same time. If dissociated cells were incubated in a solid gelatin matrix that prevents all cell to cell contact and also inhibits cell division, competence was lost at the normal time. When cyclohexamide was used to inhibit protein synthesis, there was no significant effect on the timing of the loss of competence to form mesoderm.

Therefore, cell to cell contact, cell division, and protein synthesis do not appear to be important in the loss of competence to form mesoderm by ectodermal cells. Competence must therefore be lost by an autonomous process that can operate in a single nondividing cell.

A Community Effect in Animal Development
J. B. Gurdon
Nature, 336, 772—774, 1988   5-13

To study Xenopus embryonic induction by vegetal cells of animal cells to form muscle, amphibian cells were cultured in solid gels containing cytochalasin to inhibit cell division and mobility during the induction process.

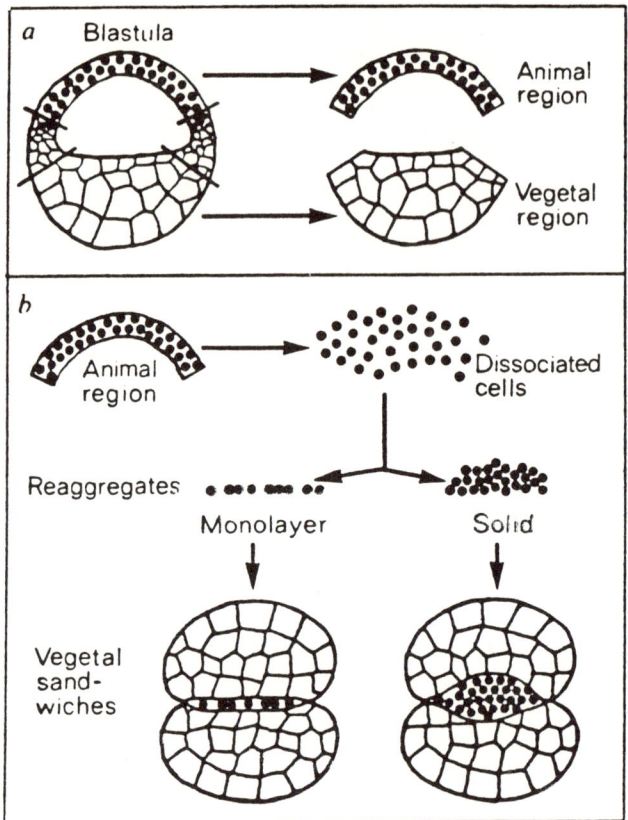

FIGURE 5-13. Diagrams illustrating the procedures used to place animal cells of a blastula into vegetal sandwiches in either monolayer or solid configurations. Vegetal cells are indicated by their outlines as large cuboidal cells; animal cells are shown diagrammatically as black circles. (a) A blastula is cut into animal and vegetal regions, discarding the equatorial cells. (b) Animal regions (radioactively labeled) are dissociated in Ca^{2+}-free medium into their component cells. These are reaggregated in Ca^{2+}-containinbg medium into monolayer or solid configurations, and then incorporated in between two undissociated vegetal pieces of unlabeled blastulae. The vegetal sandwiches are cultured until stage 28, when they are fixed, sectioned, and processed. (From Gurdon, *Nature*, 336, 772—774, 1988. With permission.)

Xenopus blastula animal regions were dissociated and then reaggregated as either monolayers or solid reaggregates. The reaggregates were sandwiched between two vegetal blastula pieces (Figure 5-13). Animal cells were prelabeled to distinguish them from vegetal cells. After the inductive period, the tissue sandwiches were fixed, sectioned, reacted with muscle-specific antibody and Hoechst nuclear stain, and autoradiographed. Each cell was scored by autoradiography for animal origin, by Hoechst stain for normal nuclear structure, and by immunofluorescence for induction.

Animal cells reaggregated as monolayers were not induced to express muscle-specific actin by the vegetal tissue sandwich. However, solid reaggregates were induced to express muscle-specific genes in the same assay.

This result is interpreted as evidence of a community effect, in which the ability of a cell to respond to induction is increased by the similar response of surrounding cells. This effect may be important in the creation of sharp boundaries between tissue types during induction.

♦The reason for grouping these papers is that each shows that the formation of mesoderm, muscle in particular, is a process involving multiple steps with different requirements for cell-cell communication. Grainger and Gurdon found that they could block cell communication, as well as protein synthesis, without stopping the time-dependent loss of competence of ectoderm to respond to vegetal mesodermalizing signals. Thus, competence loss, like expression of epidermal keratins, is a characteristic of ectoderm that is programmed in a cell-autonomous manner. The duration of the competence period is an important parameter because, in concert with cell movement and the generation of multiple inducing signals during gastrulation, it allows for correct spatial organization of induced tissues. These investigators also noted a coincidence between the loss of competence and reduction in the tendency of ectodermal and endodermal cells to adhere to each other. It is known that ectoderm does not have to adhere to endoderm for mesoderm induction to occur (the signal can pass through a porous barrier), so the loss of adhesiveness should note *cause* the loss of competence. Nevertheless, this correlation is interesting.

Symes et al. confirmed earlier findings that cell-cell communication prior to gastrulation is required for the development of muscle and extended this to include other mesodermal derivatives such as notochord. They show that exposure of dissociated embryonic blastomeres to XTC-derived mesoderm-inducing factor (MIF) does not substitute for this cell-cell communication; such cells still fail to make mesodermal markers after reaggregation. However, another effect of XTC-MIF does occur under these conditions: the suppression of epidermal differentiation was observed if the cells were treated prior to the end of gastrulation (see Figure 7 from their paper). Perhaps XTC-MIF triggers the first step in a multistep mesoderm-induction process.

The existence of secondary signal(s) is also revealed as the "community effect" described in the paper by Gurdon in *Nature:* muscle formation will not take place when the mass of responding ectodermal cells is too small or thinly spread (see Figures 2 and 3). Gurdon argues that this effect could act to sharpen the boundaries of an induced tissue by requiring large blocks of cells in order to complete the induced response. See the papers by London et al. and Savage and Phillips for discussion of a possibly related phenomenon in neural induction and the paper by Rosa for discussion of one other ectodermal response to MIF that occurs without the need for further cell communication, the expression of the Mix.1 gene. *Thomas D. Sargent*

Effects of Altered Expression of the Neural Adhesion Molecule, N-CAM, on Early Neural Development in Xenopus Embryos
C. Kintner
Neuron, 1, 545—555, 1988 5-14

Xenopus embryos the neural cell adhesion molecule, N-CAM, is expressed at low levels in the early embryo. Expression of N-CAM increases in ectoderm as the neural plate and tube are formed. This suggests that N-CAM may be involved in the regulation of early neural morphogenesis. To assess the role of N-CAM expression in early neural development, N-CAM transcripts were injected into Xenopus embryos such that injected embryos expressed high levels of N-CAM on the surface of both induced and noninduced ectodermal cells throughout gastrulation and nuerulation.

Abundant indiscriminate expression of N-CAM throughout neurulation in Xenopus embryos did not effect the formation of the neural tube. However, ectopic expression of N-CAM in epidermis cuased thickening of the epidermal layer. Ectopic expression of N-CAM in somitic mesoderm caused kinking of the body axis and disorganization of the somites.

Overexpression of N-CAM in the early Xenopus embryo does not affect the segregation of cells into different germ layers nor the normal movements of gastrulation and neurulation. Differential N-CAM expression does not appear to control morphogenesis in the early embryo.

♦The neural cell adhesion molecule, N-CAM, is prominently expressed on neuronal cells, as well as some other cell types. During embryogenesis, N-CAM is expressed in the neural plate shortly after neural induction. Because of its distribution and known ability to promote neuronal adhesion, it has been proposed that N-CAM may be involved in formation of the nervous system. However, this proposition has never been directly tested.

To study this question, Kintner has utilized the approach of "overexpression" of an adhesion molecule in early embryos. Frog embryos represent a particularly good system for this type of experiment since there is essentially no growth until the feeding larval stage. By injecting N-CAM RNA into the fertilized egg or one cell of the 2-cell stage embryo and allowing further development, one obtains an embryo with bilateral or unilateral, respectively, overexpression of N-CAM. The extra injected RNA transcripts produced at least a 50-fold increase in N-CAM expression.

Even after marked overexpression of N-CAM, neurulation occurred properly in the injected embryos. Although the neural tube appeared to form properly, some defects in the structure of the epidermis and somites were observed. These experiments show that proper regulation of N-CAM is not necessary for nervous system formation. It remains to be shown, however, whether the absence of N-CAM causes improper neurulation. This study presents a novel approach for studying the role of adhesion molecules by overexpression. *Marianne Bronner-Fraser*

Timing in the Regulation of Neural Crest Cell Migration: Retarded "Maturation" of Regional Extracellular Matrix Inhibits Pigment Cell Migration in Embryos of the White Axolotl Mutant
J. Lofberg, R. Perris, and H. H. Epperlein
Dev. Biol., 131, 168—181, 1989   5-15

In embryos of the Mexican axolotl, pigment cells derived from the neural crest (NC) migrate during development. In the white mutant, this migration does not occur normally, due to the inability of the embryonic epidermis to support this migration. To analyze whether the subepidermal extracellular matrix (ECM) is defective for migration, subepidermal ECM was adsorbed onto membrane microcarriers and transplanted from and to white and normal embryos with and without ECM.

Control carriers without ECM and subepidermal ECM from white donors did not affect NC cell migration in white or normal embryos. Subepidermal ECM from normal donors triggered NC cell, including pigment cell, migration in both white and normal embryos (Figure 5-15). Subepidermal ECM from older white embryos also stimulated this migration.

It appears that the subepidermal ECM from white embryos is transiently defective as a substrate for pigment cell migration during the period when pigment cells are competent to migrate. Therefore, the action of the *d* gene retards ECM maturity in certain regions. The restricted pigment in these mutant axolotl larva is generated by this heterochrony.

♦Because neural crest cells migrate along pathways that contain abundant extracellular matrix molecules, it has been assumed that the matrix is important for cell migration. However, few studies have addressed the role of cell-matrix interactions in cell migration directly in the embryo. The axolotl provides a particularly good experimental system to approach

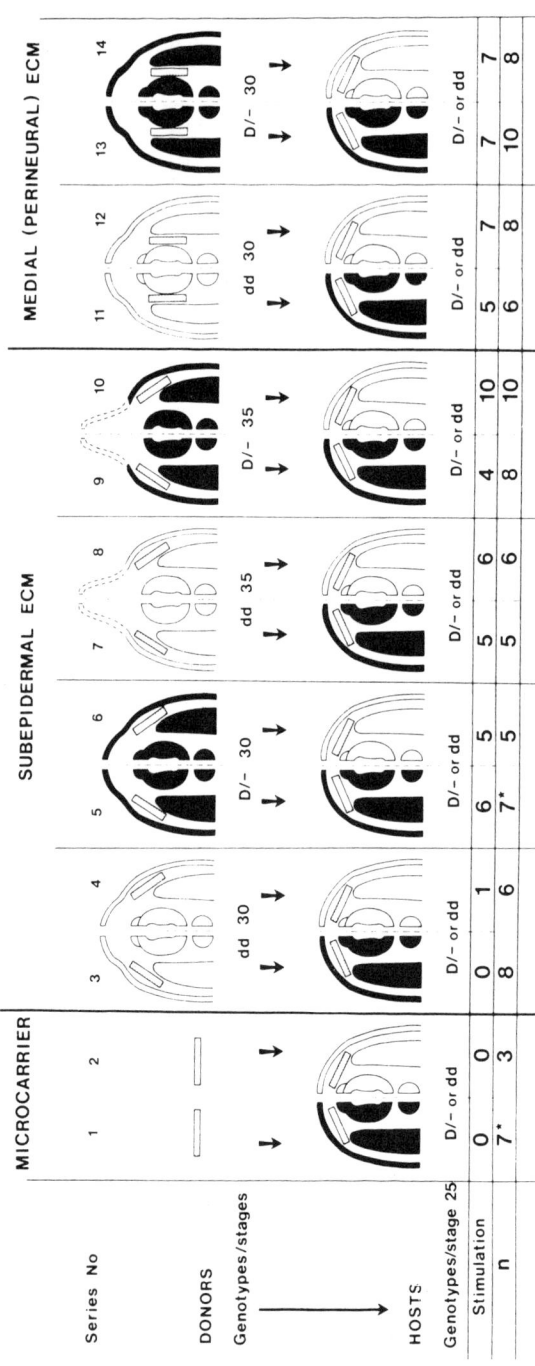

FIGURE 5-15. Diagram summarizing the results of the transplantation experiments, in which ECM was transplanted on microcarriers between axolotl embryos. The figures show cross-sectioned midtrunk regions of dark (*D/-*, solid) and white (*dd*, open) donor and host embryos. The upper row represents white and dark donor embryos of different stages with microcarriers (rectangles) implanted for adsorption of subepidermal or medial (perineural) ECM. The lower row represents white and dark host embryos into which donor ECM was implanted on microcarriers into the subepidermal space. In series 1 and 2, microcarriers without ECM were implanted as controls. In the horizontal columns, "stimulation" denotes the number of individual cases in a series in which NC cell migration was stimulated in the region of the carrier. *n* is the total number of individual cases in a series. Asterisks indicate those series which contain some individual cases from an earlier study (Löfberg et al., 1985). In series 7 to 10, the dorsal fin is outlined by dashed lines, indicating that it was not formed after the neural crest cells had been removed. (From Löfberg et al., *Dev. Biol.*, 131, 168–181, 1989. ©Academic Press.)

these questions because of the ease of manipulating this amphibian embryo and the existence of mutants in this species. This study utilizes the white axolotl mutant, in which pigment cells fail to migrate underneath the epidermis.

Previously, Lofberg and colleagues developed a technique for "transplanting" extracellular matrices and associated molecules in the living embryo. "Microcarriers" or filters were implanted into a region of an embryo and left in place for several hours, allowing the endogenous matrix to become adsorbed to the filter. These microcarriers could then be removed and transplanted to another embryonic region. In this study, this microcarrier transplantation paradigm is applied to studying neural crest migration in the white mutant.

The authors found that microcarriers conditioned with the extracellular matrix derived from under the epidermis of the white mutant was unable to support migration of pigmented neural crest cells. In contrast, microcarriers conditioned from the subepidermal region of normal embryos, from the ventral neural crest migratory pathway in the mutant, and from older subepidermal regions of the mutant *all* support pigment cell migration when implanted under the mutant epidermis. This shows that a matrix or matrix-associated molecule present in the subepidermal space of normal embryos is absent from the mutant epidermis, but is present in other regions of the mutant and even occurs in the subepidermal region, but at relatively late times.

These results are significant for a number of reasons. They demonstrate that the white mutation does not result from the simple presence or absence of a molecule. Rather, the *site* of expression and the *timing* of expression of extracellular matrix or associated molecules are abnormal in the mutant. *Marianne Bronner-Fraser*

Neural Fold Formation at Newly Created Boundaries between Neural Plate and Epidermis in the Axolotl
J. D. Moury and A. G. Jacobson
Dev. Biol., 133, 44—57, 1989   5-16

The cortical tractor model proposes that neural fold and crest formation occurs at the boundary between neural plate and epidermis because random cell movements become organized at this boundary. This model predicts that a neural fold would form at any boundary between epidermis and neural plate. To test this model, pieces of neural plate and epidermis that would not participate in fold formation in the normal axolotl embryo were juxtaposed to create novel boundaries. These boundaries were morphologically and histologically examined for the presence of neural folds.

At each and every novel boundary between epidermis and neural plate, neural folds were detected. At the interesection of two folds, a "neural tube" of neural tissue covered by epidermis formed. These folds and tubes were morphologically similar to neural folds and tubes in unmanipulated embryos.

When neural plate was transplanted into epidermis, nodules of neural tissue with central lumens and peripheral nerve fibers formed. When epidermis was transplanted into neural plate, the neural tube and dorsal fin were bifurcated at the graft site. Neural crest cells formed at the novel boundaries.

As predicted by the cortical tractor model, any boundary between neural plate and epidermis will form a neural fold in the axolotl embryo. Therefore, local interactions appear to be responsible for neural fold and neural crest formation.

◆ During development, a portion of the ectoderm becomes specialized during neural induction to form the nervous system. This neuroectoderm is morphologically distinguishable from the surrounding ectoderm well before the neuroectoderm folds to form the neural tube. This paper demonstrates that the neural folds form at any border, natural or artificial, between ectoderm (epidermis) and neuroectoderm (neural plate). Moury and Jacobson further show that multiple neural folds can develop, reflecting the number of epidermis/neural plate borders. With subsequent development, experimentally created neural folds roll up to form neural tube-like structures and give rise to neural crest cells (as shown in a later paper).

Although little is known about the mechanisms underlying neural induction, it is thought that the boundary between the epidermis and neural plate is defined during neural induction under the influence of the chordamesoderm. One possibility is that distance away from the strongest inducer, presumably the chordamesoderm at the midline, defines components of the neural crest and neural plate. Thus, the cells of the neural plate furthest from the inducer would form the most dorsal structures — i.e., neural fold and neural crest.

The experiments in this paper argue against such a possibility. Moury and Jacobson's results suggest that neural folds (and by extension, neural crest) can form at any boundary between epidermis and neural plate, regardless of the original position of either tissue relative to the midline. Thus, local interactions appear to define the sites of neural fold formation rather than interactions with a putative diffusible neural inducer.
Marianne Bronner-Fraser

A Laminin-Like Adhesive Protein Concentrated in the Synaptic Cleft of the Neuromuscular Junction
D. D. Hunter, V. Shah, J. P. Merlie, and J. R. Sanes
Nature, 338, 229—234, 1989   5-17

Regenerating motor axons preferentially reinnervate original synaptic sites on denervated muscle fibers. This topographic specificity of synapse formation is mediated by the basal lamina (BL) of the synaptic cleft. A protein involved in this process has been identified.

Monoclonal antibodies that stained BL were generated and recognized a polypeptide, s-laminin. In muscle, s-laminin was concentrated at synaptic sites. It was also abundant in perineural BL and glomerular BL. Those BLs which contain abundant s-laminin were apposed on both sides by cellular membranes.

Kidney fractions were analyzed by immunoblotting. Under nonreducing conditions, a protein of greater than 1000k was detected. After a reduction, a 190k protein was recognized. The protein was not soluble, which was expected for an extracellular matrix protein. Extraction with a chaotrope and a reducing agent released s-laminin. Collagenase did not affect the size of the protein. Concanavalin A-agarose and wheat germ agglutinin-agarose bound s-laminin, which could be released by hapten sugars. Therefore, s-laminin is a non-collagenous glycoprotein attached to the BL by hydrophobic and disulfide bonds. Chick ciliary neurons adhered to a 2000-fold purified preparation of s-laminin.

A rat kidney cDNA library was screened for s-laminin cDNAs. The predicted protein contained seven potential N-glycosylation sites and a signal sequence characteristic of membrane or secreted proteins. The predicted sequence demonstrated striking homology to laminin, especially the B1 subunit.

A component of synaptic BL has been identified. This protein is similar to laminin and interacts with motoneurons. It is possible that s-laminin present at original synaptic sites might signal axons to stop growth and differentiate into nerve terminals.

♦ Many extracellular matrix molecules contain complex and multiple binding domains for cells and other matrix molecules. For example, laminin is known to bind tumor cells, neurites, and other extracellular matrix molecules including nidogen/entactin and collagens. There has been great interest in the laminin molecule because of its pronounced and unique ability to promote neurite outgrowth *in vitro*.

The neuromuscular junction has been a particularly attractive system to study neurite ingrowth and synaptogenesis because of the ability of regenerating motor axons to reinnervate their original synaptic sites. The extracellular matrix of the synaptic region has been thought to be responsible for this selective regrowth and synaptogenesis. In this report, Hunter and colleagues have identified a molecule uniquely distributed in the synaptic region. By molecular cloning, they have found that the molecule is highly homologous to laminin and have named it s-laminin. *In vitro*, s-laminin promotes neuronal adhesion as efficiently as laminin.

This work is particularly striking because it provides evidence for a laminin gene family. Other matrix molecules and cell surface receptors, most notably collagens and integrins, have shown to exist as gene families. Because laminin is a major neurite-outgrowth promoting factor, this line of research may lead to the identification of a family of molecules involved in axonal behavior. *Marianne Bronner-Fraser*

Distinct Roles for Adhesion Molecules during Innervation of Embryonic Chick Muscle
L. Landmesser, L. Dahm, K. Schultz, and U. Rutishauser
Dev. Biol., 130, 645—670, 1988   5-18

Chick lumbosacral motor neurons project to their target muscles in a very precise manner. It is likely that cell-cell interactions involving adhesion between axons — or between axons and muscle — have an important role in the cellular events involved. *In vitro* studies have suggested that the cell adhesion molecules NCAM and G4/L1 contribute to neural development.

In this study the role of these molecules in initial nerve ingrowth and ramification in the embryonic chick iliofibularis muscle was examined directly through *in ovo* injections of specific adhesion-blocking antibodies. The resultant nerve branching patterns were analyzed in muscle whole mounts. In addition, some muscles were studied electrophysiologically before immunostaining.

Antibodies against both NCAM and G4/L1 produced axonal defasciculation and an enhanced transverse projection to the fast region of the muscle. With anti-G4/L1 antibody, there also was a large increase in side branches forming from nerve trunks in the slow region as well as enhanced nerve branching in the fast region. Anti-NCAM antibody led to a marked reduction in the number and length of side branches in the slow region and less nerve branching in the fast region. Nerve branching was similarly reduced after injection of an endosialidase, which removes sialic acid from NCAM. This material enhances fiber-to-fiber apposition, presumably by promoting cell adhesion.

These adhesion molecules appear to have complementary roles in muscle innervation, influencing axon-axon fasciculation so as to regulate the pattern of nerve branching. NCAM-mediated axon-myotube interactions likely are required to achieve the normal stereotyped pattern of nerve branching in both fast and slow regions of the muscle. The branching pattern is an important determinant of the overall pattern of synapses, and it probably influences the trophic interaction between neuron and muscle.

♦ Axon-axon, axon-cell, and axon-matrix interactions have been suggested to play important roles in neurite outgrowth and pathfinding in both the peripheral and central nervous system. Most data supporting this hypothesis come from *in vitro* assay in which various types of neurites are exposed to extracellular matrices, muscle cells, Schwann cells, or astrocytes. Such *in vitro* experiments are informative about potential interactions and surface properties of axons but do not necessarily relate to mechanisms of neurite outgrowth in the embryo.

This study by Landmesser and colleagues represents the first direct assay of the role of the cell adhesion molecules N-CAM and L1 (also called Ng-CAM) *in vivo*. Both molecules are Ca^{++} independent, with the former binding in a homophilic manner and the latter in a heterophilic manner. The authors examined the projections of motor neurons to the ileofibularis muscle in the limb. As motor axons project into the limb, they have N-CAM and L1 immunoreactivity. The muscle cells have little N-CAM immunoreactivity until later stages, and the axons appear to "pause" until muscle N-CAM immunoreactivity increases. The N-CAM in the limb appears to be primarily embryonic in form, containing a high content of sialic acid which decreases its binding affinity.

Function-perturbing antibodies against N-CAM and L1 were injected into the limb region in order to disturb neurite interactions. Some similarities in the effects of injected antibodies were noted; for example, both caused defasciculation and neither prevented neurite ingrowth. However, the major effects of the antibodies were different, but complementary. While N-CAM antibodies decreased side branch formation, anti-L1 antibodies increased side-branch formation. Another way to alter N-CAM binding efficacy in the limb is to remove the sialic residues using and endosialidase. In so doing, the authors produced a decrease in the number of side branches, similar to that achieved by anti-N-CAM antibodies. This was presumably caused by tighter axon-axon associations.

These studies sum to indicate important roles for N-CAM and L1 in axon-axon and axon-muscle interactions in the limb. Furthermore, they suggest that glycosylation of N-CAM is also important for the pattern of innervation. Interestingly, the antibodies did not appear to affect nerve-muscle function, suggesting that synaptogenesis occurred normally. The most important message is that multiple cell adhesion molecules function in a developing system *in vivo* to achieve proper formation of the innervation pattern. A side lesson is that removal of either interaction did not block neurite ingrowth completely, suggesting the presence of other or redundant mechanisms for neurite guidance. *Marianne Bronner-Fraser*

Construction of Epithelioid Sheets by Transfection of Mouse Sarcoma Cells With cDNAs for Chicken Cell Adhesion Molecules

R.-M. Mege, F. Matsuzaki, W. G. Gallin, J. I. Goldberg, B. A. Cunningham, and G. M. Edelman

Proc. Natl. Acad. Sci. U.S.A., 85, 7274—7278, 1988

During epithelial-mesenchymal transformation epithelia are converted to mesenchyme in a reversible process in which cell adhesion molecules (CAMs) have a role. These molecules decrease in prevalence as mesenchyme is produced and reappear when it condenses. In this study, tumorigenic mouse sarcoma cells that express enither the Ca-dependent liver CAM (L-CAM) or the Ca-independent neural CAM (N-CAM) were singly or doubly transfected with the cDNAs for these CAMs. For transfection, the L-CAM coding sequence was inserted into the *Bgl* II site of pkSV-10 vector downstream of its simian virus 40 early promoter.

Transfected cells expressed the appropriate CAMs at their surfaces. Those expressing L-CAM changed from spindle or round shapes to a closely linked epithelioid sheet. Cells transfected with cDNA for N-CAM expressed this molecule on their surfaces and bound brain vesicles containing N-CAM, but did not change phenotypically to an epithelioid state. L-CAM was concentrated in areas of cell contact and was codistributed with cortical actin. The cells in epithelioid sheets were polygonal in shape, but, in contrast to true epithelium, they lacked a basement membrane and polar structure. Tight junctions and desmosomes also were absent. Compared with nontransfected cells, those in epithelioid sheets exhibited large increases in adherens junctions and gap junctions, and the dye coupling studies showed the gap junctions to be functional. Expression of both types of junction was much reduced by treatment with anti-L-CAM Fab' fragments.

These findings are evidence that linkage of cells through CAMs is necessary for the extensive expression of junctional structures in cells. Transfection of cells with CAM cDNA coupled with appropriate inducible promoters should permit changes in the interactions between cell adhesion, growth, and movement.

♦ Cell adhesion molecules are thought to have long-ranging effects on cell morphology, cell interactions, and the formation of intercellular junctions. In this study, Mege and colleagues tested the effects of N-CAM and L-CAM on the morphology, cell-cell adhesions, and junctional formation in transfected cells containing ectopic levels of one or both of these molecules. Cells transfected with either N-CAM or L-CAM cDNA displayed homophilic cell interactions. Cells transfected with L-Cam, but not N-CAM, cDNA underwent phenotypic changes and appeared to form epithelioid sheets. L-CAM became accumulated at regions of cell contact in these sheets. Further, the epithelioid sheets were disrupted by antibodies against L-CAM. Ultrastructural and dye-coupling experiments demonstrated an increase in adherent junctions and gap junctions in regions of cell-cell contact in L-CAM transfected cells.

These results have several important messages. First, similar to the study described above (Yamaguchi and Ruoslahti), they show that expression of additional molecules can drastically affect cell morphology. Second, they demonstate that cell-cell interactions mediated by L-CAM appear to precede and be important for subsequent formation of intercellular junctions. Thus, cells first need to establish intimate contacts via adhesion molecules prior to formation of adherens and communicating junctions. *Marianne Bronner-Fraser*

Thymic Major Histocompatibility Complex Antigens and the $\alpha\beta$ T-Cell Receptor Determine the CD4/CD8 Phenotype of T Cells
H. S. Teh, P. Kisielow, B. Scott, H. Kishi, Y. Uematsu, H. Bluthmann, and H. von Boehmer
Nature, 335, 229—233, 1988   5-20

Thymus-derived lymphocytes, or T cells, recognize antigen on the surfaces of antigen-presenting cells in the context of class I or class II major histocompatibility complex (MHC) molecules, through the heterodimeric $\alpha\beta$ T-cell receptor (TCR). CD4 and CD8 molecule — expressed on the T-cell surface — bind to nonpolymorphic parts of class II and class I MHC molecules, respectively. In addition, this binding may help produce signals leading to T-cell activation. It has been suggested that selection of the antigen-specific T-cell repertoire involves the supression or deletion of autospecific T cells. Alternately, T cells may be positively selected by thymic MHC antigens.

Both aspects of T-cell repertoire selection were examined by constructing TCR transgenic mice expressing, on a large proportion of their RT cells, a receptor that binds to H-Y antigen in the xontext of class I H-2D^b molecules. Monoclonal antibodies identifying the expressed transgenic receptor served to analyze negative selection in male $\alpha\beta$ transgenic H-2^b mice, which express both the H-Y antigen and H-2D^b molecules. Analysis of female $\alpha\beta$ transgenic mice should indicate the influence of MHC molecules on T-cell selection in the absence of (nominal) H-Y antigen.

The results confirm that the specific interaction of the TCR on immature thymocytes with thymic MHC anmtigens determines the differentiation of CD4$^+$8$^+$ thymocytes into either mature CD4$^+$8$^-$ or CD4$^-$ 8$^+$ T cells. Selection of CD$^-$ 8$^+$ T cells takes place in the absence of the nominal antigen in $\alpha\beta$ transgenic female mice. It appears that both positive and negative selection can occur at the same stage of T-cell development. Negative selection by nominal self-antigen need not occur following positive selection by thymic MHC antigens. The signals leading to positive and negative selection presumably differ.

Positive Selection of Antigen-Specific T Cells in Thymus by Restricting MHC Molecules
P. Kisielow, H. S. Teh, H. Blutmann, and H. von Boehmer
Nature, 335, 730—733, 1988   5-21

Studies of T-cell development in T-cell receptor (TCR) transgenic mice indicate that the DC4/CD8 phentype of the cells is determined by the interaction of the $\alpha\beta$ TCR expressed on immature $CD4^+8^+$ thymocytes with polymorphic domains of thymic MHC molecules in the absence of nominal antigen. Direct evidence now is available that the positive selection of antigen-specific class I MHC-restricted $CD4^-8^+$ T cells in the thymus requires a specific interaction of the $\alpha\beta$ TCR with the restricting class I MHC molecule.

An increased proportion of $CD4^-8^+$ thymocytes results from positive selection secondary to the interaction of $CD4^-8^+$ thymocytes expressing transgenic TCR with MHC molecules encoded by the $H-2^b$ but not the $H-2^d$ haplotype. Thymus repopulation studies confirmed the specificity of the postiive selection exerted by MHC molecules. The repopulation capacity of donor stem cells correlated with the degree of compatibility within the H-2 complex. Cells expressing the highest level of transgenic TCR were virtually absent from D^b-positive recipients.

The specific low-affinity interaction of TCR with thymic MHC antigens may be adequate to select immature T cells, but not to activate mature T cells. This interaction may be complemented by accessory molecules present on immature thymocytes and/or selecting thymic cells. Such complementation could result in the positive selection of T cells having low affinity for self-MHC.

The Generation of Mature T Cells Requires Interaction of the $\alpha\beta$ T-Cell Receptor With Major Histocompatibility Antigens
B. Scott, H. Bluthmann, H. S. Teh, and H. von Boehmer
Nature 338, 591—593, 1989   5-22

In transgenic mice bearing genes for T-cell receptor (TCR) of defined specificity, T cells expressing the introduced genes form the main part of the mature T-cell population only in mice expressing the appropriate MHC product. This study utilized TCR transgenic mice homozygous for the severe combined immunodeficiency (SCID) mutation. The mutation was introduced into αβ TCR transgenic mice in order to produce animals having only a very limited choice of TCRs. Mice homozygous for the SCID mutation are grossly defective in productive rearrangements of both immunoglobulin and TCR genes.

Mature thymocytes developed only in transgenic mice expressing the MHC product that restricts the specificity of the transgenic TCR. The peripheral lymph nodes of these mice were underdeveloped. In contrast, thymus size was nearly normal in αβ-transgenic SCID mice.

Interaction of αβ TCR with thymic MHC antigen is necessary for mature T cells to develop. The peripheral expansion of mature T cells may depend on both immunoglobulin-positive lymphocytes and $CD4^+8^-$ T cells having a variety of receptors.

♦ The TCR recognizes a fragment of antigen (usually a small peptide) in association with a MHC class I (if it is a $CD4^-CD8^+$ T cell) or class II (if it is a CD^+CD8^- T cell) on the surface of an antigen-presenting cell. The T cell repertoire is generated through several steps. First of all, as seen with B cells, there is the random selection of variable region genes followed by the positive selection of T cells in the thymus determined by the products of the MHC expressed by cells of the thymus, and finally the deletion and/or inactivation of self-reacting cells. Due to the advent of transgene systems, over the past several years there has been dramatic progress in our understanding of these crucial events which occur within the thymus and control the repertoire of T cells released into the periphery.

The concept of positive selection is based upon the understanding that T cells only recognize antigen when it is presented in the context of self-MHC. Therefore, at some point in the selection of T cells within the thymus, a process must exist for selecting only those T cells expressing TCR which can bind to self-MHC. Experiments utilizing transgene animals have recently provided strong support for this occurring as an active process within the thymus. *Charles Snow*

Role of CD4 in Thymocyte Selection and Maturation
J. C. Zuniga-Pflucker, S. A. McCarthy, M. Weston, D. L. Longo, A. Singer, and A. M. Kruisbeek
J. Exp. Med., 169, 2085—2096, 1989   5-23

The selective forces guiding T-cell differentiation and maturation are not well understood. The T-cell receptor (TCE) clearly is involved in the process of negative selection or induction of tolerance. Tolerance induction — at least for certain self-antigens expressed in the thymus — occurs through clonal delection of T cells with self-reactive TCRs. One view of positive selection is that only T cells with low affinity for self-MHC glycoproteins expressed in the thymus are able to become functional and to be exported peripherally.

In this study, pregnant mice were treated with anti-CD4 monoclonal antibody to assess the interaction between CD4 and its ligand in generating the T-cell repertoire. In addition, the fetal thymus organ culture system was used to analyze the function of CD4 during intrathymic selection *in vitro*, comparing the effects of monovalent and divalent anti-CD4 monoclonal antibodies.

The fetal thymus exposed to intact anti-CD4 monoclonal antibody failed to generate CD4 single-positive T cells. Identical results were obtained using $F(ab')_2$ and Fab anti-CD4 monoclonal antibody. The generation of $CD4^+/CD8^+$ T cells was unchanged. An increase in expression of TCR/CD3 was noted after treatment with divalent or monovalent anti-CD4 monoclonal antibody.

These findings confirm that CD4 is an avidity-enhancing and signal-transducing molecule in mature T cells and also has a critical role in the differentiation of immature T cells and selection of the T-cell repertoire. One explanation of the CD3 upregulation observed is compensation for the decrease in avidity or signaling that normally is provided by interaction of the CD4 accessory molecule and its ligand.

♦ The cell surface molecules CD4 and CD8 are thought to participate as secondary adhesion molecules during the process of T cell binding to antigen presenting cells. In other words, the TCR binds antigen fragment in association with an MHC molecule, and this binding is strengthened by CD4 interacting with a nonpolymorphic determinant on class II (for $CD4^+$ T cells) or CD8 interacting with a nonpolymorphic determinant on class I (for $CD8^+$ T cells). A recent report provided the first evidence implicating these secondary adhesion molecules being operative during thymic positive selection. *Charles Snow*

Differential Exon Usage Involving an Unusual Splicing Mechanism Generates at Least Eight Types of NCAM cDNA in Mouse Brain
M. J. Santoni, D. Barthels, G. Vopper, A. Boned, C. Goridis, and W. Wille
EMBO J., 8(2), 385—392, 1989   5-24

The neural cell adhesion molecules (NCAMs) are a group of related cell-surface glycoproteins involved in cell-cell contact formation. NCAM-mediated adhesion may be important in regional segregation of cells,

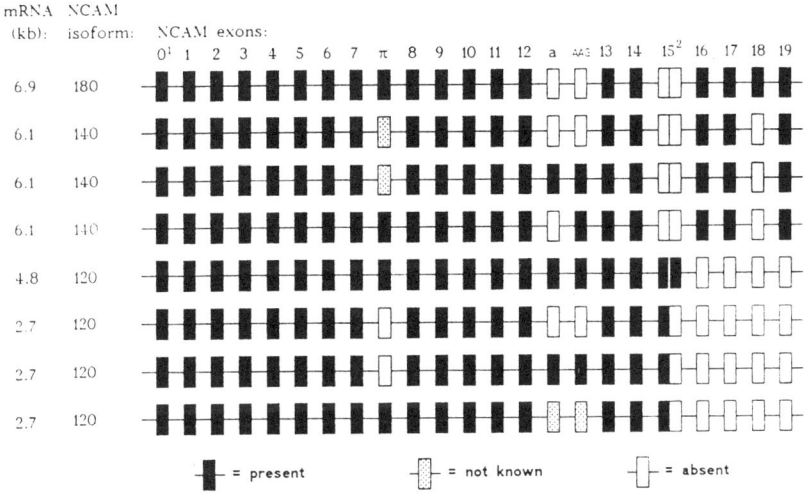

FIGURE 5-24. The theoretical exon organization of the eight different spliced NCAM transcripts of mouse brain as deduced from cDNA clones and Northern blot analyses. The related transcript sizes and the NCAM isoforms encoded by these mRNAs are given on the left-hand side. [1]Exons 1 to 11 are numbered according to Owens et al. (1987); exon 0 encodes the 5' noncoding region in the mouse NCAM messenger (Barthels et al., 1987), the complete genomic organization of the mouse NCSAM gene will be published elsewhere; [2]the double box represents the NCAM-120-specific exon 15 containing alternative polyadenylation signals. (From Santoni et al., *EMBO J.*, 8(2), 385—392, 1989. ©IRL Press.)

orderly axonal outgrowth, formation of junctional communications, and the regulation of neurotransmitter enzymes. It is not clear whether NCAM has a role in determining the selectivity of adhesion exhibited by different cell types.

The murine NCAM exists in three isoforms of differing size, which are coded for by four transcripts differing in length from 2.7 to 6.9 kb. The differences between the isoforms result from alternative splicing in the coding region for the transmembrane and cytoplasmic domains. The extracellular N-terminal part of NCAM is shared by all three protein forms. In this study, the coding region for the N-terminal domains of NCAM was found to contain at least two sites of alternative splicing. Short added sequences can be introduced at these sites, which are between the Ig-like domains and the membrane attachment site and within the Ig-like domain IV, respectively. Sequencing and S1 nuclease protection assays of independent cDNA clones, along with Northern blot analyses, demonstrated that at least 8 different mRNAs exist (Figure 5-24).

If most combinations of the splice patterns so far identified in mouse brain actually occurred, 24 different mRNAs could be generated, coding for 18 different proteins. The complexity of splicing patterns of the NCAM gene in brain and muscle is much greater than expected. Studies transfecting specific cDNAs and employing site-specific antibodies will show how the diverse NCAM transcripts, and probably proteins, modulate the function of NCAM in specifying cell interactions.

Identification of Two Protein Kinases that Phosphorylate the Neural Cell-Adhesion Molecule, N-CAM
K. Mackie, B. C. Sorkin, A. C. Nairn, P. Greengard, G. M. Edelman, and B. A. Cunningham
J. Neurosci., 9, 1883—1896   5-25

Changes in levels of expression of the neural cell adhesion molecule (NCAM) may be critically important at sites of embryonic induction. Since protein phosphorylation is an important mechanism regulating protein function in many systems, phosphorylation sites in NCAM were characterized with respect to their sites within the ld and sd polypeptides, two of the three related polypeptides generated by alternative splicing of a single gene. They are the larger polypeptides that are phosphorylated *in vivo* on several common phosphorylation sites. The smallest of the polypeptides lacks a cytoplasmic domain and is not phosphorylated. Chick, mouse, and rat NCAMs were used in the study.

Two protein kinases that phosphorylate rodent and chicken NCAM were isolated from mammalian and avian brain. They were identified as glycogen synthase 3 (GSK-3) and casein kinase I (CK I). The kinases rapidly phosphorylated NCAM to a high stoichiometry. Both phosphorylated the larger NCAM polypeptides *in vitro* in the cytoplasmic domain on threonyl residues that are phosphorylated to a low level *in vivo*. These are close to seryl residues that are phosphorylated to a high level *in vivo*. Prior phosphorylation at the *in vivo* sites was necessary for phosphorylation by GSK-3 and CK I.

NCAM may be physiologically phosphorylated on two sets of interrelated sites, one demonstrable *in vivo* and the other *in vitro*. Phosphorylation on the *in vivo* sites resists dephosphorylation and may be constitutive, while that on the *in vitro* sites is relatively quite labile.

♦ One important family of molecules involved in vertebrate development are the different structural forms of the neural cell adhesion molecule (NCAM). The NCAMs are membrane-associated sialoglycoproteins that serve as ligands in the formation of homophilic cell-cell bonds. The NCAMs are known to result from a primary message which is encoded by a single gene. Structural heterogeneities have been found both in the protein and the carbohydrate portions of the molecule. The polypeptide itself exists primarily in three molecular weight forms, the expression of

which is also dependent on tissue source and age. The recent paper from Santoni and colleagues demonstrate that, through the process of alternative splicing, at least eight different mRNAs exist in mouse brain. Furthermore, these investigators predict that if free combination of all the splice patterns that have been identified in mouse brain actually occurs, as many as 24 different mRNAs could potentially be generated resulting in 18 distinct proteins. In addition to these types of structural heterogeneities, it is now known from the work of Mackie and co-workers that the alternative forms of NCAM are differentially phosphorylated, probably due to the activity of several different protein kinases. Such differences in phosphorylation may have profound effects on the function of the protein. For example, its surface density, binding characteristics, or interactions with other molecules at the cell surface, in the cytoskeleton, or in the cytoplasm all may be affected by the degree to which the molecule is phosphorylated. Thus, mechanisms such as alternative splicing and differential phosphorylation appear to be responsible for generating an unexpected degree of diversity of NCAM proteins which presumably is important for subtle modulation and control of NCAM-mediated adhesion.
Terry Magnuson

Cell Lineage and Developmental Fate 6

INTRODUCTION

The expression of a particular cell phenotype could result from either the restricted expression of certain genes or the increased expression of others. Nuclear transplantation experiments clearly demonstrate that the full potential of the genome becomes restricted as development progresses. On the other hand, the expression of new phenotype-specific genes is what defines the biological nature of cells within a particular cell lineage. Thus, a combination of the two processes could operate to determine specific cell lineages.

There is evidence suggesting that genetic, molecular genetic, and cellular factors can all influence cell lineage decisions. DNA structure and the interaction of multiple transcription factors can serve to regulate the expression of particular genes in a lineage-specific manner. "Who your neighbors are" can also influence the developmental fate of an individual cell or group of cells.

A critical and still unanswered question regarding the initiation of cell lineage determination is what the primary event really is. Is the restricted expression of certain gene products the cause of cell-lineage determination or simply the result of an earlier lineage decision? New molecular components are being described that could provide some clues to the answer of this fundamental question.

Lineage-Specific Development of Calcium Currents during Embryogenesis
L. Simoncini, M. L. Block, and W. J. Moody
Science, 242, 1572—1575, 1988   6-1

It is not clear how cells of differing developmental fate acquire their characteristic electrical properties during embryogenesis. The authors followed the development of the electrophysiologic properties of isolated ascidian blastomeres from the fertilized egg to the neurula. The goal was to identify the stage at which cells of differing lineage first express different functional ion channel populations. It has proved difficult to make voltage-clamp recordings from small embryonic cells. This study employed the whole-cell patch clamp technique. Identification of the final products in dissociated preparations was aided by *Boltenia villosa*, a species of ascidian in which endogenous pigment marks cells of specific developmental fates.

Muscle-lineage blastomeres developed a voltage-dependent calcium current within about 3 h after gastrulation. Surrounding blastomeres of differing lineage did not. At about the same time all the cells developed delayed outward potassium currents and lost the inwardly rectifying potassium currents that were present at earlier stages.

Development of calcium currents is an early event in the terminal differentiation of ascidian muscle. Since the calcium currents appear well in advance of their function in contractility, they may have a role in the terminal differentiation of muscle cells in the larva.

♦ The timing of certain phenotypic events is closely correlated with the time at which cell-lineage determinations occur. The ability to separate the timing of a series of such phenotypic events can facilitate the identification of those events most closely related to the determinative events leading to multiple cell lineages. The electical properties of cells during embryogenesis are such phenotypic events. In this article, the authors use a species of ascidian to successfully separate the timing of lineage-specific calcium currents from changes in potassium currents common to all cell lineages. By successfully separating these events, the authors can now focus on the relationship between calcium currents and the determination of its cell lineage specificity. These observations may have broad application for how a cell's electrical properties are generated during development. *Joel M. Schindler*

Evolutionary Modification of Cell Lineage in the Direct-Developing Sea Urchin *Heliocidaris erythrogramma*
G. A. Wray and R. A. Raff
Dev. Biol., 132, 458—470, 1989   6-2

Two Australian sea urchins of the genus *Helicidaris* provide a useful model system with which to study evolutionary change in development. *H. erythrogamma* undergoes direct development, bypassing the usual echinoid pluteus larva. In contrast, *H. tuberculata* develops by means of a typical larva. Nevertheless, adults of the two species are morphologically similar, occupy the same habitat, and breed at the same time of year. Microinjection of fluorescinated tracer dye and surface marking with vital dye served to follow the larval fates of 2-, 8-, and 16-cell blastomeres and to examine axial specification.

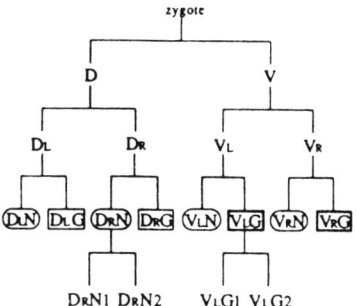

FIGURE 6-2. Cell lineage in *H. erythrogramma* to 16-cell stage. Blastomere names are modified from Cameron et al. (1987); N, animal; G, vegetal; L, left; R, right; D, adult dorsal; V, adult ventral. The four animal eight-cell blastomeres (circled) give rise exclusively to ectoderm and ectodermally derived neurons. Descendents of the vegetal quartet of eight-cell blastomeres are found in all three germ layers; fates of dorsal blastomeres (heavy boxes) and ventral blastomeres (light boxes) differ qualitatively and quantitatively. Only one example each of an animal and a vegetal fourth cleavage are shown for the sake of clarity. (From Wray/Raff, *Dev. Biol.*, 132, 458—470, 1989. ©Academic Press.)

The fates of animal cells were quite similar to those of typically developing species, but vegetal cell fates in *H. erythrogramma* were considerably altered. In the eight-cell embryo, dorsal vegetal blastomeres contributed proportionately more descendants to ectodermal fates and fewer to mesodermal fates. Ventral vegetal blastomeres exhibited a complementary bias in developmental fates. Vegetal-cell fates were more variable than usually is observed. All vegetal cells in the 16-cell embryo of *H. erythrogramma* could contribute progeny to ectoderm and to gut.

The lineage and fate of eight-cell blastomeres of *H. erythrogramma* are shown in Figure 6-2. While the fates of animal 16-cell blastomeres resemble mesomere fates in typically developing sea urchins, those of vegetal 16-cell blastomeres differ substantially (Figure 6-2A). Delayed segregation of cell lineages appears to be a general feature of *H. erythrogramma* development (Table 6-2). Available evidence suggests that at least some cell lineages are evolutionarily conserved in typically developing echinoids.

♦Sea urchin embryological studies have been dominated by those species which undergo "typical" development, that is, cleavage, gastrulation, feeding pluteus larvae, and metamorphosis into a juvenile adult; yet there is a rich diversity in the way sea urchins obtain adulthood. In addition to

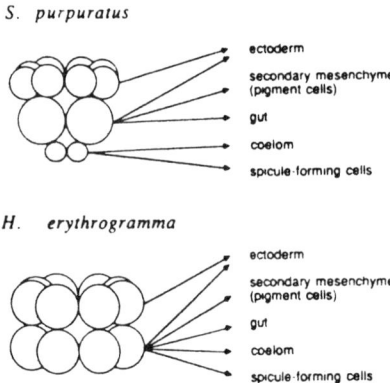

FIGURE 6-2A. Comparison of 16-cell fates. In typically developing echinoids such as *S. purpuratus*, partitioning of cell lineages is well under way by the 16-cell stage (Hörstadius, 1939). For example, the four micromeres give rise to only two cell types, spicule-forming cells, and coelomic cells. In *H. erythrogramma*, the vegetal blastomeres retain a wider range of cell fates, reflecting a general delay in cell lineage partitioning during development. (From Wray/Raff, *Dev. Biol.*, 132, 458—470, 1989. ©Academic Press.)

Table 6-2
Number of Blastomeres Contributing to Larval Cell Types

	S. purpuratus[a]		*H. erythrogramma*[b]	
	8-cell	16-cell	8-cell	16-cell
Ectoderm	8	12	8	14-16
Mesoderm	4	8	2-4	4-6
Endoderm	4	4	4	6-8

Note. In both typical and direct developing sea uchins which have been studied, larval cell types are not clonally derived. However, the number of cells which give rise to particular larval cell types has changed during the evolution of direct development in *H. erythrogramma*. These differences reflect alterations in timing of cell lineage allocations. *H. erythrogramma* resembles *S. purpuratus* more closely at the 8-cell than at the 16-cell stage.

[a] Numbers given for *S. purpuratus* from Davidson (1986).

[b] Numbers for *H. erythrogramma* embryos estimated from the percentage of cases giving rise to cells of each germ layer. Where results indicate variability between individual embryos, the estimated range of founder blastomeres is given.

From Wray/Raff, *Dev. Biol.*, 132, 458—470, 1989. ©Academic Press.

the typical developers, there are partial pluteus developers and direct developers. Partial pluteus developers gastrulate into a nonfeeding abbreviated pluteus which metamorphoses quickly into an adult. Direct developers produce nonfeeding embryos (either free swimming or brooded by their mothers) which develop directly into adults, completely bypassing the larval stage. About 20% of the described species of sea urchins obtain adulthood via such nontypical development. The loss of larval stages and accelerated development into adults are common in other echinoderms such as starfishes, brittle stars, and sea cucumbers as well as in mollusks, brachiopods, bryozoans, ascidians, and frogs. Direct developers have arisen independently in several lineages, and in some cases members of the same genus may develop by quite distinct mechanisms. Such is the case in the sea urchin genus *Heliocidaris* where *H. tuberculata* is a typical developer, and a coexisting species *H. erythrogramma* is a direct developer. Raff's laboratory has compared the developmental characteristics in these two species to understand the potential for evolutionary divergence in development. In the accompanying paper, Wray and Raff document evolutionary alterations in the patterns of lineage specification in *Heliocidaris*. When combined with the previously documented alterations in egg size, blastula formation, and larval morphology, a general conclusion arising from their studies is that developmental mechanisms impart few constraints on the evolution of direct development. This is of particular interest from a molecular perspective since gene expression in direct developers must also be modified from a program of distinctly embryonic or adult as found in typical developers. Once compared with other direct developer species, analyses such as these described by Wray and Raff should help define evolutionarily conserved features of developmental programs which, presumably, are rules which embryos must strictly follow. *Gary M. Wessel*

A Regulatory Domain That Directs Lineage-Specific Expression of a Skeletal Matrix Protein Gene in the Sea Urchin Embryo
H. M. Sucov, B. R. Hough-Evans, R. R. Franks, R. J. Britten, and E. H. Davidson
Genes Dev., 1, 1238—1250, 1988   6-3

The skeletal elements, or spicules, of the sea urchin embryo are produced by cells of four equivalent lineages deriving from the fourth-cleavage micromeres. Most known sea urchin embryo lineages require specific interactions for their normal fates to be realized.

In this study, DNA sequences derived from the 5′ region of a gene coding for the 50-kDa skeletal matrix protein (SM50) of sea urchin embryo spicules were linked with the CAT reporter gene and injected into unfertilized eggs. CAT mRNA and emzyme were synthesized from the fusion constructs in embryos derived from the injected eggs. In situ hybridization with a CAT antisense RNA probe showed that expression was confined to skeletogenic mesenchyme cells. An average of 5.5 of the 32-blastula-stage skeletogenic mesenchyme cells exhibited CAT mRNA, suggesting that the injected NDA probably is incorporated in a single blastomere during early cleavage. In vitro mutagenesis and deletion studies showed marked enhancement of CAT enzyme activity in transgenic embryos when the number of SM50 amino acids at the amino terminus of the fusion protein was reduced from 43 to 4. Lineage-specific spatial expression was promoted by cis-regulatory sequences located between −440 and +120 with respect to the transcriptional initiation site.

The lineage-specific expression of a gene needed for mesenchymal skeletogenesis in the sea urchin embryo is mediated by cis-regulatory sequences located within a few hundred nucleotides of the initiation site of the gene. It is possible that the regulative capacity of given parts of an early embryo reflects the distribution of trans-regulators that can be converted to active forms via the ectopic operation of cell signaling systems normally used for founder cell specification.

♦The primary mesenchyme cells of the sea urchin are derived fom the four micromeres resulting from the fourth cleavage division of the embryo. The descendants of these cells differentiate to produce the skeletal system of the embryo and can do so in vitro, isolated from the remainder of the embryo. Several primary mesenchyme-cell specific genes and epitopes have been identified recently; one of the best characterized genes expressed only in this lineage is SM50. Originally identified by Benson et al. (see 1989 Year Book), the SM50 protein appears to participate in the biomineralization process of spicule growth in mesenchyme cells. In the accompanying paper, Sucov et al. define and characterize an upstread domain of the SM50 gene which is sufficient to correctly regulate expression of microinjected promoter-reporter constructes in vivo both spatially and temporally. This result is important for it provides a probe to identify the transcriptional regulatory factors specific for this cell type. It is hoped that a characterization of such factors may lead to an understanding of the specification in the micromere lineage, and possibly to the maternal determinants of vegetal polarity in the early embryo. A complimentary study is in progress on the specification of ectoderm cell lineage and determinants of animal axis polarity (see Calzone et al., this volume). An understanding of SM50 transcription will also be important to explain the capacity for SM50 expression in cells other than in the primary mesenchyme cell lineage. For example, SM50 expression can be induced in animal blastomeres by LiC1 treatment (see Livingston and Wilt, this volume; see also Ettensohn and McClay, 1989 volume). This

study, therefore, provides the groundwork for an understanding of both autonomous and regulated gene expression in the sea urchin embryo.
Gary M. Wessel

Lithium Evokes Expression of Vegetal-Specific Molecules in the Animal Blastomeres of Sea Urchin Embryos
B. T. Livingston and F. H. Wilt
Proc. Natl. Acad. Sci. U.S.A., 66, 3669—3673, 1989   6-4

The way in which early embryonic cells are determined is a fundamental problem in biology. Separated animal halves of early sea urchin embryos form hollow epithelial spheres with little overt differentiation, while vegetal halves form more normal embryos. The authors developed a simple means of studying isolated blastomeres in culture. The overt differentiation of separated animal and vegetal blastomere pairs resembles that of hemispheres separated by manual dissection. Alkaline phosphatase and spicule matrix protein (SM50) RNA were used as markers to examine determination and its respecification by lithium chloride. Lithium reportedly leads to a pattern of differentiation in which archenterons are exaggerated and surface epithelium is reduced.

Treatment of animal blastomeres with LiCl led to morphologic features resembling those of isolated vegetal blastomeres. Histochemical staining and *in situ* hybridization studies indicated that gut alkaline phosphatase and SM50 RNA normally are expressed only in vegetal blastomeres, but that their expression can be elicited in animal blastomeres by LiCl treatment.

Induction of morphologic changes of vegetal differentiation in animal blastomeres by LiCl is associated with the expression of lineage-specific molecular markers in the vegetal structures. The effector molecules needed to activate expression of these makers presumably are present in both animal and vegetal blastomeres of the sea urchin embryo. Possibly, a signal at the cell surface is required to prevent the expression of vegetal markers in animal blastomeres.

♦In the sea urchin embryo, fates of the blastomeres are specified during early cleavages. This is evident after the fourth cleavage in which separated animal and vegetal blastomeres will differentiate into distinct structures. The mechanism of specification is not understood, but the classic experiments of Horstadius (1973) demonstrated that cell-cell interactions of the blastomeres are crucial for early specification (see also Hurley et al., this volume). This is particularly evident in the mesomeres of the 16-cell stage embryo which, when isolated from their vegetal cells, do not differentiate normally. Ectopic micromere apposition to the mesomeres, however, is sufficient to rescue normal embryonic development. This respecification of animal blastomeres can also be accomplished by LiCl treatment (Von Ubish, 1929). In the accompanying paper, Livingston and Wilt demonstrate that the respecification of animal blastomeres by Li treatment involves genes normally expressed only in cells derived from vegetal blastomeres. Thus, transcription of vegetal-specific gene products is possible in both the animal and vegetal blastomeres, and LiCl treatment activates a vegetal potential in animal blastomeres where it is normally silent. Thus, one step of animal blastomere specification may require a repression of vegetal-specific transcriptional events as well as an activation of animal potential. This conclusion is particularly interesting since recent evidence suggests that the site of Li action involves the second messenger system (e.g., inositol phosphate turnover and G proteins in the membrane) and that LiCl treatment produces pattern abnormalities in amphibian embryos as well. The possibility that the second messenger system participates in the regulation of blastomere gene expression provides an intriguing mechanistic clue to the blastomere transplantation results of Horstadius (1973). *Gary M. Wessel*

Lateral Inhibition during Vulval Induction in *Caenorhabditis elegans*
P. W. Sternberg
Nature, 335, 551—554, 1988   6-5

During vulval induction in *Caenorhabditis elegans*, the anchor cell of the gonad specifies a spatial set of three cell types among six multipotent epidermal cells. These vulval precursor cells (VPCs) are homologous epidermal cells formed in the first postembryoic larval stage of *C. elegans* hermaphrodites. Each VPC is multipotent, but the spatial pattern of cell types is invariant in the intact animal. There is evidence from past studies that the anchor cell emits a graded inductive signal that can directly stimulate VPCs away from the ground state (type 3). Interactions among VPCs were examined in a mutant, *lin-15*, in which VPC destinies are partially independent of the inductive signal.

Type 1 cells actively inhibited adjacent cells from becoming the same type of cells. It appeared that the fate of each VPC depends on the combined actions of two intercellular signals, a graded inductive signal from the anchor cell and a lateral inhibitory signal from its neighbors.

Pattern formation among the VPCs in the *lin-15* mutant is analogous to the establishment of a neuroblast/dermatoblast pattern in early insect neurogenesis. Similarities in molecular structure of the *lin-12* and *Notch* gene products, which are involved in these pattern formation processes, might extend to functional similarities.

♦ This is one paper in a series of genetic and cell ablation experiments aimed at understanding the mechanisms of vulval induction in *C. elegans*. This induction results in pattern formation and requires both long-range interactions (between a cell in the gonad called the anchor cell and vulval precursor cells or VPCs) and short-range interactions between the VPCs. The anchor cell produces a graded signal that induces VPCs near it from a ground state (3° fate) to a 1 or 2° fate. VPCs far away from the signal adopt a 3° fate. By studying mutants of *lin-15* in which the VPCs fates are partially independent of the anchor cell signal, Sternberg was able to show that the choice between 1 and 2° fates occurs by 1° cells actively inhibiting neighbors from also becoming 1° cells. Thus, the fate of each VPC depends on the combined action of two kinds of intercellular signals, one from the anchor cell and one from its neighbors. The molecular mechanisms involved in both signaling pathways involve the *lin-12* receptor as well as VUL (vulvaless) and MUV (multivulva) genes. This work provides an important component in the genetic model of vulval induction in *C. elegans*, one which should facilitate our understanding of how combinations of developmental choices involving intercellular interactions are executed at the cellular and molecular levels to bring about pattern formation. *Joseph G. Culotti*

The Combined Action of Two Intercellular Signaling Pathways Specifies Three Cell Fates during Vulval Induction in *C. elegans*
P. W. Sternberg and H. R. Horvitz
Cell, 58, 679—693, 1989

Each of the six vulval precursor cells (VPCs) of *C. elegans* has three possible fates (1, 2, 3). The fate of each VPC depends on a graded inductive signal acting at a distance, and also on a short-range lateral signal among the VPCs (Figure 6-6). The authors examined interactions among mutations that cause differing misspecifications of VPC fates. In particular, the interactions of *lin-12* mutations with vulvaless and multivulva mutations were studied.

Particular combinations of mutations lead all six VPCs to have a single fate independent of their position. Specification of the three possible fates is achieved through two binary decisions, each affected by one of the two signaling pathways. The gene *lin-12* acts in the lateral signaling pathway and specifies fate 2. The vulvaless and multivulva genes act in the inductive signaling pathway and specify fate 1 independently of *lin-12* and fate 2 via *lin-12*.

A model for the regulation of VPC determination during vulval induction (Figure 6-6A) incorporates a role for *lin-12* in both autocrine and

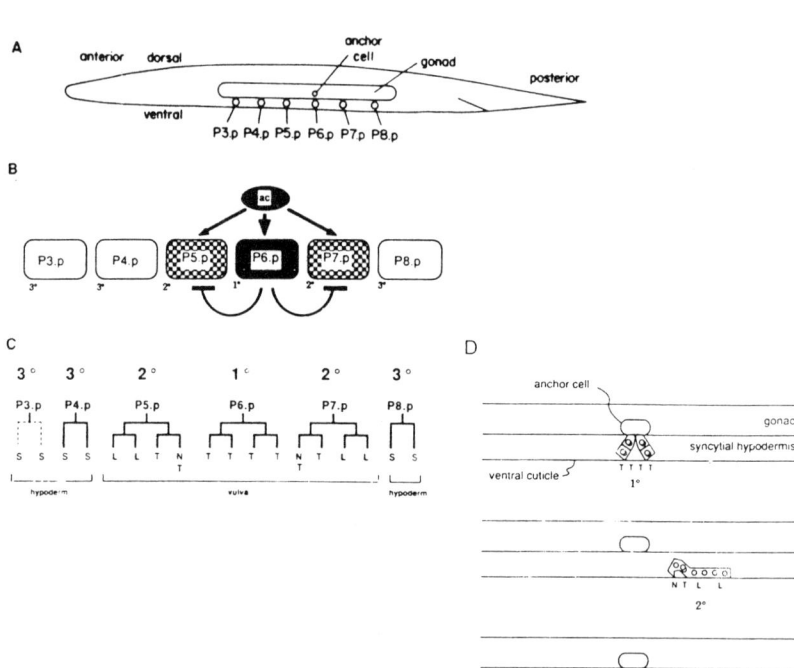

FIGURE 6-6. Development of the vulva of the wild-type hermaphrodite. (A) Six multipotent hypodermal cells, the VPCs P3.p, P4.p, P5.p, P6.p, P7.p, and P8.p, are located just ventral to the gonad. (B) According to our current model (Sternberg and Horvitz, 1986; Sternberg, 1988), the anchor cell induces P5.p, P6.p, and P7.p to generate vulval sublineages as opposed to a nonvulval sublineage. The potency of the signal produced by the anchor cell is graded and causes the six cells to adopt fates in a precise pattern. 1° VPCs inhibit neighboring VPCs from also becoming 1° via a lateral signal. (C) Each fate is to generate a particular sublineage, an apparently intrinsically determined pattern of cell divisions producing a characteristic set of progeny cell types. The vulval cell types are defined after two rounds of precursor cell divisions by two criteria: the axis of the third round nuclear divisions (L, longitudinal axis; T, transverse axis; N, no division), and adherence to the ventral cuticle indicated by boldface type (e.g., L). The types of progeny cells derived from a 2° sublineage are distributed in an asymmetric pattern, e.g., (**LL**TN), while the progeny of a 1° sublineage are distributed in a symmetric pattern, e.g., (**TTTT**). (D) The fate of a VPC is correlated with its position. (From Sternberg/Horvitz, *Cell*, 58, 679—696, 1989. ©Cell Press.)

paracrine VPC signaling. The formation of patterns of multiple cell types may, in general, involve the superposition of simple binary decisions. Each of these decisions involves an intercellular signaling system distinguishing two alternative cell states.

♦ The molecular mechanisms for specifying patterns of different cell type

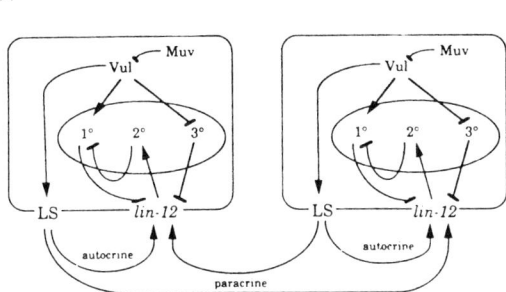

FIGURE 6-6A. Model for gene interactions during vulval induction. (A) Regulatory circuitry and inherent metastability of the presumptive 1° and 2° VPCs. The two cells are shown; the interior ovals represent the nuclei. Each line represents a proposed regulatory interaction; an arrowhead represents activation, a bar represents inhibition. 1°,1°-specific functions; 2°,2°-specific functions; 3°,3°-specific functions; LS, lateral signal. Induced VPCs (whether they receive inductive signal or are in a Muv animal) are poised between becoming 1° or 2°. *lin-12* is indicated as the receptor for the lateral signal. One source of the metastability might be the antagonsim between the 1° and 2° cell types: in this example, 1°-specific functions inactivate *lin-12* and thus indirectly 2° functions, while 2° functions inactivate 1° functions. The coupling between VPCs occurs via the lateral signal, which might act by paracrine stimulation of *lin-12*. Since the inductive signal can specify a 2° cell directly, *lin-12* must also be activated in a single VPC; the simplest hypothesis is that *lin-12* is stimulated by an autocrine mechanism. (It is also possible that *lin-12* is not stimulated in an autocrine fashion, but rather that it is constitutively active but its effect is antagonized by 3°-specific functions; thus, in an isolated 2° cell, Vul inhibits 3° which then allows *lin-12* to activate 2° functions.) (From Sternberg/Horwitz, *Cell*, 58, 679—696, 1989. ©Cell Press.)

in multicellular organisms by intercellular signaling are just beginning to be understood. Intercellular signaling can occur by diffusible factors acting at a distance (e.g., retinoic acid in the chick limb bud) or by short-range interactions via cell surface molecules (e.g., Drosophila R7 photoreceptor cell determination). By studying a number of mutants affecting vulval development in *C. elegans*, Sternberg and Horvitz were able to derive a model for gene interactions used to induce vulva formation. This pathway involves short-range lateral interactions between the vulval precursor cells (VPCs) (Sternberg, *Nature*, 335, 551—554, 1988) as well as a longer range inductive signal from a gonadal cell called the anchor cell onto the VPCs. The activity levels of products of three kinds of genes determine vulval cell fates: MUV (multivulva) genes, VUL (vulvaless) genes, and *lin-12*. *lin-12* and the VUL and MUV genes act in two separable pathways to specify the fates of the VPCs. One pathway, involving *lin-12*, distinguishes between 2° vs. 1° or 3° dates. A second pathway, involving the VUL and MUV genes, distinguishes between 1° or 2° vs. 3° fates, the combined action of the two pathways can force each VPC to adopt a unique fate. The VUL and MUV products mediate the response to the anchor cell signal, and the *lin-12* pathway mediates the response to the lateral pathway. This model should provide a framework for understanding the molecular circuitry of vulval fate determination in *C. elegans*, one which involves both long- and short-range interactions and a host of genes. Similarities in the structure and function of the *lin-12* gene product with other cell autonomously acting gene products, such as *Drosophila Notch* and *C. elegans glp-1*, suggest that the molecular mechanisms involved will be conserved in other organisms. Moreover, as stated at the conclusion of this paper, "the formation of patterns of multiple cell types may in general involve the superposition of simple binary decisions, each of which involves an intercellular signalling system that distinguishes two alternative cell states." Analysis of these decision-making processes at the molecular level in *C. elegans* may lead to a better understanding of these processes in vertebrates. *Joseph G. Culotti*

Cell Autonomy of *lin-12* Function in a Cell Fate Decision in *C. elegans*
G. Seydoux and I. Greenwald
Cell, 57, 1237—1245, 1989   6-7

The *lin-12* gene of *C. elegans* encodes a predicted transmembrane protein which controls a decision by two cells, Z1.ppp and Z4.aaa, to follow the anchor cell (AC) and ventral uterine (VU) precursor cell fates (Figure 6-7). In this study laser ablation was used to determine whether specification of the VU fate of these cells depends on a AC-to-VU signal. Genetic

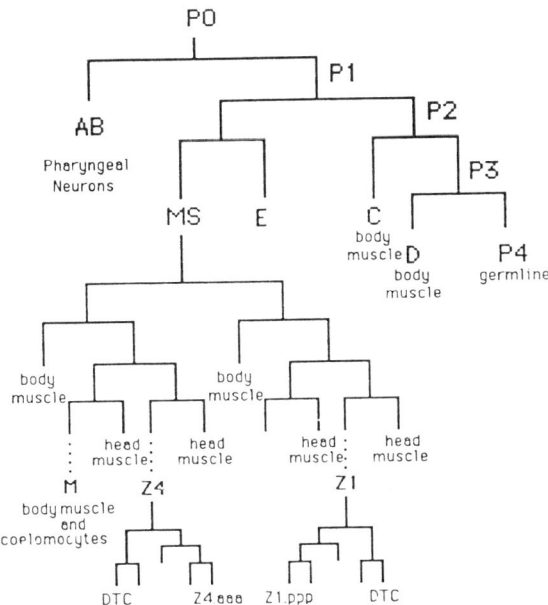

FIGURE 6-7. Abbreviated *C. elegans* lineage. Selected embryonic cell divisions (Sulston et al., 1983) and selected postembryonic cell divions of Z1 and Z4 (Kimble and Hirsh, 1979) are diagrammed. P_0 is the zygote. The embryonic blastomeres AB, MS, E, C, and D are precursors for the soma, while P_4 gives rise exclusively to the germline. Z1 and Z4 give rise to the somatic structures of the gonad. Note that Z1.ppp and Z4.aaa are descendants of Z1 and Z4, respectively. Under each precursor, we have indicated the cell type that was scored for that lineage in our genetic mosaics. DTCs: distal tip cells. A dotted line indicates three rounds of cell division. (From Seydoux/Greenwald, *Cell*, 57, 1237—1245, 1989. ©Cell Press.)

mosaics were produced in which defined cells lacked *lin-12* activity. Correlating the fates of Z1.ppp and Z4.aaa with the *lin-12* genotype of nearly all cells in the mosaics indicated that *lin-12* function is VU cell autonomous.

The VU fate adopted by Z1.ppp or Z4.aaa cells depends on a signal emanating from the presumptive AC. The function of *lin-12* apparently is autonomous, indicating a role for *lin-12* acting downstream of, and perhaps in the receiving mechanism for the AC-to-VU signal. The simplest explanation is that *lin-12* is the receptor for the AC-to-VU signal and is activated by binding its ligand, the signal itself (Figure 6-7A). The gene is not involved in both the signaling and receiving mechanisms. Even if it does not function biochemically as a receptor, *lin-12* functions cell autonomously to specify the VU fate of Z1.ppp and Z4.aaa.

♦ This study is an elegant use of genetics and cell ablation to reveal the

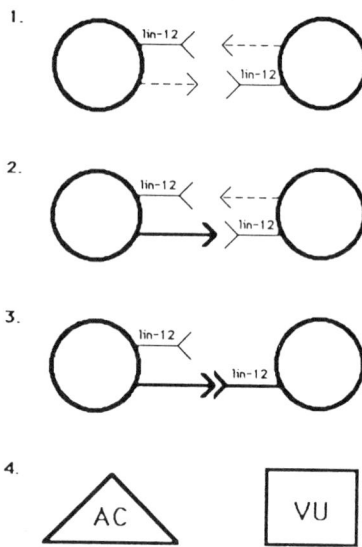

FIGURE 6-7A. Proposed model for the specification of the AC and VU fates. Step 1, uncommitted Z1.ppp and Z4.aaa (circles) express the receptor (*lin-12*). The dashed arrows represent either the potential of both cells to signal or the presence of subthreshold level of signal (i.e., does not activate *lin-12*). Step 2, one cell produces enough AC-to-VU signal (solid arrow) to activate *lin-12* in the other cell. Step 3, activated *lin-12* (bold) in the receiver causes the decision to be a VU and results in reduced signaling potential or level. Step 4, the committed AC and VU. Our data do not address the possibility that the committed AC and VU continue to interact to maintain their fates. (From Seydoux/Greenwald, *Cell*, 57, 1237—12465, 1989. ©Cell Press.)

cellular and molecular mechanisms of a cell fate determination in *C. elegans*. During development of the wild-type gonad, two cells of equivalent potential (Z1.ppp and Z4.aaa) interact, one (Z1.ppp or Z4.aaa with equal probability) becomes an anchor cell (AC) and the other becomes a ventral uterine cell (VU) as a result of this interaction. *lin-12* activity is necessary and sufficient for Z1.ppp and Z4.aaa to become VUs since in *lin-12* (dominant) mutants which have elevated *lin-12* activity, both cells become VUs and in *lin-12* (0) mutants which lack *lin-12* activity both cells becomes ACs. By studying mosaic animals in which Z1.ppp is mutant and Z4.aaa wild type (or vice versa) the authors were able to show (1) that *lin-12* function is VU cell autonomous, suggesting that *lin-12* acts downstream of, perhaps in the receiving mechanism for, the AC to VU signal and (2) that the relative level of *lin-12* activity in Z1.ppp and Z4.aaa is assessed in the initial decision of one of the cells to become AC. These results suggest a surprising model in which Z1.ppp and Z4.aaa both have receptor (*lin-12*), and sometime in the second larval stage, when AC and

VU fates are determined, the presumptive AC cell produces enough signal to activate *lin-12* in the receiver. Somehow, activation of *lin-12* in the receiver must lower the probability of activating *lin-12* in the signaler. Exactly how this happens may be a common mechanism used in fate determination by cell-cell interaction and should be the focus of future research. *Joseph G. Culotti*

The *Caenorhabditis elegans* Heterochronic Gene *lin-14* Encodes a Nuclear Protein that Forms a Temporal Developmental Switch
G. Ruvkun and J. Giusto
Nature, 338, 313—319, 1989   6-8

In *C. elegans* the temporal pattern of the cell types generated during development is regulated by heterochronic genes. Mutations of these genes lead particular cells in various lineages and tissues to adopt fates ordinarily associated with cells at an earlier or later stage of development. The heterochronic gene *lin-14* is central in controlling the temporal pattern of the postembryonic cell lineage of *C. elegans*. The gene has been cloned, and in this study antibodies to *lin-14* protein served to document the cellular and subcellular location of the protein and its accumulation during wild-type and *lin-14* mutant development.

The protein product of *lin-14* is localized in nuclei of specific somatic cells in embryos and early larvae of *C. elegans*, but is absent in the late larval and adult soma stages. Normally down regulation of the *lin-14* protein level encodes a temporal switch between early and late cell fates. Gain-of-function *lin-14* mutations caused the level of *lin-14* protein to remain high throughout development. Developmental reinterations of early cell lineages resulted.

Other heterochronic genes might regulate developmental timing of the *C. elegans* cell lineage by participating in generation of interpretation of the *lin-14* temporal switch. Many of the developmental control genes identified to date seem to be components of binary or multistate switches like *lin-14*. It is possible that mutations like the *lin-14* gain-of-function and loss-of-function mutations are the main cause of the variation required for evolutionary change to take place. The *C. elegans* heterochronic mutations are analogous to the heterochronic variation between species observed in phylogenetic studies.

♦Analysis of mutations in the heterochronic gene *lin-14* suggests that it is playing a central role in determining the temporal pattern of cell lineages in *C. elegans*. Loss-of-function *lin-14* mutations cause a precocious appearance of late larval lineages in earlier larval stages. Conversely, gain-of-function *lin-14* mutations cause an opposite transformation in cell fate, i.e., early larval lineages are repeated during later larval states (Ambros and Horvitz, *Science,* 226, 409—416, 1984). These results indicated that a transition from high *lin-14* activity to low *lin-14* activity as development progresses controls a switch from early to late cell fates in all of the cells affected by *lin-14* mutations.

The *lin-14* gene has been cloned by RFLP mapping (Reuvkun et al., *Genetics,* in press). This paper examines the spatial and temporal regulation of *lin-14* protein accumulation using specific antibodies and provides the first insights into the molecular mechanisms underlying heterochrony. The first surprise is that *lin-14* encodes a nuclear protein. Not surprisingly, all postembryonic blast cells affected by *lin-14* mutations accumulate the *lin-14* protein at the expected time, and those postembryonic blast cells not affected by *lin-14* mutations do not accumulate the *lin-14* protein. Staining in all of these cells fades following the period of known *lin-14* activity, but persists in *lin-14* gain of function alleles. This is consistent with the genetic data which suggest that a decrease in *lin-14* activity controls the temporal change in the fates (lineages) of these blast cells. A second surprise is that certain embryonic hypodermal, muscle, and neuronal cells not previously shown to be affected by *lin-14* mutations, nevertheless accumulate the *lin-14* protein. The fates of these cells were not examined in the mutants because only postembryonic blast cell divisions were scored, so it is still not known whether the *lin-14* expression observed in these differentiated cells is in fact functional. This paper goes a long way toward understanding heterochrony in *C. elegans* and it should be interesting to see if similar mechanisms operate during vertebrate development. *Joseph G. Culotti*

Ubiquitous Expression of *sevenless*: Position-Dependent Specification of Cell Fate
K. Basler and E. Hafen
Science, 243, 931—934, 1989   6-9

Cellular interactions are entirely responsible for the specification of cell fate in the compound eye of *Drosophila*. The *sevenless* (*sev*) gene is required for correct determination of one of the eight photoreceptor cells (R7) in each ommatidium. The gene encodes a transmembrane protein having a tyrosine kinase domain. It is expressed transiently on a population of ommatidial precursor cells including the precursors of R7 cells. The goal of this study was to determine whether the restricted distribution of the wild-type *sev* protein is necessary for the correct specification of photoreceptor cell fate.

Indiscriminate expression of a *sev* cDNA throughout development, induced by heat shock, correctly specified R7 cell identity without influencing the development of other cells. The discontinuous supply f *sev* protein during development of the eye led to the formation of mosaic eyes containing stripes of *sev*$^+$ and *sev*$^-$ ommatidia, indicating that R7 cell fate is specifiable only within a relatively brief time during ommatidial assembly.

In this system, specification of cell fate by position depends on the interaction of a localized signal with a receptor present on undifferentiated cells. The specificity of cell fate determination comes from a localized presentation of the signal, not restricted expression of the receptor. The presence of the receptor is not in itself adequate to specify cell fate.

Ommatidia in the Developing *Drosophila* Eye Require and Can Respond to *sevenless* for Only a Restricted Period
D. D. L. Bowtell, M. A. Simon, and G. M. Rubin
Cell, 56, 931—936, 1989   6-10

The eye of *Drosophila* serves well for studies of how distinct patterns of gene expression contribute to development of a highly ordered structure. The *sevenless (sev)* gene encodes a putative cell-surface receptor that is expressed in a complex and highly specific manner in the developing *Drosophila* eye. Mosaic analysis indicates that wild-type *sev* protein is required only in the R7 cells for normal development to take place. Is the transience of *sev* expression in the R7 precursor required for placing R7 differentiation in the correct temporal sequence with respect to other cells in the ommatidium?

The complex pattern of expression of *sev* was found not to be necessary for the development of R7 cells or any other cells in the retina. The time during which *sev* protein is required for formation of the R7 photoreceptor is brief. Since the specificty of cell-to-cell signaling mediated by *sev* protein is not a result of a particular distribution of the protein, it must reside in some aspect of the *sev* signaling pathway.

It is possible that the wild-type *sev* pattern functions as a redundant mechanism ensuring specificity of signal transmission in the developing *Drosophila* eye. Alternately, the complex pattern may reflect evolutionary adoption of a pattern that allows expression of *sev* in the presumptive R7 cells during the required time of ommatidial development. As always, caution is in order when drawing inferences about the role of a given pattern of genic expression during development.

♦ About 800 ommatidia compose the adult *Drosophila* eye. Each ommatidium includes eight photoreceptor cells termed R (1-8). Differentiation of the ommatidia occurs as a morphogenetic furrow moves in a posterior to anterior direction during a period of 36 h, "sweeping" the monolayer of cells that makes up the eye imaginal disc. At all times there is a precise spatial arrangement and patterns of cell-cell contacts among the photoreceptor cells. These cells differentiate in the following order: R8, R2 and R5, R3 and R4, R1 and R6, and finally R7. Inactivation of the *sevenless* gene results in the inability of the R7 cell to differentiate. The gene has been cloned and shown to code for a putative transmembrane protein with a tyrosine kinase domain, typical of a signal-transducing protein. Even though mosaic analysis had shown that *sevenless* function is required only in R7 for proper differentiation of the eye, the protein is also expressed in R3 and R4, but not in R1, R2, R5, R6, and R8.

The two articles describe similar experiments and reach similar conclusions. The *sevenless* gene was fused to a heat shock promoter and the consequences of ubiquitous expression of the gene at different developmental times was examined. The questions that were addressed were: is the time of *sevenless* expression important for proper differentiation of the R7 cell? Can indiscriminate expression of *sevenless* affect the differentiation of other photoreceptor cells?

Expression from the (heat shock-*sevenless*) gene construct could completely rescue the *sevenless* mutation. Transient expression of the *sevenless* gene in a *sevenless*⁻ fly during eye morphogenesis had only a localized rescuing effect. That is, although expression of *sevenless* occurs in all cells of the developing eye, rescue was only observed in a "stripe" of ommatidia that were at the position of the morphogenetic furrow at the time of *sevenless* expression. This suggests that the gene product is required at a precise time of development (when the cell is localized in the vicinity of the morphogenetic furrow) and that its activity falls to below-threshold concentrations in a relatively short period of time.

Expression of *sevenless* in any tissue and at any time of development had no deleterious effects. This suggests that a generalized expression of *sevenless* is of no consequence as long as it is present in R7. Most likely, the specificity of cell fate is determined not by the distribution of the receptor, but by the localized production of a *signal* that is recognized by the *sevenless* protein. The product of the *bride-of-sevenless* gene (see separate report by Reinke and Zipursky in this volume) may well constitute the signal or be intimately connected with its production. Marcelo Jacobs-Lorena

---

Cell Lineage Analysis Reveals Multipotency of Some Avian Neural Crest Cells
M. Bronner-Fraser and S. E. Fraser
Nature, 335, 161—164, 1988   6-11

In most developmental systems it is not clear whether precursor cells can form many or all cell types, or only a single type. The question is central to the neural crest, which gives rise to many and diverse derivatives including peripheral neurons, glial and Schwann cells, pigment cells, and cartilage. In this study the vital dye LRD (lysinated rhodamine dextran) was microinjected iontophoretically into individual dorsal neural tube cells of he chick embryo in order to identify their descendants.

Many of the labeled clones consisted of multiple cell types, judging from both their location and morphology. Sensory neurons, presumptive pigment cells, ganglionic supportive cells, adrenomedullary cells, and neural tube cells all were found within individual clones. Descendants of single cells were seen in as many as four neural crest derivatives. A majority of clones giving rise to neural crest cells were distributed in multiple neural crest sites. Even those clones with no members remaining in the neural tube differentiated into multiple cell types.

At least some neural crest cells are multipotent before leaving the neural tube and are able to give rise to both sensory and autonomic cell lineages. The findings do not preclude the existence of some predetermined cells in the premigratory neural crest. Restriction may be initiated by the emigration of neural crest cells from the neural tube. Alternately it may be environment dependent and mediated by interactions along their migratory path of cells or at their final position.

♦ The neural crest is a migratory population of cells that gives rise to many diverse derivatives, including neurons and glia of the peripheral nervous system, pigment cells, and cartilage and bone of the face. A long-standing question concerns the developmental potential of individual neural crest cells: are they able to give rise to numerous derivatives (multipotent) or are they restricted to a single fate (predetermined)? Previous studies designed to answer this question have relied upon examining cell fate using large populations of cells *in vivo* or upon culture techniques to isolate single cells *in vitro*. However, populations of cells may not reflect individual cell fate and the behavior of the cells *in vitro* may not reflect their behavior *in vivo*. Recent approaches for *in situ* studies of cell lineage have employed either microinjection of large, inert lineage tracer dyes or infection by a recombinant retrovirus. The use of retroviral lineage markers, in which clones are typically defined as contiguous cells, is not readily applicable to the neural crest system because of the extensive dispersion of neural crest cells.

This study represents the first direct analysis of neural crest cell fate *in situ*. A lineage tracer dye was microinjected into individual premigratory neural crest cells within the trunk neural tube of chick embryos. Because the dye is only passed to progeny of the labeled cell by cell division, all labeled cells can be identified as members of the same clone, even after the cells disperse. Bonner-Fraser and Fraser found that about half of the clones contributed cells to numerous neural crest derivatives, such as the dorsal root ganglia, sympathetic ganglia, adrenal medulla, and/or pigment cells. This represents the first evidence that many neural crest cells are multipotent prior to their migration. Thus, environmental influence rather than an inherent program may be important for determining the differentiated phenotype arising from some neural crest cells. This does not, however, rule out the possibility that other neural crest cells are developmentally restricted. *Marianne Bronner-Fraser*

Radial Arrangement of Clonally Related Cells in the Chicken Optic Tectum: Lineage Analysis with a Recombinant Retrovirus
G. E. Gray, J. C. Glover, J. Majors, and S. R. Sanes
Proc. Natl. Acad. Sci. U.S.A., 85, 7356—7360, 1988   6-12

In many parts of the vertebrate brain neurons are organized both in a series of laminae paralleling the pial surface and radially, perpendicular to the laminae. Radial organization may be most evident in the visual system. It is of interest to ask whether clonally related cells are restricted to — and potentially involved in establishing — individual laminae or radial arrays. A method of lineage analysis developed in rodents has now been applied to the chick optic tectum. A recombinant retrovirus serves to insert a foreign gene into the genome of a precursor cell, and the gene product is detected histochemically in the progeny of the infected cell.

The descendants of a single precursor formed narrow, radially oriented columns spanning the thickness of the developing tectum. Study of embryos injected with virus at different stages of development suggested that early-born progeny are displaced laterally, and late-born porgeny, radially (Figure 6-12). Periods of lateral and radial displacement were largely sequential, not simultaneous. Mixing of cells from neighboring clones was observed but was limited in extent.

Neurons descended from a single precursor are widely interconnected and functionally related. Many cells in a vertically arrayed clone likely are functionally related. Similarly, radially displaced cells in different laminae have similar visual receptive fields, and some receive somatosensory or auditory inputs from corresponding regions of body space. Possibly a progenitor cell is programmed to generate a specific ensemble of neurons that form a functional unit. Alternately cellular phenotypes may be determined totally by epigenetic factors. An intermediate view is that epigenetic factors predominate in determining phenotype, while regular mitotic and migratory patterns combine to make sure that cells are delivered at appropriate times and sites to produce the cellular diversity and interconnections required for a functional circuit.

♦Two techniques have surfaced for the analysis of cell lineage in intact

Cell Lineage and Developmental Fate 201

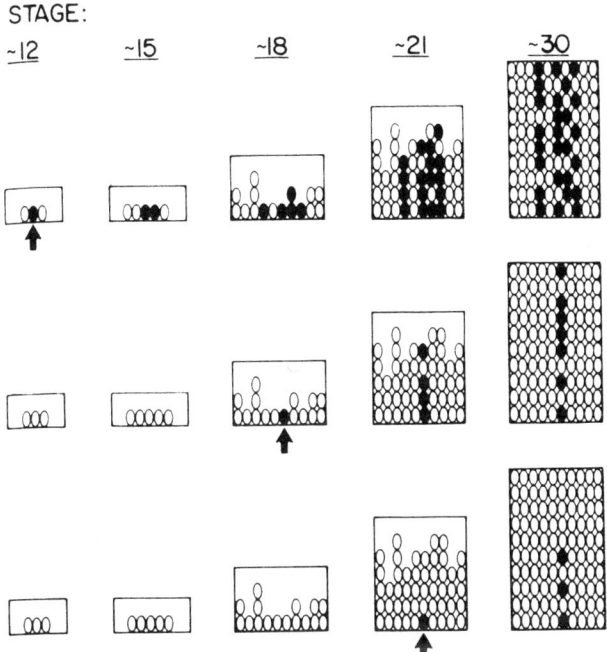

FIGURE 6-12. A simplified scheme of tectal neurogenesis that summarizes results obtained by varying the stage of virus injection. Arrows indicate time of infection and solid ovals indicate clonally related cells. In this scheme, early-born progeny are displaced laterally, later-born progeny are displaced vertically, and cell mixing is possible but limited at all stages. (From Gray et al., *Proc. Natl. Acad. Sci. U.S.A.*, 85, 7356—7360, 1988. With permission.)

embryos: one is direct microinjection of lineage tracer dyes (as discussed above), and the second is infection of cells with a replication-incompetent retrovirus. The latter technique has been particularly useful in rodents, which are difficult to manipulate, and for examination of the central nervous, which is often inaccessible to microinjection. The retroviral techniques have the advantage of introducing a heritable marker gene, but the disadvantage of it being difficult to establish clonality in a labeled group of cells.

In this study, Gray and colleagues have applied the retroviral approach to studying cell lineage in the chick optic tectum. They have shown that each group of contiguously labeled cells is likely to be a clone based on results of infecting with limiting dilutions of virus. They found that the number of cells per clone remained constant, indicating that each labeled group resulted from one "hit", whereas the total number of clusters decreased with decreasing virus concentration. This is an important consideration since only in this way can clonality be inferred.

From their lineage analysis, the authors find that early cell divisions of tectal cells give rise to laterally displaced daughter cells. This is followed by a second stage in which progeny are radially displaced as they arise. Cells labeled at early times are multipotent and contribute clonally related cells to multiple layers. Furthermore, there is only limited cell mixing between neighboring clones. The radially arranged cells give rise to different classes of neurons that are known to be functionally interconnected. This suggests that clonally related cells form functional units.—Marianne Bronner-Fraser

Clone-Forming Ability and Differentiation Potential of Migratory Neural Crest Cells
A. Baroffio, E. Dupin, and N. M. Le Dourin
Proc. Natl. Acad. Sci. U.S.A., 85, 5325—5329, 1988   6-13

Development of the neural crest in the vertebrate embryo involves the segregation of various cell lineages from the initial migratory cells. The microenvironment is an important determinant of the phenotypes expressed, but the cells themselves are heterogeneous in their developmental potential. In this study neural crest cells were taken from quail embryos at the level of the mesencephalon and metencephalon and cultured as single cells on 3T3 feeder layers. Nearly 250 clones, some containing more than 20,000 cells, were analyzed after 7 to 10 d in culture.

Cells with very diverse developmental potency were observed. Both committed neuron progenitors and pluripotent cells were present; the latter differentiated into several types of neurons, nonneuronal cells, and melanocytes. Cells of intermediate developmental potency also were seen. In all, at least 11 types of colonites were seen; a majority contained more than one type of cell.

The neural crest is very heterogeneous with regard to the proliferative and developmental potential of its component cells. The precursors of several types of clones are more or less developmentally restricted members of the same lineage, implying that the cells become progressively committed during their migration.

♦This work uses tissue culture analyses to examine the developmental potential of individual cranial neural crest cells. *In vivo*, the cranial neural crest gives rise to neurons and glia of some cranial ganglia, cartilage and bone of the face, connective tissue, and hormone-secreting cells. One problem that has limited the utility of cell lineage studies *in vitro* is the need for cell contacts for the differentiation of some neural crest derivatives. In this study, individual cranial neural crest cells were grown on feeder layers of 3T3 cells. By providing heterologous cell contacts, these culture conditions help to promote differentiation of cloned neural crest cells without the necessity of abundant homologous cell contacts. The authors found numerous types of clones arising under these conditions, ranging from clones containing only neurons to clones which contained numerous phenotypes including neurons, nonneuronal cells, and pigment cells. Interestingly, clones containing cartilagenous cells never had neuronal cells and vice versa. These results suggest that many cranial neural crest cells are multipotent under the culture conditions used. In addition, the finding of clones containing only a few neurons is consistent with predetermined of some cells. However, this cannot be taken as firm evidence of predetermination until the prospective fate of these cells is challenged by altering the local environment. Another interesting conclusion from these studies is that neural crest cells are heterogeneous, based on the observation of numerous classes of clones. Whether this heterogeneity reflects an early lineage restriction or simply a probablistic component to prospective cell fate remains unclear. *Marianne Bronner-Fraser*

Cloning of cDNA for the Major DNA-Binding Protein of the Erythroid Lineage through Expression in Mammalian Cells
S.-F. Tsai, D. I. K. Martin, L. I. Zon, A. D. D'Andrea, G. G. Wong, and S. H. Orkin
Nature, 339, 446—451, 1989
6-14

Regulation of the globin gene family in the erythroid lineage of hemopoietic cells is an excellent system with which to study tissue and developmental stage-specific gene expression. Genes expressed in erythroid cells contain binding sites for a cell-specific factor thought to be an important regulator for this lineage. The authors isolated cDNA encoding the murine protein by high-level transient expression in mammalian cells.

The cell-specific factor, termed GF-1, is a new member of the zinc-finger family of DNA-binding proteins; murine GF-1 contains two Cys-Cys zinc fingers. GF-1 is limited to the erythroid lineage at the level of RNA expression. The murine and human homologues of GF-1 are very closely related at the level of the amino acid sequence, at least in a region flanking the putative DNA-binding domain. This occurs despite seemingly weak cross-hybridization of nucleotide sequences.

The sites to which GF-1 binds in human beta-globin gene clusters are associated with DNAse I hypersensitivity limited to erythroid progenitors. Since DNAse I sensitivity appears to precede active globin gene transcription *in vivo*, the potential role of GF-1 in promoting structural changes in chromatin warrants investigation. A better understanding of the molecular basis for cell-restricted expression of GF-1 RNA should help elucidate the early events in commitment of stem cells to the erythroid lineage.

♦The appearance of a particular cell lineage results from the expression of a set of genes that confer phenotypic specificity. How those genes are recognized and coordinately expressed to produce such specificity remains an unanswered question. This article offers a possible explanation. The authors report the application of a new strategy for the isolation of DNA-binding proteins and discuss a specific one unique to the red blood cell lineage. Genes expressed in this lineage seem to contain a specific binding site recognized by this binding protein. It is obvious that the lineage specificity is conferred by the restricted expression of the binding protein itself. Therefore, understanding how the binding protein is regulated should provide important insights into the mechanism of cell lineage determination. *Joel M. Schindler*

Separate Lineages of T Cells Expressing the $\alpha\beta$ and $\gamma\delta$ Receptors
A. Winoto and D. Baltimore
Nature, 338, 430—432, 1989

The T-cell antigen receptor is a heterodimer consisting of either $\alpha\beta$ or $\gamma\delta$ chains. The $\alpha\delta$ receptor molecules are expressed chiefly in helper and killer T cells, while the $\gamma\delta$ receptor molecules are expressed mainly in $CD4^-CD8^-$ T cells. Since helper ($CD4^+CD8^-$) and $CD4^-CD8^-$ cells arise from a class of $CD4^-CD8^-$ cells during thymic development, the question arises of whether cells rearranging their receptors later give rise to $\alpha\beta$ T cells through further rearrangements of the receptor genes — or whether rearrangements and expression of the receptor genes occur in separate lineages.

The δ-chain gene is between the $V\alpha$ (variable) and $J\alpha$ (joining) gene segments. When the rearrangements allowing α- and β-receptors take place, the DNA between these segments is deleted as small circles that may be isolated from developing thymocytes. This study found that the δ-chain gene in the T-cell receptor α-circles has a germline configuration, indicating that $\alpha\beta$ and $\gamma\delta$ T cells are distinct lineages.

These findings indicate that a majority of α-producing T cells never rearrange their δ-chain locus. The split of $\alpha\beta$- and $\gamma\delta$-producing T-cell lineages might occur after the initial γ-chain rearrangements at gestation. of 2 weeks. Cells with a productive γ-chain gene rearrangement then would go on to rearrange their δ-chain gene, while others would differentiate to the $\alpha\beta$ lineage. More likely, the T-cell lineage decision is made independently of the rearrangement process. One subset of T cells then would commit to α-chain rearrangement and another to δ-chain rearrangement.

Surface Expression of Only γδ and/or αβ T Cell Receptor Heterodimers by Cells With Four (α, β, γ, δ) Functional Receptor Chains
T. Saito, F. Hochstenbach, S. Marusic-Galesic, A. M. Kruisbeek, M. Brenner, and R. N. Germain
J. Exp. Med., 168, 1003—1020, 1988   6-16

Previous studies of T-cell receptor gene expression in cloned cell lines have indicated that γδ- and αβ-expressing T cells represent distinct lineages derived from a common precursor. Expression of a receptor presumably precludes further receptor gene rearrangement, and therefore αβ expression. Failure to rearrange either γ or δ allows continued rearrangement and the eventual expression of an αβ receptor. In this study, functional α and/or β genes were introduced into γδ$^+$ cells, and the resultant cell surface expression of CD3-associated heterodimers was analyzed.

Cells containing intact γ, α, and β chains had only γδ dimers on their surfaces. In PEER cells from a human tumor line, addition of a functional α chain led to a loss of γδ dimer expression and to the expression of only αβ dimers. This was not a result of transcriptional down regulation of the γ or δ loci. In murine cells expressing all four chains, both γδ and αβ dimers were found on a single cell. No other chain combinations were demonstrated.

It appears that cells express the two types of receptor dimer in a mutually exclusive manner. Since the effector functions of T cells bearing either receptor type are similar — if not identical — the two receptor types probably function to provide their respective cells with diverse, minimally overlapping sets of specificities. This allows a larger antigenic universe to elicit immune effector activity.

Intestinal Intraepithelial Lymphocytes are a Distinct Set of γδ T Cells
M. Bonneville, C. A. Janeway Jr., K. Ito, W. Haser, I. Ishida, N. Nakanishi, and S. Tonegawa
Nature, 336, 479—481, 1988   6-17

In addition to CD4$^+$ T cells and CD8$^+$ T cells, a third rearranging T-cell specific gene called γ is responsible for a new class of T cells bearing a new receptor type, CD3 γδ. These cells are the main lymphocytes of epidermis and might therefore be important in surveillance of all epithelia. In this study, intraepithelial lymphocytes isolated from the murine small bowel were found to predominantly or exclusively express CD3 γδ receptors. In contrast to epidermal lymphocytes, the cells also express CD8, and they use a different $V\gamma$ gene to form their receptor.

The intraepithelial lymphocytes of the small intestine are T cells having γδ receptors. The fact that both epidermis and intestinal epithelium are sites of predominant γδ T-cell populations indicates that these cells have a critical role in epithelia. Probably the cells are specialized for homing to intestinal epithelium, but the mechanism of homing or selective retention of the cells in particular epithelial sites remains to be determined.

♦ The search for the T cell receptor (TCR) α gene resulted in the discovery of a new receptor gene locus now referred to as the TCR γ (Saito et al., Nature, 309, 757, 1984). This initial discovery was extended to show that between 1 and 10% of peripheral T cells express the TCR γ (Brenner et al., Adv. Immunol., 34, 133, 1988). This class of T cells apparently precedes those cells which express TCRαβ during thymic development (Raulet, Annu. Rev. Immunol., 7, 175, 1989). Also, the TCRγδ T cells are mainly CD4⁻CD8⁻ (Davis and Bjorkman, Nature, 334, 395, 1988). Since the T cells which express the TCRαβ are mainly CD4⁺CD8⁻ or CD4⁻CD8⁺ and these cells arise from CD4⁻CD8⁻ in the thymus (von Boehmer, Annu. Rev. Immunol., 6, 309, 1988), it has been speculated that the TCRγδ cells give rise to the TCRαβ cells. A recent communication, employing a very nice trick, demonstrated that the αβ and γδ T cells probably represent distinct lineages of cells.

In addition, it has been formally shown that T cells do not cross utilize these four independent chains to generate additional diversity within the T cell repertoire.

These types of studies have lead investigators to wonder about the function of these γδ T cells. At this time it seems that these T cells localize into sites which represent potential portals of entry for pathogens. The γδ T cells are the most predominant CD3⁺ cells found in the epithelium of the skin (Kuziel et al., Nature, 328, 263, 1987). This lead to the prediction that the γδ T cells represent a major line of defense against pathogens which enter either through the skin or across an epithelial lining covering a mucosal surface (Janeway et al., Immunol. Today, 9, 73, 1988). Supporting this idea is the observation that intraepithelial cells of the gut are members of this class of T cells. *Charles Snow*

Cytodifferentiation — Cell- and Tissue-Specific Gene Expression and Maintenance

7

INTRODUCTION

The terminal phenotype of any individual cell is the result of cell- and tissue-specific gene expression. Once all the appropriate signals are received, integrated, and interpreted, the existing molecular genetic machinery responds by expressing the correct battery of specific genes. The gene products encoded by those genes confer functional specificity to each individual cell.

Unique DNA sequences have been identified that are responsible for that cell specificity. It is clear that multiple sequences, and multiple factors that interact with them, will be necessary to completely define the molecular mechanisms that support cell- and tissue-specific gene expression.

There are several different organ systems that display unique aspects of cell- and tissue-specific gene expression. These are the nervous, immunological, and gonadal systems. Each system includes a number of different cell phenotypes, all of which are unique. The expression of their collective function is primarily cellular. They do not necessarily secrete a specific product. Instead, they base their function on the activity of a specific cell phenotype — neurons, B- or T-cells, sperm, or oocytes. Cytodifferentiation in each of these organ systems involves a complex interaction between several different cell phenotypes. Thus, cell- and tissue-specific gene expression not only involves the integration of signals at the cellular level, but collectively at the tissue level as well.

The regulation of this collective response is not understood well. It includes genetic, cytoplasmic, cell surface, and extracellular components. Determining how these various components operate together in multiple cell phenotypes to generate an accurate and appropriate cellular response should become the focus of much interesting research.

Evidence That Elevated Intracellular Cyclic AMP Triggers Spore Maturation in *Dictyostelium*
R. R. Kay
Development, 105, 753—759, 1989   7-1

Spore maturation is a normal part of development of *Dictyostelium*, when environmental effects lead a migrating slug to transform to a fruiting body. A burst of enzyme accumulation accompanies the conversion of amoeboid prespore cells to refractile spores. The authors found that the process is triggered by an increased intracellular cyclic AMP concentration. The phenotypes of certain mutants with advanced spore maturation suggested this idea. Permeant cAMP analogues were used to induce spore formation in wild-type strains.

A number of rapidly developing mutants are able to form spores in submerged monolayers, while wild-type strains are not capable of this. The mutant phenotypes are best explained by derepression of the signal transduction pathway utilizing intracellular cAMP. Permeant cAMP analogues — but not cAMP itself — rapidly induced spore differentiation in wild-type amoebae incubated in submerged monolayers. The analogues also promoted accumulation of UDP-galactose epimerase in slug cells transferred to shanken suspension. This is one of the enzymes that increases during spore maturation.

Spore differentiation may be induced by permeant cAMP analogues at very low cell densities, indicating that neither cell contact nor additional soluble inducers are required. In other instances an additional inducer may be necessary. Spore differentiation is inhibited by DIF-1, a stalk-specific inducer, reflecting probable inhibition of a target downstream of intracellular cAMP in the signal transduction pathway leading to spore differentiation. This is further evidence that cell-type differentiation is regulated by interactions between cells rather than being predetermined by intrinsic differences between them.

A *Dictyostelium* Prespore-Specific Gene is Transcriptionally Repressed by DIF *In Vitro*
A. E. Early and J. G. Williams
Development, 103, 519—524, 1988   7-2

Cells in the anterior fifth of the slug *Dictyostelium* normally differentiate to form dead, vacuolated stalk cells, while a majority of posterior cells differentiate into spores. One important role of DIF, the stalk cell-specific inducer of *Dictyostelium*, may be to divert cells away from the spore-cell path of differentiation. The D19 gene encodes an mRNA that is highly enriched in prespore cells in the migratory slug. The gene encodes PsA, a cell surface glycoprotein of prespore cells. The authors used this gene to determine who DIF acts to repress specific prespore gene expression.

Studies with a mutant defective in DIF accumulation showed that the concentration of D19 — and of several other prespore mRNA sequences — declines in the presence of exogenous DIF. Both transcriptional and posttranscriptional controls appear to operate to regulate expression of these genes. *In vitro* nuclear transcription and mRNA half-life analyses indicated that DIF acts at the transcriptional level to repress the accumulation of D19 mRNA.

Since DIF does not cause a rapid turnover of prespore protein, its regulation of gene transcription likely is of primary importance in mediating prespore repression. DIF and adenosine act at different stages in the cAMP signal transduction pathway, and they may function synergistically to divert cells from the prespore path of differentiation.

♦ During *Dictyostelium* development, the cells of the multicellular aggregates differentiate into spatially separated sets of prestalk and prespore cells. Only at culmination do these cells finally differentiate into mature stalk and spore cells. Despite a considerable body of knowledge of the biochemical and molecular events associated with sporulation, the specific trigger of this process has remained unknown. Two new studies now offer considerable insight into this important morphogenetic process.

The first paper draws upon earlier observations that certain sporogenous mutants can differentiate into spores at low cell density in the presence of cAMP and that rapidly developing mutants have elevated cAMP levels. Kay uses 8-Br cAMP to elevate the levels of cAMP and induce spore differentiation in low density cultures of wild-type cells. These results support the idea that cell-cell interactions are not involved in the terminal stage of spore differentiation. Interestingly, DIF-1 which is the morphogen responsible for stalk cell differentiation can overcome the effect of the 8-Br cAMP and cause the entire population to become spores. This indicates that DIF inhibition of spore differentiation does not occur by blocking the cell surface cAMP receptors. DIF has been shown to induce the transcription of prestalk-specific genes. Early and Williams now demonstrate that DIF also acts to repress the transcription of several prespore-specific genes. There is no effect of DIF on the stability of the mRNAs. Thus, DIF has both positive and negative regulatory roles, at the level of transcription, in inducing stalk cell differentiation. *Stephen Alexander*

Characterization of an Antigenically Related Family of Cell-Type Specific Proteins Implicated in Slug Migration in *Dictyostelium discoideum*

S. Alexander, E. Smith, L. Davis, A. Gooley, S. B. Por, L. Browne, and K. L. Williams

Differentiation, 38, 82—90, 1988   7-3

The monoclonal antibody *MUD50* recognizes a group of proteins expressed almost exclusively by prespore cells in developing aggregates of *Dictyostelium discoideum*. Some of the antigens are closely associated with the cell membrane. The MUD50 proteins are largely prespore-specific at the slug stage and are posttranslationally modified. One of the products is a common unique non-N-linked carbohydrate which, with other carbohydrates, may have a significant role in slug migration and morphogenesis.

The *MUD50*-reactive proteins are glycosylated and some of them are phosphorylated. The common antigenic feature recognized by the antibody in these cell type-specific proteins is posttranslational modification. Flow cytometric studies showed that the MUD50 antigenic determinant is exposed at the surface of prespore but not prestalk cells (Figure 7-3). A glycosylation-defective mutant, DL118, fails to express the MUD50 epitope but does express the MUD52 epitope, which is found on another group of glycoproteins.

The authors believe that the common *MUD50* determinant on the proteins of sheath and prespore cells has functional significance in slug motility and differentiation. Further knowledge of how these proteins function in morphogenesis awaits their biochemical characterization.

Glycoproteins that Exhibit Extensive Size Polymorphisms in *Dictyostelium discoideum*
E. Smith, A. A. Gooley, G. C. Hudson, and K. L. Williams
Genetics, 122, 59—64, 1989   7-4

Electrophoretic variants arising from amino acid substitutions lead to charge differences between proteins, which have been extensively used in genetic analysis. Polymorphisms in protein size may be similarly used.

The authors studied a group of glycoproteins sharing a common carbohydrate epitope which vary in size in different isolates of the cellular slime mold *Dictyostelium discoideum*. One of them, PsA, is a developmentally regulated prespore-specific surface glycoprotein which exists in three size forms due to allelic variation at the *pspA* locus on linkage group I. A second glycoprotein, PsB — which also is prespore-specific and is found within prespore cells — maps to linkage group II. It has at least four different size forms. PsB appears to be a cytoplasmic, not a membrane-associated protein.

The size polymorphisms of PsB probably are the result of allelic variation at the *pspB* locus, reflecting differences in the number of repeat units. In cell contact molecules the functional contact domain may be held some distance from the cell membrane by a spacer domain. Nonsurface

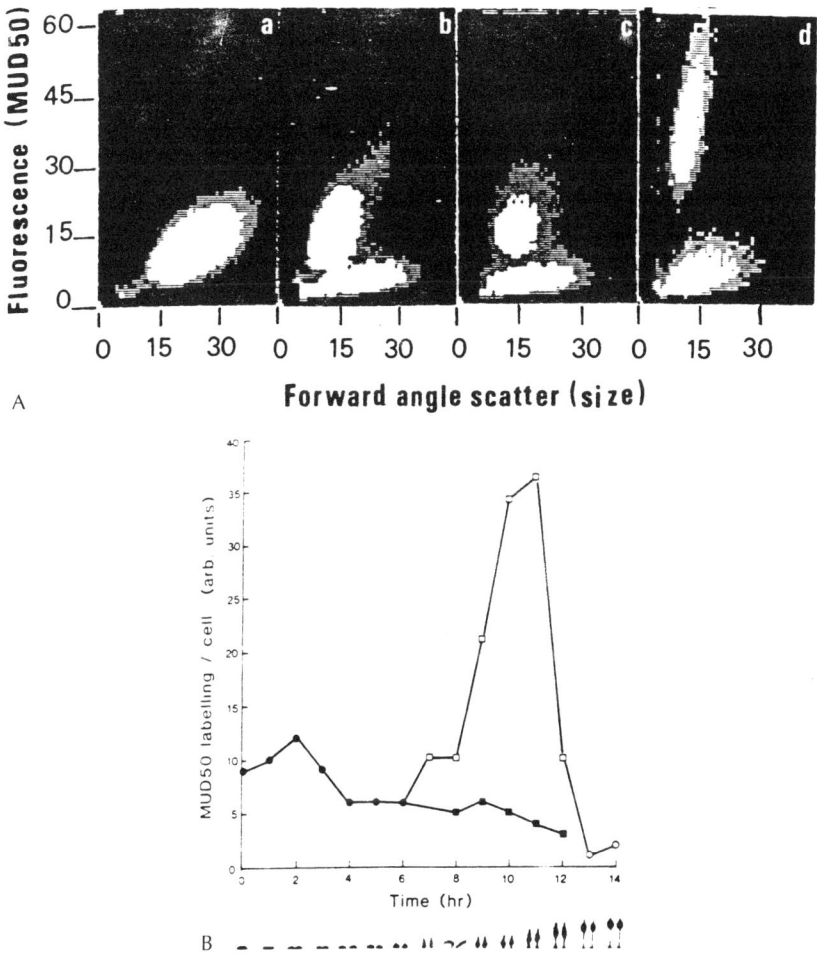

FIGURE 7-3. A,B. Developmental time course of MUD50 binding to the surface of WS380B cells measured by flow cytometry. (A) Profiles of MUd50 labeling at 2 h (a), 9 h (b), and 10 h (d). The profile shown in (c) is of disaggregated cells from slugs migrating on water agar. In profiles b, c, and d, the two populations are unlabeled prestalk cells and labeled prespore cells. (B) The level of MUD50 labeling of cell types at each point of the developmental time course. Amebae (closed circles), prespores (open circles), prestalks (closed squares — note thsat since prestalk cells could not be identified as a well--separated population at 7 h, no point is given), and mature spores (open circles). (rom Alexander et al., *Differentiation*, 38, 82—90, 1988. ©Springer-Verlag.)

molecules such as PsB may have two functional domains separated by a similar spacer domain.

♦ A number of monoclonal antibodies have been produced to probe the mechanism of *Dictyostelium* slug migration. Some of these antibodies are directed toward carbohydrate modifications and recognize a number of different glycoproteins. Other antibodies are directed towards the polypeptide backbone of glycoproteins and recognize only a single species. Many antibodies show crossreactivity. These two reports add valuable new information on the biochemical nature and function of these antigens.

The paper by Alexander et al. describes the characterization of the proteins detected by the MUD50 antibody. They show that the antibody reacts with an amazingly diverse group of prespore-specific proteins including membrane-associated and soluble proteins, that the MUD50 epitope is an unusual O-linked oligosaccharide, and that this oligosaccharide is involved in the migration of the slug. One of the MUD50 reactive proteins is PsA which has been shown to have different polymorphic forms (as judged by molecular weights on SDS gels). The data in this paper also clearly show that these differences are due to differences in the protein backbone rather than differential numbers of carbohydrate sidechains.

Smith et al. extend the studies of the size polymorphisms of the MUD50 reactive proteins by analyzing another member of the antigen family with the use of another protein-specific antibody. MUD102 recognizes an antigen (PsB) of between 80 and 89 kDa depending on the strain examined. Four different polymorphic forms exist differing by approximately 3000 Da. Unlike PsA it is not membrane bound and is encoded by a different gene. The authors suggest that this family of proteins has a common feature of repeat amino acid sequences, that the exact number of repeats is not critical to the function of the protein. Deletion of one or more repeats has resulted in the appearance of the polymorphisms.
Stephen Alexander

A Genetic Pathway for the Development of the *Caenorhabditus elegans* HSN Motor Neurons
C. Desai, G. Garriga, S. L. McIntire, and H. R. Horvitz
Nature, 336, 638—646, 1988   7-5

A total of 38 genes define the pathways for development of the hermaphrodite-specific neurons (HSNs) in *C. elegans,* which are required for egg laying by the hermaphrodites. The two bilaterally symmetrical HSNs are generated midway through embryogenesis, and shortly thereafter they begin migrating anteriorly from the tail. In the hermaphrodite the HSNs migrate about half way along the length of the embryo. Developmental steps affected in HSN-defective mutants include sex-specific HSN survival, HNS migration, hood formation, axonal outgrowth, serotonin expression, and HSN function. Some mutants have multipe HSN defects in which HSN migration, serotonin expression, and axonal outgrowth are altered.

HSN development occurs not as a series of dependent steps but, to a considerable degree at least, in parallel, independent steps (Figure 7-5). The independence of the various steps in development shows the ability of the HSN to adapt to its environment. Some of the genes may be expressed not within the HSNs themselves but within other cells that control some aspects of HSN development. Acquisition of identity is the first cell type-specific step in HSN development. Subsequently a transition from an early, immature HSN cell state to a later mature state takes place. An additional regulatory step termed "specialization" controls the expression of certain specialized HSN traits, both early and late traits, expressed intermittently during development.

The genetic pathway of HSN development can serve as a model for the determination and differentiation of any cell type. Initially a decision is made on whether or not a given cell identity is to be acquired. Genes controlling the acquisition of cellular identity regulate other control genes determining the expression of overlapping subsets of cell-specific traits. These genes, in turn, regulate those that decode the products responsible for specific differentiated cell characteristics.

♦ This paper describes a genetic pathway for the development of a small subset of neurons in *C. elegans* derived through the phenotypic analysis of mutations in 35 genes that affect these neurons. The HSN neurons of *C. elegans* are required for egg laying in *C. elegans* which provides a convenient assay for selection of HSN mutants and for function of HSN neurons. These cells are born in the posterior of the animal then migrate anteriorward before undergoing differentiation. Differentiation includes hood formation (detectable by Nomarski optics), axon outgrowth, serotonin expression, and function. (Mutants with normal HSNs in every respect except function can be detected because egglaying muscles are sensitive to serotonin agonists but insensitive to imipramine, a drug which in the wild-type blocks serotonin reuptake by the presynaptic HSN neurons.)

Some of these genes affect only one HSN trait, demonstrating that HSN migration, axonal outgrowth, and serotonin expression are independent events. Other genes affect more than one HSN and may be regulatory. Nearly all of the mutations that affect HSN development are pleiotropic, that is, they also function in the development of other cell types. This is reminiscent of the genes that affect development of the mechanosensory neurons of *C. elegans*, mutations in most of these genes also have pleiotroic effects (Chalfie and Au, Science XXXXX). Together, these results suggest that there probably are not many gene products that are specific to the development of particular nerve types, rather, nerve types are probably specified by different combinations of developmental genes. Some nerve traits in *C. elegans*, e.g., axon morphology and choice of

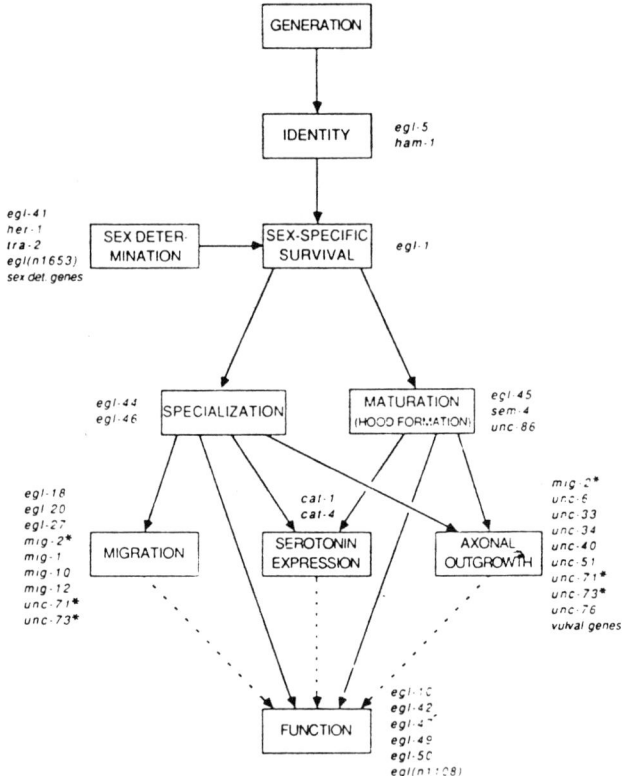

FIGURE 7-5. A genetic pathway for the development of the HSN neurons. Boxes represent steps in HSN development. Mutations in the genes adjacent to each box affect but may not prevent that step. This pathway was constructued from the HSN phenotypes of the mutants and from a limited number of gene interaction studies. Mutants defective in a given step are defective in all steps in the pathway subsequent to that step. Hood formation has been placed with maturation because we have no mutations that distinguish this morphologically visible event from the decision step of maturation. That the migration, serotonin expression, and axonal outgrowth steps are independent is supported by the following evidence: (1) migration is normal in mutants defective in serotonin expression or axonal outgrowth; furthermore, in the wild type, migration occurs long before axonal outgrowth and expression of detectable serotonin, (2) serotonin expression is normal in migration mutants even when the HSN cell bdies are located in the postnatal region of the tail (data not shown); serotonin expression is also normal in mutants with severely attenuated axonal outgrowth, such as *unc-51* animals, (3) axonal outgrowth is completed in animals with total failures in HSN migration (data not shown) and also in cat-1 mutants with very low serotonnin levels; furthermore, axonal outgrowth occurs in the wild type before detectable HSN serotonin appears. The dotted lines indicate that none of the mutants abnormal in HSN migration, axonal outgrowth, or serotonin expression displays the serotonin-sensitive imipramine-resistant Egl phenotype characteristic of animals with impaired HSN function. Although some of these mutants are Egl, they probably have defects in other components of the egg-laying system (data not shown). Therefore, we cannot be sure if the steps of migration, axonal outgrowth, and serotonin expression are required for HSN function. *vulval genes* refers collectively to genes involved in vulval development. *sex det. genes* refers to other sex-determining genes, which control whether or not the presumptive HSNs undergo programmed cell death. It is likely that *mig-2, unc-71*, and *unc-73* (denoted *), which affect both HSN migration and axonal outgrowth, function separately in these two processes. (From Desai et al., *Nature*, 336, 638—646, 1988. With permission.)

synaptic partners, may be determined in part by position (White et al., *Cold Spring Harbor Symp. Quant. Biol.,* 48, 633—640). *Joseph G. Culotti*

Sex-lethal, a Drosophila Sex Determination Switch Gene, Exhibits Sex-Specific RNA Splicing and Sequence Similarity to RNA Binding Proteins
L. R. Bell, E. M. Maine, P. Schedi, and T. W. Cline
Cell, 55, 1037—1046, 1988   7-6

Somatic sex development in Drosophila is regulated by a network of genes operating in a hierarchical manner. *Sex-lethal (Sxl)* is a central element in this process and the only gene known to be required for all aspects of sex development: signaling, determination, and differentiation. It must be active in females and inactive in males throughout development. In addition to interacting with downstream genes, *Sxl* exhibits positive autoregulatory activity. This study examined the molecular basis for the function of *Sxl* in both maintenance and expression of the male/female decision.

Analysis of *Sxl* cDNAs showed that on/off regulation may be explained by differential RNA splicing. Only female transcripts encode functional products. All male transcripts contained an exon that truncates the open reading frame. The functional female product exhibits sequence similarities with ribonucleoproteins, suggesting that it is an RNA binding protein.

It is suggested that *Sxl* encodes a factor that interacts with both its own pre-mRNA, explaining positive autoregulation, and that of downstream genes to confer female-specific splicing. An action of *Sxl* on its own pre-mRNA is suggested by the similarity of the *Sxl* gene product to RNA binding proteins (Figure 7-6). The presence of functional *Sxl* product continuously directs the *Sxl* splicing pattern into the active form, perpetuating the female state. In males, where no functional *Sxl* product is produced, the male state is maintained by the default splicing pattern. In this way *Sxl* can maintain a stable commitment to the female or male developmental pathway. The question of how *Sxl* enters the female or male RNA processing mode remains to be answered.

The Sex-Determing Gene *tra-2* of Drosophila Encodes a Putative RNA Binding Protein
H. Amrein, M. Gorman, and R. Nothiger
Cell, 55, 1025—1035, 1988   7-7

In the proposed scheme for sex determination in Drosophila, the differential activity of the key gene Sex-lethal *(Sxl)*, via the subordinate

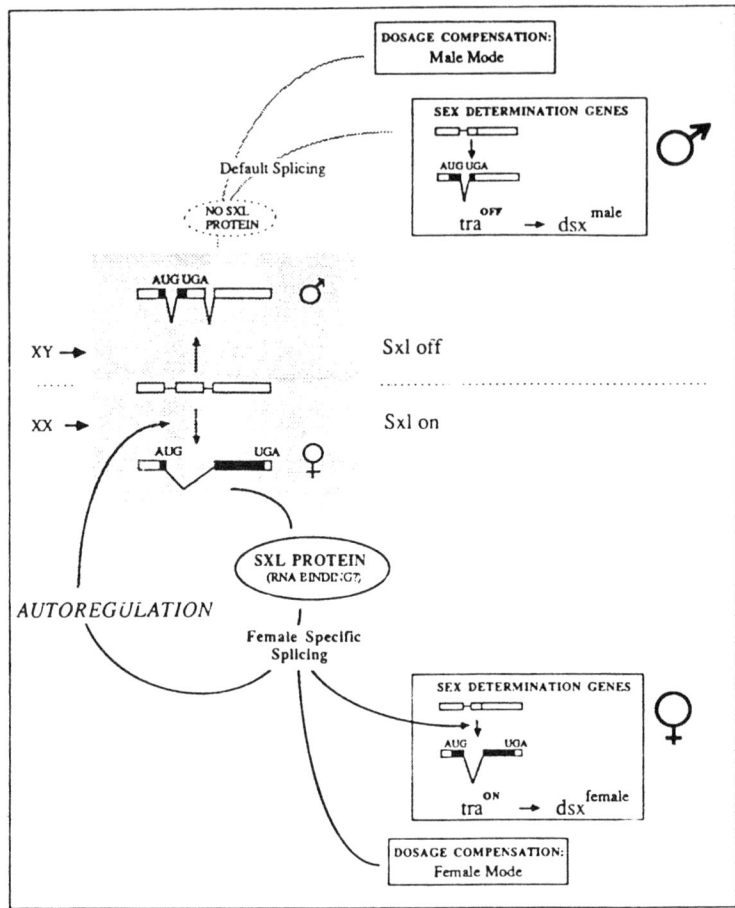

FIGURE 7-6. A model for *Sxl* regulation of the sex determination and dosage compensation pathways. *Sxl* is regulated by alternative splicing that produces transcripts encoding either a functional product in females or a truncated product in males. This is accomplished by female-specific removal of a translocation-terminating, male-specific exon. The female mode of *Sxl* activity is maintained through a positive autoregulatory activity: the *Sxl* product continuously directs female-specific splicing of its own pre-mRNA. The female *Sxl* product also regulates sexual differentiation by directing splicing of *tra* RNA in the female mode; it is postulated to regulate the dosage compensation pathway in a similar manner. In males, *Sxl* RNA is spliced in the default mode, no functional *Sxl* product is made, and as a consequence RNA of genes in the sexual differentiation and dosage compensation pathways are spliced in the default mode. To simplify the figure, only the regulatory genes downstream of *Sxl* that have been analyzed at the molecular level are shown. (From Bell et al., *Cell*, 55, 1037—1046, 1988. ©Cell Press.)

control genes *transformer (tra)*, *transformer-2 (tra-2)*, and *intersex (ix)*, achieves alternative expression of *double sex (dsx)*, the last gene in the cascade. The *tra-2* gene, which is necessary for both female sexual differentiation and normal spermatogenesis in males, now has been isolated and characterized.

The putative protein product of *tra-2* has a domain that is homologous to RNA binding proteins. This finding suggests that *tra-2* protein may achieve female-specific splicing of the transcript of *dsx*, the sex-determining gene whose mode of expression depends on *tra-2*. The protein-coding region of the *tra-2* transcript is identical in males and females. In both sexes, a low level of transcript is present in the soma and a high level in the germ line.

The transcriptional regulation of *tra-2* appears to be tissue-specific rather than sex-specific. High expression of *tra-2* is a consequence of tissue type in cells that have entered the gametogenic pathway, while low expression is observed in somatic cells.

The Sex Determination Locus *transformer-2* of Drosophila Encodes a Polypeptide With Similarity to RNA Binding Proteins
T. J. Goralski, J.-E. Edstrom, and B. S. Baker
Cell, 56, 1011—1018, 1989   7-8

Control of sex differentiation is a very useful system with which to examine gene regulation during development. The Drosophila *transformer-2 (tra-2)* gene regulates somatic sex differentiation in females and is required for spermatogenesis in males. Wild-type *tra-2* function is necessary for female-specific splicing of the pre-mRNA of *doublesex*, the next known gene downstream of *tra-2* in the sex determinative regulatory hierarchy.

The authors cloned the *tra-2* gene and, using P element-mediated transformation, found that a 3.9-kb genomic fragment contains all the sequences necessary for *tra-2* function. A 1.7-kb transcript is the product of the *tra-2* locus; it was present in reduced levels in animals containing a *tra-2* mutant allele. Sequencing of a cDNA corresponding to this transcript indicated that it encodes a polypeptide strongly resembling a family of RNA-binding proteins that includes proteins associated with hnRNPs and snRNPs.

These findings indicate that the *tra-2* product may directly regulate the processing of *double-sex* mRNA in females. Several genes utilize alternative splicing to generate RNA transcripts that can express different protein products. As is true for the sex differentiation-regulating genes *tra* and *dsx*, some of these exhibit RNA processing patterns specific to certain cell types and developmental stages. This clearly is a general mechanism for regulating gene expression.

Drosophila *doublesex* Gene Controls Somatic Sexual Differentiation by Producing Alternatively Spliced mRNAs Encoding Related Sex-Specific Polypeptides
K. C. Burtis and B. S. Baker
Cell, 56, 997—1010, 1989

The *doublesex (dsx)* gene regulates somatic sex differentiation in both Drosophila males and females. It is the only gene in the hierarchy that is required for somatic sexual differentiation in both sexes. The gene encodes two functional products; one is expressed in females and represses male differentiation, while the other is expressed in males and represses female differentiation.

The *dsx* gene is transcribed to produce a common primary transcript that is alternatively spliced and polyadenylated to produce male- and female-specific mRNAs. These mRNAs share a common 5' end and three common exons, but they possess alternative sex-specific 3' exons. As a result the mRNAs encode polypeptides having a common amino-terminal sequence and sex-specific carboxyl termini.

There is evidence that sequences including and adjacent to the female-specific splice acceptor site have an important role in regulation of *dsx* expression by the *transformer* and *transformer-2* loci. In females, the product of the *intersex* gene is required in addition to *dsx* to repress male-specific somatic sexual differentiation. The exact mechanism by which *dsx* polypeptides regulate the expression of terminal differentiation genes remains to be determined. Studies are in progress to learn whether either the male or female *dsx* polypeptides interact directly with nucleotide sequences implicated in the sex-specific regulation of yolk protein gene expression.

daughterless, a Drosophila Gene Essential for Both Neurogenesis and Sex Determination, Has Sequence Similarities to *myc* and the *achaete-scute* Complex
M. Caudy, H. Vassin, M. Brand, R. Tuma, L. Y. Jan, and Y. N. Jan
Cell, 55, 1061—1067, 1988

Embryonic lethal mutations of the *daughterless (da)* gene of Drosophila eliminate all sensory neurons and sensory structures. Apart from the need for zygotic *da* activity to ensure formation of the peripheral nervous system, maternal *da* activity is required for proper sex determination.

The authors cloned the region containing *da* and found that five recessive lethal *da* mutations map to a single transcription unit. A predicted protein product of this unit exhibits sequence similarities with the oncogene *myc*. There also are similarities to the gene *MyoDl*, which is involved in myoblast determination, and to the Drosophila *achaete-scute* complex, involved in neuronal precursor determination.

There is genetic evidence for a single functional product of the *da* gene. The same *da* protein, perhaps acting as a DNA binding transcription factor, may be required for both neuronal development and sex determination. Possibly *da* acts in conjunction with one set of factors to initiate sex determination of somatic cells and with another to initiate neuronal precursor formation.

Molecular Characterization of *daughterless*, a *Drosophila* Sex Determination Gene with Multiple Roles in Development
C. Cronmiller, P. Schedl, and T. W. Cline
Genes Dev., 2, 1666—1676, 1988   7-11

In *Drosophila* sexual phenotype and the level of X-linked gene expression depend on the number of X chromosomes relative to the number of sets of autosomes, as expressed by the X:A ratio. The role of X/A balance in activating the feminizing functions of Sx^+ depends on material expression of the gene *daughterless (da)*. In addition to its role in sex determination, embryonic expression of da^+ is required for formation of the peripheral nervous system, and possibly for proper function of genes in heterochromatic regions of the genome.

The authors cloned *da* through transposon tagging and examined the patterns of transcription for wild-type and partial-loss-of-function mutant alleles. In addition, the sequences of the wild-type coding region and of regions altered in two mutant *da* alleles were determined.

The gene was found to code for two transcripts, which are present in both sexes and t all stages of development. The nucleotide sequence of a nearly full-length cDNA predicts a protein product of 710 amino acids, the sequence of which resembles the His-Pro repeat of the *Drosophila* genes *bicoid* and *paired*. Two partial-loss-of-function *da* mutations, one of them temperature-sensitive, apparently are caused by DNA insertions in the 5'-untranslated region of the gene.

The sequence similarity between *da* and *bcd* could reflect the fact that both these genes are involved in processes by which the organism transduces quantitative differences in a developmental signal into qualitatively different cellular responses.

A New DNA Binding and Dimerization Motif in Immunoglobulin Enhancer Binding, *daughterless*, *MyoD*, and *myc* Proteins
C. Murre, P. S. McCaw, and D. Baltimore
Cell, 56, 777—783, 1989   7-12

Expression of immunoglobulin genes depends on sequence motifs in their enhancer and promoter regions including the so-called E boxes, first identified as protein- binding sites on DNA. Mutational analysis has demonstrated a role for many of the E boxes in controlling transcription. The $_kE2$ and $\lambda E2$ sites are important for immunoglobulin light chain gene and insulin gene transcription, respectively.

The authors used an oligonucleotide screening procedure to isolate cDNAs encoding $_kE2$ binding proteins. Two cDNA were isolated whose dimerized products bound specifically to $_kE2$, located in the immunoglobulin kappa chain enhancer. Both the cDNAs share a region of extensive identity with the Drosophila *daughterless* gene. The region also resembled a segment in three *myc* proteins, *MyoD*, and members of the Drosophila *achaete-scute* and *twist* gene family. The homologous regions are able to form two amphipathic helices separated by a loop. Hydrophobic residues are strictly conserved in both helices.

The homology among the helix-loop-helix proteins is so marked that they may well have a common function. Since E12 and E47 are specific DNA-binding proteins and there is similar evidence for *MyoD*, the *myc*, *twist*, and *achaete-scute* proteins probably serve a similar purpose. It is possible that *daughterless* and E12/E47 are quite homologous functionally as well as structurally.

♦ In Drosophila the primary signal for sex determination is the X-to-autosome ratio. This signal is transmitted to the key gene *Sex-lethal (Sxl)* which acts through the subordinate genes *transformer (tra)* and *transformer-2 (tra-2)* to achieve alternative expression of the *double-sex (dsx)* gene either in the male or in the female mode. These articles report on the cloning and characterization of *Sxl, tra-2,* and *dsx*.

Interestingly, two of the three genes (*Sxl* and *dsx*) have a sex-specific splicing pattern leading to the production of functional proteins in only one of the sexes (*Sxl*) or to the production of different proteins in each sex (*dsx*). Thus, if one includes the previously characterized *tra* gene, three out of four sex-determination genes that have been cloned are regulated at the level of splicing.

Perhaps surprisingly, no differential regulation of gene expression was detected in case of the *tra-2* gene. While the gene is known to be required only in female somatic tissues and in the male germ line, the same mRNA is expressed in both males and females, in all tissues, and at all times of development. It is not clear what function, if any, *tra-2* would have, for instance, in male somatic tissues. It is possible that *tra-2* simply acts as a cofactor with *tra* to regulate splicing of *dsx*. Regulation of *tra* expression alone would suffice to modulate activity of the [*tra-tra-2*] complex; in the absence of *tra*, *tra-2* would be functionless.

Another finding of interest was that the predicted protein sequence of two of the three genes (*Sxl* and *tra-2*) show sequence similarity to RNA-binding proteins. The predicted *tra* protein shares sequence similarity with a component of small nuclear ribonucleoprotein particles. Considering that these genes participate in a regulatory cascade that involves differential RNA splicing, this observation raises the possibility that the protein products of *Sxl* and *tra-2* regulate RNA splicing. Genetic evidence indicates that *Sxl* regulates its own expression. This fact plus the findings that the gene is both differentially spliced and codes for an RNA binding protein, suggests that the *Sxl* protein regulates the splicing of its own RNA.

The papers by Caudy et al., Cronmiller et al., and Murre et al. concern *daughterless*, a gene whose maternal expression is required by embryos for the regulation of sex determination and dosage compensation. *daughterless* is also required for a variety of other apparently unrelated functions, such as differentiation of the embryonic peripheral nervous system and differentiation of the adult epidermis. Significantly, *daughterless* has sequence similarity to a number of Drosophila and non-Drosophila presumed DNA-binding proteins. The possibility then arises that *daughterless* acts as a transcriptionally regulatory element. Significantly, there is a class of transcripts that is specific for the early embryo. These data and previously obtained genetic information suggest a model whereby *daughterless* is involved in the transcriptional activation of *Sxl* to generate early embryonic RNAs with functions in dosage compensation. Later in embryogenesis, the *Sxl* transcription pattern shifts to produce female-specific RNAs coding for proteins with self regulatory functions. These proteins may function by controlling the splicing of its own RNA thus perpetuating the sex-specific transcription patterns. Additional data that address the validity of this model should become available in the near future. *Marcelo Jacobs-Lorena*

Cell-Autonomous and Inductive Signals Can Determine the Sex of the Germ Line of Drosophila by Regulating the Gene *Sxl*
M. Steinmann-Zwicky, H. Schmid, and R. Nothiger
Cell, 57, 157—166, 1989   7-13

The genes that control somatic sex determination in Drosophila are not involved in determing the sex of the germ line. It remains unclear whether *Sxl*, which is necessary for proper oogenesis, is needed to determine the sex of the germ line or for some other function. In this study pole cells were transplanted with either wild-type, amorphic (loss-of-function), or constitutive (gain-of-function) alleles of *Sxl* in order to determine whether the state of activity of the gene dictates the sexual pathway to germ cells.

In ovaries, germ cells developed according to their X-chromosome/autosome (X:A) ratio: XX cells underwent oogenesis while XY cells formed spermatocytes. In testes, in contrast, both XY and XX germ cells entered the spermatogenic pathway. The findings with amorphic and constitutive mutations of *Sxl* indicated that both the genetic and somatic signals act through this gene to achieve sex determination in germ cells.

The germ cells of Drosophila appear to have an inherent genetic sex that is specified by the X:A ratio. The XX cells obey inductive somatic stimuli, and both signals act through Sxl to determine the sex of the germ line (Table 7-13). In many species the germ cells utilize an inductive signal for sex determination. In Drosophila the somatic sex-determining signal may be viewed as an evolutionary residual in a species moving towards cell-autonomous genetic sex determination of the germ line.

♦ A fair amount of information is available on mechanisms of sex determination in Drosophila somatic tissues. The primary signal, the X to autosome ratio, is read by the key gene *Sex-lethal (Sxl)* and the information is transferred through a regulatory cascade that involves the *transformer, transformer-2,* and *intersex,* to *double-sex*, the last known gene in this hierarchical set. In contrast, little is known about sex determination in the germ line. With the exception of *Sxl*, the above mentioned genes are known not to be required for germ cell sex determination.

Pole cells (the germ cell precursors), are formed very early during Drosophila embryogenesis. By transplanting genetically marked pole cells, several investigators have attempted to determine the fate of pole cells of one sex transplanted into embryos of the opposite sex. Although homotypic transplants were always successful, functional egg or sperm have never been obtained from pole cells transplanted into embryos of the opposite sex. The inference from these experiments was that germ cells determine their sex cell autonomously.

The present set of experiments resorted to a similar transplantation approach but with one important difference: pole cells were transplanted into mutant embryos that had no pole cells of their own, thus eliminating competition between resident pole cells of the correct chromosomal composition (X-to-autosome ratio) and the injected pole cells carrying an "incorrect" chromosomal composition. Perhaps not unexpectedly, germ cells were found to possess an inherent mechanism of sex determination that depends on the X-to-autosome ratio. In the ovary, XX germ cells develop as oocytes and XY cells develop as sperm. Surprisingly, however, XX cells entered the spermatogenic rather than the oogenic pathway in the testis, implying that germ cells are prone to inductive signals by the somatic tissues of the testis. Transplantation of pole cells carrying dominant gain-of-function or loss-of-function null alleles of the *Sxl* gene clearly demonstrated that this gene is crucial for the sex determination in

Table 7-13
Summary of Main Results and Conclusions

Donor Germ Cells		Sexual Differentiation of Transplanted Germ Cells		Conclusion
Karyotype	Genotype	in Testes	in Ovaries	
XY	Sxl^+	spermatogenesis	spermatogenesis	XY germ cells differentiate autonomously.
XX	Sxl^+/Sxl^+	spermatogenesis	oogenesis	Somatic induction acts on XX germ cells.
XX	Sxl^{f1}/Sxl^{f1}	spermatogenesis	spermatogenesis	Function of Sxl is required for female pathway.
XX	Sxl^{M1}/Sxl^+	oogenesis	oogenesis	Sxl^{M1} overrules somatic induction.
X0	Sxl^{M1}	spermatogenesis	spermatogenesis	Sxl^{M1} has no effect on X0 germ cells.

From Steinmann-Zwicky et al., *Cell*, 57, 157–166, 1989. ©Cell Press.

germ cells. For instance, while wild-type germ cells entered the spermatogenic pathway in the testis, germ cells carrying a gain-of-function *Sx1* allele developed along the oogenic pathway, thus overriding the inductive effects of the testis. The latter finding suggests that the inductive signals of the testicular tissues operate in the germ cells through the *Sx1* gene.

Although the nonautonomy of germ cell sex determination in Drosophila may come as a surprise, examples of inductive determination by gonadal soma exist. These include a number of lower insects, the nematode *C. elegans*, and spermatogenesis in the mouse. Marcelo Jacobs-Lorena

The CArG Promoter Sequence is Necessary for Muscle-Specific Transcription of the Cardiac Actin Gene in *Xenopus* Embryos

T. J. Mohum, M. V. Taylor, N. Garrett, and J. B. Gurdon
EMBO J., 8, 1153—1161, 1989

Earlier studies indicate that a 416-nucleotide segment of the cardiac actin gene promoter of *Xenopus* can confer muscle-specific expression on the bacterial chloramphenicol acetyl transferase transcription unit. An RNAse protection assay now has been used to measure transcription of a chimeric cardiac actin/beta-globin fusion gene injected into *Xenopus* embryos.

Promoter sequences were analyzed by micro-injecting the chimeric genes into two-cell embryos. Transcription was monitored during the differentiation of embryonic muscle and non-muscle tissues. The effects of various mutations, including internal deletions and linker scan mutations, were examined. The mutations lay between -64 and -396 of the cardiac actin promoter, a region containing four copies of a conserved motif, the CArG box, that is common to vertebrate striated-muscle acting gene promoters.

The most proximal motif, CArG box 1, located at -80, proved essential for muscle-specific transcription. Other CArG motifs could substitute functionally for CArG box 1 when placed at that position. CArG boxes 3 and 4 bound the same activity in a neurula embryo nuclear extract as CArG box 1. The degree of binding activity remained constant through early development.

The most proximal CArG box may be responsible for muscle-specific transmission. Alternately, the CArG box motif may be necessary for transcription from the muscle actin gene promoter in conjunction with other *cis*-acting sequences, while not itself determining the muscle-specific pattern of expression of the cardiac actin gene. CArG box binding activity is not confined to embryonic muscle tissue; it also is present in both oocytes and developing embryos.

Muscle-Specific (CArG) and Serum-Responsive (SRE) Promoter Elements Are Functionally Interchangeable in *Xenopus* Embryos and Mouse Fibroblasts

M. Taylor, R. Treisman, N. Garrett, and T. Mohun

Development, 106, 67—78, 1989

The cardiac actin gene of *Xenopus* contains four copies of a promoter element, the CArG box, that is conserved in striated muscle actin genes and is necessary for tissue-specific expression in the developing embryo. The authors have attempted to identify embryo and muscle proteins that interact with the CArG box. The CArG box has some sequence similarity to the SRE, which mediates the transcriptional activation of genes such as c-*fos* and cytoskeletal actin.

The most proximal cardiac actin CArG box is recognized by the same binding activity as cytoskeletal actin SRE in nuclear extracts from both *Xenopus* embryos and mammalian muscle cells. The activity is indistinguishable from HeLa cell SRE-binding activity (serum response factor, SRF). The finding that the CArG box and SRE are functionally interchangeable in both *Xenopus* embryos and murine fibroblasts imply that they can bind the same protein *in vivo* as well as *in vitro*.

A key sequence in muscle-specific transcription can also mediate growth factor-dependent transcription in fibroblasts. It may be that a number of different proteins that have the same or similar sequence specificity recognize the SRE/CArG motifs. Alternately, a single SRE/CArG box-binding activity, SRF, may be involved in both serum-responsive and muscle-specific transcription.

♦Muscle is the largest derivative of mesoderm in *Xenopus*, so the regulation of muscle-specific α-actin genes is important in understanding the molecular basis of mesoderm specification. Mohun and colleagues showed previously by exonuclease deletion experiments that α-actin genes transferred into embryos depended upon 5′ flanking sequences from around −400 to the initiation site for correct expression in muscle.

From these papers comes the interesting conclusion that there appears to be only one vital element in this promoter region, a Serum response element (SRE, also designated CArG box) located at −80. There are a total of four SREs in the regulatory region, but the distal three are not required for α-actin gene expression. It is intriguing that except for those disrupting the proximal SRE, none of the small deletions or linker scan mutations in the −1 to −400 interval has any discernable effect on expression. The apparent discrepancy between this observation and the earlier exonuclease deletion data has not been explained yet. Another important point is that none of the mutations leads to ectopic α-actin expression: either nothing happens or the gene becomes completely inactive.

The α-actin SRE is demonstrated to be functionally equivalent to the

SREs found in other genes, including those encoding cytoskeletal (i.e., non-muscle) actins and c-*fos*. All of these SREs, in *Xenopus* as well as in mammalian cells, bind a single identifiable protein factor, serum response Factor (SRF). The SRE/SRF regulatory mechanism is involved in the rapid transcriptional response to serum, and it is possible that this may point to some connection to the mesoderm-inducing role of growth factors. However, muscle differentiation and the concomitant activation of the α-action genes is a process which is at least one step removed from mesoderm induction *per se,* so this coincidence may be fortuitous. The authors discuss models whereby the widely distributed SRF might nonetheless control the tissue-specific expression of α-actin. The most likely alternative is perhaps the involvement of additional transcription factors that interact with SRF, and possibly also as yet unidentified protein-DNA interactions elsewhere in the 5' regulatory region. *Thomas D. Sargent*

Cholinergic Phenotype Developed by Noradrenergic Sympathetic Neurons after Innervation of a Novel Cholinergic Target *In Vivo*
R. J. Schotzinger and S. C. Landis
Nature, 335, 637—639, 1988
7-16

Mammalian sympathetic neurons can express either a noradrenergic or cholinergic phenotype and target-appropriate expression of neurotransmitter obviously is critical. Development of a cholinergic phenotype in the sympathetic innervation of sweat glands is characterized by production of acetylcholinesterae, choline acetyltransferase (ChAT), and cholinergic transmission. Sweat gland innervation differs from other sympathetic projections that retain noradrenergic features throughout development.

A transplantation paradigm was used to examine the role of the rat sweat gland in the cholinergic differentiationof its innervation. The study employed hairy skin which normally receives noradrenergic sympathetic innervation and glabrous skin of the footpad, which normally receives cholinergic sympathetic innervation.

Sympathetic neurons innervating the novel cholinergic target altered their neurotransmitter properties and developed a cholinergic phenotype. ChAT activity increased markedly in footpad skin transplanted to the thorax. ChAT activity was not found at any time in hairy skin. Acetylcholinesterase staining also appeared during development of the sweat gland innervation. The neurons responsible for the increased cholinergic expression in transplants were sensitive to 6-hydroxydopamine shortly after birth. The transplant procedure itself was not responsible for altered neurotransmitter properties.

This study shows that a cholinergic sympathetic target, the sweat gland, promotes cholinergic differentiation in sympathetic neurons that ordinarily would undergo noradrenergic differentiation. Apparently postnatal sympathetic neurons *in vivo* are not irrevocably committed to a particular neurotransmitter phenotype. Neurotransmitter plasticity is demonstrable in cell culture. An alternative explanation is that the transplanted target organ leads to selective survival of cholinergic sympathetic neurons that normally would die during development.

♦ The autonomic nervous system contains both sympathetic and parasympathetic neurons, both of which are derived from the neural crest. In general, sympathetic neurons are adrenergic, utilizing the neurotransmitter norepinephrine, whereas parasympathetic neurons are cholinergic and utilize acetylcholine as a neurotransmitter. The exception is the cholinergic sympathetic innervation to sweat glands.

In vitro, there is now ample evidence that sympathetic neurons can change their neurotransmitter from adrenergic to cholinergic under the influence of environmental factors. Landis and colleagues have shown that a similar transition is likely to occur in the innervation to the sweat gland in the rat footpad. This innervation is derived from sympathetic neurons which originally project adrenergic processes to the target; these processes subsequently acquire cholinergic properties.

In this study, Schotzinger and Landis have examined the ability of the sweat gland target tissue to cause an adrenergic to cholinergic transition in sympathetic neurons that would not normally innervate the sweat gland. By transplanting the foot pad containing sweat glands into the epidermis of newborn rats, the authors found that the glands become reinnervated by adrenergic sympathetic fibers. With time, the sympathetic innervation develops cholinergic properties. These results demonstrate that a factor produced by the target is able to induce phenotypic plasticity in postnatal sympathetic neurons in an intact animal. *Marianne Bronner-Fraser*

A Laminin-Like Adhesive Protein Concentrated in the Synaptic Cleft of the Neuromuscular Junction
D. D. Hunter, V. Shah, J. P. Merlie, and J. R. Sanes
Nature, 338, 229—234, 1989

Reinnervation of vertebrate skeletal muscle is topographically specific in that axons preferentially form new junctions at original synaptic sites. In some instances more than 95% of regenerated synapses form on the 0.1% of the muscle fiber surface that originally was synaptic. Some of the factors associated with original synaptic sites that axons recognize are related to the basal lamina ensheathing the muscle fiber, which passes through the synaptic cleft at the neuromuscular junction.

The authors have identified a glycoprotein, s-laminin, that is selectively associated with synaptic basal lamina and is recognized by motoneurons. Molecular cloning studies showed s-laminin to be a novel homologue of laminin, which strongly promotes neurite outgrowth.

The finding of s-laminin is evidence for the existence of a laminin gene family extending beyond the genes for the three laminin subunits. Laminin itself may be one of a group of related molecules that act together to regulate axonal behavior. In muscle, laminin-rich paths could provide a favorable site for regenerating axons to follow, and s-laminin at original synaptic sites might direct axons to stop growing and/or to differentiate into nerve terminals.

♦This paper reports a totally unexpected finding in the field of basal lamina/extracellular matrix work. Laminin, a major component of the basal lamina, comprises three tightly associated and structurally related chains called A, B1, and B2. Until this report it was believed that there was one gene encoding each laminin chain in mammals. The cloning and characterization of an additional laminin B1 gene from the rat suggests that there may be families of laminin A, B1, and B2 genes. The specificity of cell matrix interactions may result in part from the differential expression of individual members of such families. The s-laminin gene was cloned because its product is localized to neuromuscular junctions. Perhaps its role there is to tell the axon where to stop and/or form neuromuscular synapses.

Laminins have been demonstrated to hae nerve growth-promoting activity in tissue culture and many axons grow along laminin-rich pathways *in vivo*. Certain laminin gene products may provide guidance cues which direct axons to their targets, and once they reach those targets, other laminins, like s-laminin, may tell them to stop growing and/or to differentiate into nerve terminals. *Joseph G. Culotti*

Establishment of a Leukemic Cell Model for Studying Human Pre-B to B Cell Differentiation
B. Wormann, J. M. Anderson, J. A. Liberty, K. Gajl-Peczalska, R. D. Brunning, T. L. Silberman, D. C. Arthur, and T. W. LeBien
J. Immunol., 142, 110—117, 1989   7-18

There are no good reproducible models for studying the early stages of human B cell differentiation. The authors established and characterized a novel human leukemic cell line, BLIN-1 (B lineage 1), which recapitulates the pre-B to B cell stage of differentiation.

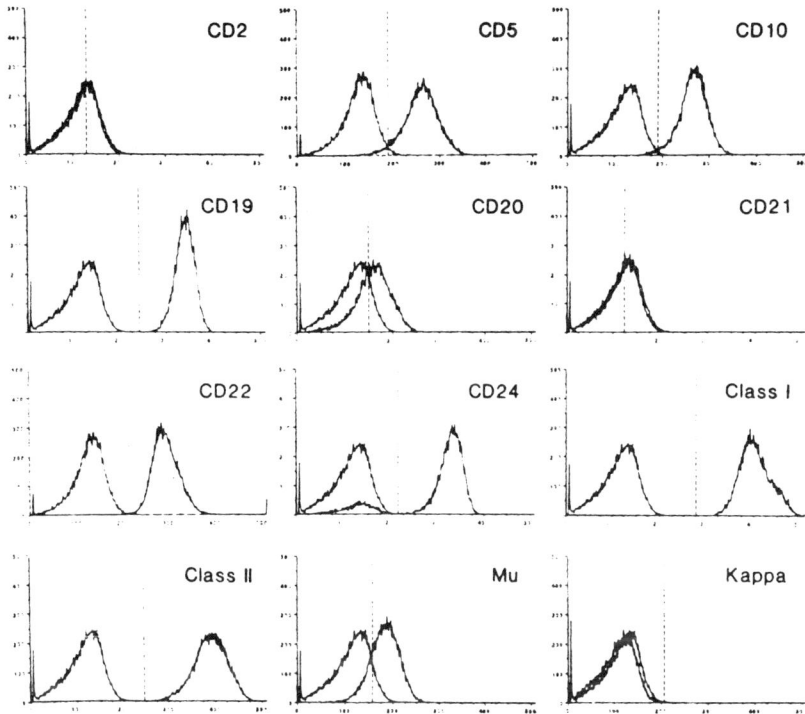

FIGURE 7-18. Flow cytometric analysis of cell surface Ag on BLIN-1 detected by indirect immunofluorescence and analyzed in a FACs IV. The vertical dashed line defines positive and negative cells. Background (negative) staining is shown by the left curve, with specific antibody (positive) staining shown by the right curve. Fluorescence intensity (log scale) increases from left to right, and cell number from bottom to top. The designation in the upper right corner of each profile represents the cell surface Ag recognized by a mAb. (From Wörmann et al., *J. Immunol.,* 142, 110—117, 1989. With permission.)

The BLIN-1 line was established from the marrow aspirate of a child with acute leukemia, in tissue culture medium containing low-molecular-weight B cell growth factor. BLIN-1 cells have a 9_p chromosomal abnormality identical to that present in leukemic blasts from the original marrow aspirate. Their immunologic phenotype is consistent with a cell arrested at the pre-B cell stage of development (Figure 7-18). Studies of immunoglobulin gene rearrangement and immunoglobulin expression in BLIN-1 subclones showed that the cells spontaneously rearrange κ light chain genes, leading to the differentiation of surface κ-negative pre-B cells into surface κ-positive B cells.

This is the first defined human model with which to examine pre-B to B cell differentiation, a critical stage in B cell ontogeny. The model should prove useful to determine what factors influence the onset of κL chain gene rearrangement and expression.

♦ The differentiation of pro-B cells into mature, immunocompetent B cells requires the successful rearrangement of both immunoglobulin heavy chain and light chain genes. As one might expect, this process is quite difficult to study in the laboratory, and the many important advancements in our understanding of this process has occurred following the development of appropriate model systems. In the mouse, the Whitlock-Witte system, (Whitlock and Witte, *Proc. Natl. Acad. Sci. U.S.A.,* 79, 3608, 1982) has allowed the long-term cultivation of B cell progenitors, and the treatment of bone marrow cells with the Abelson murine leukemia virus has allowed for the transformation of B cell precursors (Baltimore et al., *Immunol. Rev.,* 48, 3, 1979). In the past year, a defined model system for the analysis of human pre-B cell to B cell development has been described.

This system involves the subcloning of an *in vitro*-adapted, human leukemic cell line which displays all of the characteristics of pre-B cells. The subclones spontaneously differentiate into B cells by completing the requisite DNA rearrangements. *Charles Snow*

Tolerance in T-Cell-Receptor Transgenic Mice Involves Deletion of Nonmature CD4$^+$8$^+$ Thymocytes
P. Kisielow, H. Bluthmann, U. D. Staerz, M. Steinmetz, and H. von Boehmer
Nature, 333, 742—746, 1988
7-19

Mature T lymphocytes usually do not respond to self-MHC molecules presenting self-antigens, but it remains unclear whether immunologic tolerance involves the deletion of autospecific lymphocytes. Tansgenic mice offer a means of analyzing mechanisms of self-tolerance. In this study, transgenic mice were constructed that express in a large fraction of their T cells an αβ TCR (T cell receptor) specific for a minor histocompatibility antigen (H-Y) present only on male cells.

It was found that autospecific T cells are deleted in male mice. The deletion affects only transgene-expressing cells having a relatively high surface density of CD8 molecules, including nonmature CD4$^+$CD8$^+$ thymocytes. Anti-idiotype cells are not responsible for the deletion.

Both CD4 and CD8 accessory molecules are involved in the deletion of autospecific cells in this model. It is likely that at least some doubly positive CD4$^+$8$^+$ thymocytes act as precursors for functional singly positive cells. The number of peripheral T cells — which is normal in transgenic males — probably can be adjusted independently of the export of newly formed cells from the thymus. This would allow the accumulation of cells having rare phenotypes in the periphery of male mice.

The T-Cell Repertoire is Heavily Influenced by Tolerance to Polymorphic Self-Antigens
A. M. Pullen, P. Marrack, and J. W. Kappler
Nature, 335, 796—801, 1988

The repertoire of $\alpha\beta$ receptors carried by mature T cells is shaped by interactions of developing T cells with self-antigens and the products of particular MHC alleles expressed in the host. The shaping process occurs partly in the thymus where T cells whose receptors can interact with the self-MHC expressed on thymic epithelial cells are positively selected for further differentiation. In addition, T cells that are reactive to self-antigens complexed to self-MHC are deleted to achieve a repertoire that is tolerant of host antigens. The extent to which clonal deletion limits the repertoire during induction of self-tolerance has proved difficult to measure.

In examining a number of $V\beta$ elements, the elimination of T cells bearing a particular $V\beta$ element by self-tolerance to a particular polymorphic self-antigen/MHC ligand appeared to be the rule rather than the exception. Definitive evidence for this was obtained in the case of $V\beta 3$. Expression of $V\beta 3$ varies widely in mice; it is controlled chiefly by tolerance to the combination of polymorphic self-antigens and particular alleles of the murine MHC, H-2.

Polymorphisms in the MHC and/or the self-antigens that cause massive deletion of T cells using particular $V\beta$ elements may reflect the need to balance the desirability of a diverse T-cell repertoire against potential involvement of those elements in autoimmune disease. There is evidence that induced autoimmunity can be driven by T cells having quite restricted repertoires. This type of theory could explain how certain human HLA alleles protect against the onset of juvenile diabetes.

♦ The next stage of T cell ontogeny within the thymus must result in either the elimination (clonal deletion) or inactivation (clonal anergy) of self reacting T cell clones. This past year has seen support for both types of mechanisms. Two years ago saw the first strong evidence of negative selection occurring within the thymus by the actual deletion of various T cell clones (Kappler et al., *Cell,* 49, 273, 1987). Two additional papers have been released over the past year which extend this initial report.
Charles Snow

Limited Diversity of γδ Antigen Receptor Genes of Thy-1⁺ Dendritic Epidermal Cells

D. M. Asarnow, W. A. Kuziel, M. Bonyhadi, R. E. Tigelaar, P. W. Tucker, and J. P. Allison
Cell, 55, 837—847, 1988
7-21

While T cells bearing γδ antigen receptors are a minor population in most peripheral lymphoid tissues, they are the major T cell population in some epithelial including epidermis. The striking lack of diversity seen in dendritic epidermal cell (dEC) clones implies strict regulation of gene expression and somatic diversification in dEC pregentors. It is possible that dECs are the earliest emigrants from the pool of maturing T cells in the fetal thymus.

This study demonstrated that murine dEC clones express Vγ and Vδ gene segments, which are rare in adult T cells but predominate in fetal thymocytes. Analysis of the junctions of the rearranged γ and δ genes showed marked homogeneity among the receptors of five dEC clones.

These findings are consistent with a model in which dECs are one of perhaps several waves of emigrants from the early fetal thymus. It would seem that dECs have a role in immune surveillance distinct from that of αβ- and other γδ-bearing T cells. An earlier stage of thymocyte differentiation involves the generation of cells bearing specific V gene segments that have highly restricted diversity. The cells may seed epithelial tissues and act there to recognize common antigens expressed on damaged cells. Subsequently additional Vγ and Vδ elements and the full complement of α and β genes are available for use.

Isolation of CD4⁻CD8⁻ Mycobacteria-Reactive T Lymphocyte Clones From Rheumatoid Arthritis Synovial Fluid

J. Holoshitz, F. Koning, J. E. Coligan, J. De Bruyn, and S. Strober
Nature, 339, 226—229, 1989
7-22

A majority of peripheral T cells express a heterodimeric α/β T-cell receptor that recognizes specific antigenic peptides bound to self MHC molecules, and either the CD4 or CD8 surface markers. A subset of T cells — whose function is uncertain — express a distinct CD3-associated receptor consisting of γ and δ chains. This population includes cells lacking both CD4 and CD8 surface markers. To date the γ/δ receptor has been defined only in a murine cell line exhibiting MHC-linked specificity.

The authors isolated CD4⁻CD8⁻, γ/δ T-cell receptor-bearing T-cell clones from the synovial fluid of a patient with early rheumatoid arthritis. Synovial fluid cells but not peripheral blood cells responded to the acetone-precipitable fractions of mycobacterium tuberculosis. The T-cell clones responded specifically to mycobacterial antigens without MHC restriction.

Doubly negative γ/δ cells are able to proliferate in response to specific soluble antigen. Since the clones were from a site of autoimmune inflammation, they may be pathogenically relevant. T cells bearing the double negative phenotype are implicated in a range of murine autoimmune conditions.

Lymphocytes Bearing Antigen-Specific γδ T-Cell Receptors Accumulate in Human Infectious Disease Lesions
R. L. Modlin, C. Pirmez, F. M. Hofman, V. Torigian, K. Uyemura, T. H. Rea, B. R. Bloom, and M. B. Brenner
Nature, 339, 544—548, 1989

A majority of T cells bear the T-cell receptor (TCR) αβ complex that recognizes foreign antigen peptides only in the context of self MHC molecules. A small subset of T cells, however, bears a distinct TCR consisting of γ and δ subunits. These cells comprise about 5% of $CD3^+$ cells in organized lymphoid organs and in skin- and gut-associated lymphoid tissues. Extensive junctional variation, especially in the δ gene, provides great diversity for this receptor.

The frequency of γδ-bearing cells was increased 5- to 8-fold in granulomatous leprosy reactions, compared with normal lymphoid tissues, peripheral blood, and skin. TCR $γδ^+$ lymphocyte lines from cutaneous lesions of leprosy proliferated *in vitro* specifically in response to mycobacterial antigens. Culture supernates from activated γδ T lymphocytes induced adhesion and aggregation of bone marrow monocytes in the presence of granulocyte-monocyte colony stimulating factor. TCR γδ lymphocytes also accumulate in lesions of cutaneous leishmaniasis.

These lymphocytes accumulate in reactive granulomatous lesions and, on stimulation with antigen, secrete lymphokines that lead to macrophage adhesion and aggregation. At least some TCR γδ cells specific for foreign antigens may have a significant role in resistance to certain infectious pathogens and in the tissue damage associated with autoimmune disorders.

Activation of γδ T Cells in the Primary Immune Response to *Mycobacterium tuberculosis*
E. M. Janis, S. H. E. Kaufmann, R. H. Schwartz, and D. M. Pardoll
Science, 244, 713—716, 1989

Little is known about the function of T cells bearing the γδ T-cell receptor (TCR). The authors examined the role of γδ T cells in the immune response to *M. tuberculosis* (MT). Mice were immunized with MT in their front and hind limbs and the draining axillary and popliteal lymph nodes were collected.

The number of TCR γδ cells in the draining nodes of immunized mice was greatly increased compared with the number of TCR αβ cells. Three biochemically distinct γδ TCRs were observed. Cell cycle analysis and measures of interleukin-2 receptor expression and IL-2 responsiveness indicated that a large proportion of the γδ T cells were activated *in vivo*. The TCR γδ cells responded to solubilized MT antigens *in vitro* but, in contrast to MT-specific αβ T cells, their response did not require MHC class II recognition.

These findings indicate that γδ T cells may have a role in generating a primary immune response to certain microorganisms such as *M. tuberculosis*. It is not likely that expansion of the γδ T cell population is due to nonspecific proliferation of IL-2 receptor cells in response to production of IL-2 by γβ T cells. It is clear that γδ T cells and αβ T cells recognize antigen differently, and they probably have distinct roles in the primary immune response. Studies of *in vivo* responses of γδ T cells to other bacterial pathogens will be of interest.

Stimulation of a Major Subset of Lymphocytes Expressing T Cell Receptor γδ by an Antigen Derived from *Mycobacterium tuberculosis*

R. L. O'Brien, M. P. Happ, A. Dallas, E. Palmer, R. Kubo, and W. K. Born

Cell, 57, 667—674, 1989   7-25

Current studies are focusing on the function of lymphocytes expressing the γδ antigen receptor. The authors previously found that many γδ$^+$ hybridomas derived from C57BL/10 newborn thymuses, but no αβ$^+$ hybridomas from the same source, produced interleukin-2 (IL-2) in the absence of any other cell type. It then was learned that this apparently sponaneous IL-2 production is triggered through the receptor and probably results from recognition of an antigen on, or produced by the hybridoma itself.

Spontaneous production of IL-2 was quite frequent among γδ T-cell receptor (TCR) surface-positive hybridomas — especially those expressing a certain Vδ gene or gene family (VδM23). Production of IL-2 was triggered via TCRγδ. Each of the spontaneously reactive γδ$^+$ hybridomas was further stimulated by PPD from *M. tuberculosis*, possibly through cross-reaction with a bacterial antigen homologous to eukaryotic heat shock proteins.

These findings indicate that γδ TCR-bearing lymphocytes can recognize antigen and that a major subset of γδ⁺ cells can recognize mycobacterial antigens. The same subset of γδ⁺ lymphocytes may recognize antigens produced by eukaryotic cells. Recognition of such antigens would be consistent with a role for γδ⁺ cells in immune surveillance.

♦ In terms of function, some very intriguing observations have been made over the past year. First of all, it has been shown that the Thy–1⁺ dendritic epidermal cells of the skin represent a subpopulation of γδ T cells which resemble cells found in the fetal thymus.

These cells may function as a primitive line of defense against pathogens and/or transformed cells. A developing concept is that these types of T cells may represent a major line of defense against mycobacteria. The isolation of T cell clones, reactive with *Mycobacterium tuberculosis*, from arthritic joints yielded mainly γδ T cells.

The frequency of γδ T cells increases five- to eightfold in the granulomas induced *M. leprae*, suggesting that this subpopulation of T cells may be actively involved in the development of various types of granulomas. Another mycobacterium associated with the formation of granulomas is *M. tuberculosis*. The isolation of T cells from animals immunized against this mycobacterium also yielded a high percentage of antigen-reactive γδ T cells.

Finally, this subpopulation of cells exhibits many members who respond to stress or "heat-shock" proteins. These heat-shock proteins may represent a class of proteins, associated with both bacteria and transformed cells, which function as a major target for the γδ T cells. *Charles Snow*

Location of the Genes Controlling H-Y Antigen Expression and Testis Determination on the Mouse Y Chromosome
A. McLaren, E. Simpson, J. T. Epplen, R. Studer, P. Koopman, E. P. Evans, and P. S. Burgoyne
Proc. Natl. Acad. Sci. U.S.A., 85, 6442—6445, 1988

In the mouse, expression of the male-specific antigen H-Y is controlled by the Sxr (sex reversed) region. A variant region, Sxr', retains the genetic information needed for testis determination *(Tdy)* but has lost *Hya*, the gene responsible for expressing H-Y antigen. An X/X male fathered by an X/Y Sxr' male unexpectedly was positive for H-Y antigen. This apparently can be explained as the product of meiotic pairing and recombination between the two testis-determining regions of the father's Sxr'-carrying Y chromosome, as long as the region of *Hya* and *Tdy* in the normal mouse Y chromosome is on the short arm — not the long arm as has been assumed.

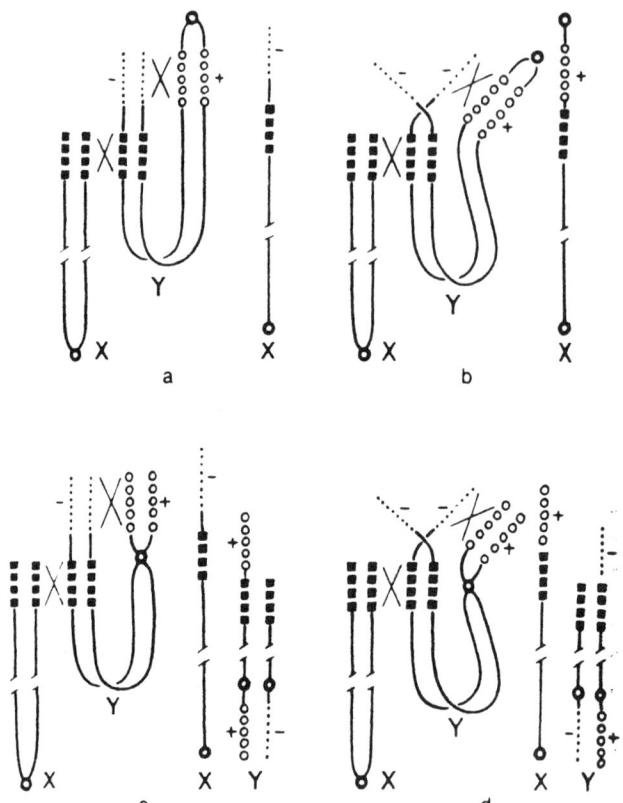

FIGURE 7-26. Possible meiotic configuration in an X/Y Sxr' testis if pairing occurs between the two homologous testis-determining regions of the Y chromosome as well as between the X and Y chromosome. Four additional configurations are possible, with the Y chromatids not crossed; the recombinant X chromosomes generated are identical to those illustrated. In a and b, only the recombinant X chromosomes have been drawn; in c and d, the recombinant Y chromosomes have also been included. Chromosomes are not drawn to scale. (From McLearn et al., *Proc. Natl. Acad. Sci. U.S.A.*, 85, 6442—6445, 1988. With permission.)

DNA fingerprinting showed that the banding pattern typcial of Sxr' had been replaced by the pattern associated with the native testis-determining region of the normal Y chromosome. Pairing between the two ends of a Y Sxr' chromosome has been observed by electron microscopy. To obtain the result observed, the native homologue of the Sxr region would have to be located on the very small short arm of the Y chromosome. In this case an X chromosome carrying *Tdy* and *Hya*$^+$ could be generated without transferring a centromere from the Y (Figure 7-26).

These findings indicate that the linked genes *Tdy*, which controls testis determination, and *Hya*, which controls H-Y antigen expression, are located on the short arm of the mouse Y chromosome.

Molecular and Cytogenetic Evidence for the Location of *Tdy* and *Hya* on the Mouse Y Chromosome Short Arm
C. Roberts, A. Weith, E. Passage, J. L. Michot, M. G. Mattei, and
C. E. Bishop
Proc. Natl. Acad. Sci. U.S.A., 85, 6446—6449, 1988   7-27

Sex reversed *(Sxr)* is a small fragment of the murine Y chromosome that has transposed to the distal pairing/recombination region of the Y chromosome in XY*Sxr* mutant mice. The authors obtained direct cytologic and molecular evidence indicating that *Sxr* is derived from the normal Y chromosome short arm. The study employed a combination of *in situ* mapping and DNA analysis with recombinant DNA probes specific for the *Sxr* region of the mouse Y chromosome.

The gene controlling sex determination, *Tdy*, and that controlling expression of the male-specific antigen H-Y, *Hya*, both are located on the short arm of the Y chromosome. These studies showed that the H-Y⁻ variant of *Sxr (Sxr')* arose through a partial deletion within the *Sxr* region. Intrachromosomal recombination between the Y short arm and *Sxr'* can occur during male meiosis, restoring the deleted DNA sequences and resulting in an H-Y⁺ mouse.

Generation of the *Sxr'* mouse from *Sxr* is not due to a point mutation of the *Hya* gene, but involves a partial deletion within *Sxr* itself. The entire *Sxr* region or part of it apparently is relocated to the distal pseudoautosomal region. Either a translocation or a single recombination event might be responsible; the latter is thought to be more likely.

Sequences Homologous to *ZFY*, a Candidate Human Sex-Determining Gene, Are Autosomal in Marsupials
A. H. Sinclair, J. W. Foster, J. A. Spencer, D. C. Page, M. Palmer,
P. N. Goodfellow, and J. A. M. Graves
Nature, 336, 780—783, 1988   7-28

Since marsupials diverged from placental mammals at least 130 million years ago, comparing gene arrangements between these groups might cast light on the evolution of sex chromosomes and sex determination. In placental mammals sexual differentiation results from the action of a testis-determining gene encoded by the Y chromosome. A candidate gene is *ZFY* (Y-borne zinc-finger protein). The *ZFY* probe detect a Y-linked sequence in DNA from a range of eutherian mammals as well as an X-linked sequence *(ZFX)* mapping to the human X chromosome.

Studies with the *ZFY* probe showed that *ZFY* homologous sequences are not on either the X or the Y chromosome in marsupials, but instead map to the autosomes. the *ZFY* probe maps to the same location as *DMD*, with a secondary site near the *OTC* and *SYN1* genes. All of these genes are close to *ZFX* on the human X chromosome.

If testis determination proves to be achieved by unrelated genes in eutherian mammals and marsupials, it will be of interest to know which was the ancestral system. There is some evidence that *ZFY*-homologous sequences may be autosomal in birds, suggesting that the testis-determining function of *ZFY* has evolved since divergence of the eutherian lineage. A previously secondary step in a sex-determining pathway may have taken on a primary function in eutherians.

Duplication, Deletion, and Polymorphism in the Sex-Determining Region of the Mouse Y Chromosome
G. Mardon, R. Mosher, C. M. Disteche, Y. Nishioka, A. McLaren, and D. C. Page
Science, 243, 78—80, 1989   7-29

The *ZFY* gene in the sex-determining region of the human Y chromosome encodes a so-called zinc-finger protein that may be the testis-determining factor, *TDF*. Homologues of human *ZFY* are found on the Y and X chromosomes in a wide range of placental mammals. In the mouse — in contrast to other mammals — two homologues are present, both in the sex-determining *(Sxr)* region. *Zfy-1* alone may be sufficient to determine males. *Zfy-2* was deleted in an *Sxr* variant that retains sex-determining function but has lost other genes, and so is dispensable. Both loci map near the centromere of the mouse Y chromosome. The Y chromosome of the subspecies *Mus musculus musculus* and *M. m. domesticus* were distinguished by a *Zfy-1* restriction fragment polymorphism.

It may be that *Zfy-1* alone is sufficient for testis determination, while *Zfy-2* encodes a functionally redundant testis-determining gene product. Alternately *Zfy-2* may carry out a function unrelated to sex determination, or it maybe a pseudogene of *Zfy-1*.

Chromosome Mapping and Expression of a Putative Testis-Determining Gene in Mouse
C. M. Nagamine, K. Chan, C. A. Kozak, and Y.-F. Lau
Science, 243, 80—83, 1989   7-30

Recently a human Y-chromosomal DNA fragment was isolated that is a candidate for the testis-determining factor *(TDF)* gene, the human equivalent of *Tdy*. A homologous sequence was found on the X chromosome. The present study isolated and mapped a mouse cDNA sequence (mouse Y-finger) that encodes a multiple, potential zinc-binding finger protein homologous to the candidate *TDF*.

Four similar sequences were identified in Hind III-digested mouse genomic DNA. Two of them (7.2 kb and 2.0 kb) were mapped to the Y chromosome. Only the 2.0-kb fragment correlated with testis determination. Polymerase chain reaction studies indicated that both Y loci are transcribed in the adult testis. A 3.6-kb fragment was mapped to the X chromosome between the T16H and T6R1 translocation breakpoints. A fourth fragment of 6.0 kb was mapped to chromosome 10.

These findings show that mYfin sequences have been duplicated several times in the mouse, while they are not duplicated in humans.

The Sex-Determining Region of the Mouse Y Chromosome Encodes a Protein with a Highly Acidic Domain and 13 Zinc Fingers
G. Mardon and D. C. Page
Cell, 56, 765—770, 1989

The *ZFY* gene in the sex-determining region of the human Y chromosome encodes a zinc-finger protein. In contrast to most placental mammals, the mouse Y chromosome carries not one but two homologues of the human *ZFY* gene as the result of an intrachromosomal duplication that occurred during evolution. One or both of these genes *(Zfy-1* and *Zfy-2)* may serve as the primary sex-determining signal in mice.

Both *Zfy-1* and *Zfy-2* are transcribed in their adult testis. Nucleotide sequence analysis of a *Zfy-2* cDNA indicated that it encodes a 783-amino acid protein having two domains. The amino-terminal portion is highly acidic (Figure 7-31); 25% of its residues are glutamic or aspartic acid. The carboxy-terminal domain contains 13 zinc fingers. The presence of an acidic domain combined with a putative nucleic acid binding domain suggests that *Zfy-2* activates transcription in a sequence-specific manner.

The *Zfy-1* and *Zfy-2* genes are very similar. Although both are in the sex-determining region of the mouse Y chromosome, no sex-determining function has yet been demonstrated for either gene. Testis determination

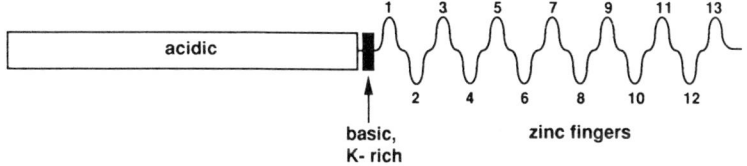

FIGURE 7-31. Schematic diagram of the putative mouse *Zfy-2* protein. Almost half the protein — an amino-terminal domain of about 369 residues — is highly acidic. A tandem array of 13 putative zinc fingers, with alternating fingers of two types (i.e., a two-finger repeat), comprises the carboxy-terminal half. Between these two large domains is a small, highly basic region. (From Mardon/Page, *Cell,* 56, 765—770, 1989. ©Cell Press.

does not require both genes. Their similarity suggests that they may encode functionally redundant proteins.

ZFX Has a Gene Structure Similar to ZFY, the Putative Human Sex Determinant, and Escapes X Inactivation
A. Schneider-Gadicke, P. Beer-Romero, L. G. Brown, R. Nussbaum, and D. C. Page
Cell, 57, 1247—1258, 1989   7-32

The authors have cloned the human X homolog of the *ZFY* gene, *ZFX*, which apparently encodes a protein containing a zinc finger domain closely related to that found in the ZFY protein.

The human *ZFX* gene was cloned by cross-hybridization to *ZFY* (Figure 7-32). Similarities in the intron/exon organization and exon DNA sequences show that *ZFY* and *ZFX* diverged from a common ancestral gene. The carboxy-terminal exons of *ZFY* and *ZFX* both encode 13 zinc fingers. All but 10 of 393 amino acid residues are identical, and there are no insertions or deletions. The ZFY and ZFX proteins therefore may bind to the same nucleic acid sequences. The genes are transcribed in a wide range of XY and — in the case of *ZFX* — XX cell lines. Transcription analysis of human-rodent hybrid cell lines containing "inactive" human X chromosomes indicated that *ZFX* escapes X inactivation.

The finding that *ZFX* escapes X inactivation contradicts the dosage/X-inactivation model, which postulates that sex determination depends on the total amount of functionally interchangeable ZFY and ZFX proteins. *ZFX* is not the only human X-chromosomal gene to escape X inactivation. A few genes that map to the extreme distal short arm of the X chromosome do so (Figure 7-32A). The posulated sex-determining role of *ZFY* remains to be established.

♦ Studies by Page and co-workers have localized an essential portion of the testis-determining factor to a 140-kb region of the short arm of the human Y chromosome (discussed in the *1989 Year Book of Developmental Biology).* This region contains a gene that encodes a protein (the putative testis-determining factor) with multiple zinc-finger domains and a highly acidic amino-terminal domain. In mouse, this gene (as well as the gene encoding male-specific H-Y antigen) also maps to the short arm of the Y chromosome (previously thought to be located on the Y-long arm) (McLaren et al.; 1988; Roberts et al., 1988). Surprisingly, homologous sequences in marsupials are autosomal in location even though the Y chromosome is required for testis determination (Sinclair et al., 1988). These results suggest that the primary sex-determination signal may be accomplished by unrelated genes in eutherians vs. metatherians. Alterna-

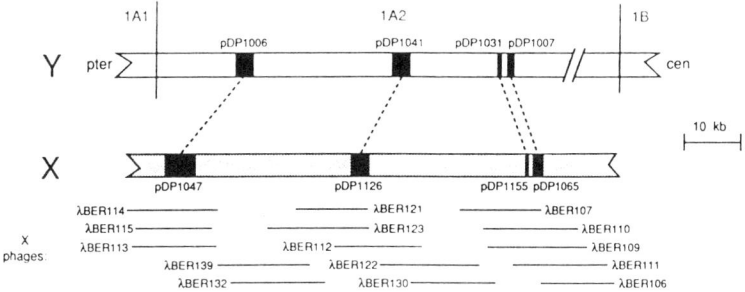

FIGURE 7-32. Cloning of the human *ZFX* gene by cross-hybridization to *ZFY*. At top is a schematic representation of interval 1A2 and adjoining intervals on the short arm of the human Y chromosome. Interval 1A2, which measures 140 kb, contains at least an essential portion of the sex-determining function. The orientation with respect to the short-arm telomere (pter) and centromere (cen) is shown. Within interval 1A2, four segments (in black) contain DNA sequences that were highly conserved during evolution, apparently contain exons of the *ZFY* gene, and cross-hybridize to sequences on the X chromosome. From left to right, these four segments represent the inserts of plasmids pDP1006, pDP1041, pDP1031, and pDP1007. Shown below is a nearly 90 kb block of DNA cloned from the human X chromosome by cross-hybridization to these four plasmids at high stringency (47°C, 50% formamide, 0.75 M NaCl). Eleven X-derived phages were identified in the initial screen: three with pDP1006, three with pDP1041, four with pDP1031 and pDP1007, and one with pDP1007 alone. The four other X phages were isolated by chromosomal walking. Dotted lines connect strongly cross-hybridizing restriction fragments (in black) within these *ZFY* and *ZFX* loci. It is likely that each blackened segment contains but does not consist entirely of one or more exons with a high degree of X-Y sequence similarity. Apart from the blackened segments, there may exist additional, smaller regions of X-Y similarity not yet detected. As judged by nucleotide sequencing of pDP1007 and its X counterpart, pDP1065, the direction of transcription is from left to right in both *ZFY* and *ZFX*. CpG islands are found in both pDP1006 and its X counterpart, pDP1047. It is possible, as is the case with many CpG islands, that transcription is initiated within these regions. (From Schneider-Gädicke et al., *Cell*, 57, 1247—1258, 1989. ©Cell Press.)

tively, the testis-determining gene in humans and mice may not be the primary-sex determining gene. This latter possibility, however, would seem unlikely given all of the data generated from genetic and molecular analyses of the sex-determining region of the human Y. Nonetheless, in the absence of direct evidence for sex-determining function, the human and mouse putative testis-determining genes are now referred to as *ZFY* (for zinc-finger protein on the Y) and *Zfy*, respectively. Interestingly, a homolog of this gene has been found on the X chromosome in both humans and mouse. In humans, *ZFX* has been cloned and sequence data suggest that the gene encodes a protein very similar to *ZFY*. Furthermore, *ZFX* apparently escapes X inactivation. The existence of an autosomal homolog (chromosome 10) and a second Y homolog have also been shown in mouse. The function of the X and autosomal loci relative to the Y loci is unclear. The two mouse Y-specific genes are now called *Zfy-1*

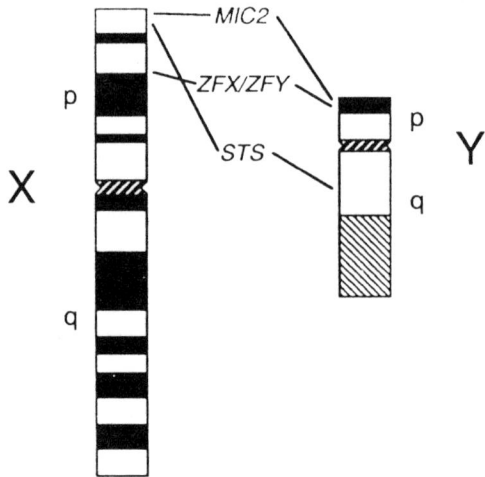

FIGURE 7-32A. Three cloned human genes known to escape X inactivation have homologs on the Y chromosome. MIC2 maps to the pseudoautosomal region of Xp22.3 and Yp and escapes X inactivation. The *STS* gene is located in Xp22.3 and exscapes X inactivation; a pseudogene is found on Yq. (Mouse *Sts*, presumably the homolog of human *STS*, is functional on both X and Y and escapes X inactivation.) *ZFX* and *ZFY* map to Xp21.3/22.1 and Yp, respectively. Though not cloned, the human *XG* gene also escapes X inactivation, and there may be a related gene on the Y chromosome. (From Schneider-Gädicke et al., *Cell,* 57, 1247—1258, 1989. ©Cell Press.)

and *Zfy-2*. Evidence using sex-reversed mice (*Sxr'* and *Sxr"*) indicates that *Zfy-1* alone is sufficient for testis determination. Nonetheless, both *Zfy-1* and *Zfy-2* appear to be expressed in adult testes (embryonic gonads apparently have not been tested). The predicted amino acid sequence for the proteins encoded by these two genes are nearly identical, suggesting that they might encode a functionally redundant testis-determining gene product. Transgenic experiments using both genes will clarify this possibility. *Terry Magnuson*

Genetically Haploid Spermatids are Phenotypically Diploid
R. E. Braun, R. R. Behringer, J. J. Peschon, R. L. Brinster, and
R. D. Palmiter
Nature, 337, 373—376, 1989   7-33

Spermatids differ genetically because chromosomal homologues segregate from one another during meiosis. Postmeiotic gene expression could lead to gametic differences and to the preferential transmission of certain alleles over others. Nevertheless in both insects and mammals, all cells derived from a single spermatogonial cell develop within a common syncytium formed through incomplete cytokinesis at each mitotic and meiotic cell division (Figure 7-33). The intercellular bridges connecting

Cytodifferentiation 243

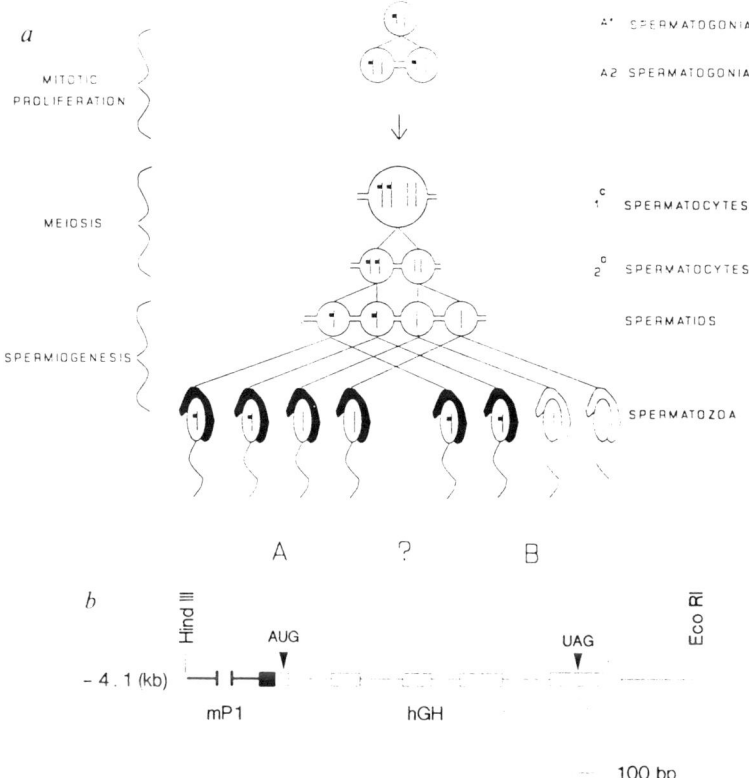

FIGURE 7-33. a, Illustration of the syncytial nature of spermatogenesis and the three general stages of development. The spermatogonial cells are drawn with a pair of homologous chromosomes inside them with one chromosome containing the introduced transgene (shown by a solid box). After the commitment of A1 spermatogonial cells to differentiate there are five or six rounds of mitotic proliferation. The first cell division generating A2 spermatogonial cells is shown. Primary spermatocytes contain a 4N complement of DNA, that is, each chromosome is associated with a sister chromatid. Each primary spermatocyte gives rise to four spermatids, two that contain a wild-type chromosome and two that carry a chromosome with the transgene. During spermiogenesis, round spermatids are transformed into mature spermatozoa. The question is whether all of the spermatozoa contain the product of the transgene (in this case the hGH protein localized to the acrosome), as shown in A, or whether only those spermatozoa that contain the transgene will have the product, as shown in B. b, Plasmid mP1-hGH contains approximately 4.1 kb of mP1 5′ sequence, including the promoter and 91 base pairs (bp) of 5′ untranslated sequence fused to a genomic clone for the hGH gene. Boxes represent exons and lines represent either nontranscribed DNA or introns. Bold lines and the solid box represent mP1 DNA whereas thin lines and open boxes represent hGH DNA. The break in the line that corresponds to mP1 sequence denotes that portion of the transgene that is not drawn to scale. The positions of the translational start and stop codons in the first and last exon, respectively, are shown above the gene structure. Transgenic mice were constructed by microinjecting a 6.3-kb HindIII-EcoRI fragment containing the mP1-hGH gene, free of plasmid vector sequences, into the pronuclei of fertilized C57BL/6 × SJL F2 mouse eggs. Two lines of transgenic mice, 314-4 and 315-6, were generated and used for this study. Both lines contain multiple copies of the transgene integrated on one chromosome. (From Braun et al., Nature, 337, 373—376, 1989. With permission.)

the cells may allow sharing of cytoplastic constituents, thereby ensuring the synchronous development of a clone of cells and gametic equivalence between haploid spermatids.

This study was intended to show whether RNA or protein can pass through the intercellular bridges connecting spermatids, which are about 1 μm in diameter. The product of a transgene expressed only in post-meiotic germ cells in hemizygous transgenic mice was analyzed. Genetically distinct spermatids were found to share the transgene product, human growth hormone protein.

These findings show that genetically distinct spermatids can be phenotypically equivalent, and they provide evidence against functional differences between sperm due to post-meiotic gene expression. The products of most genes may well be distributed equally among spermatids.

♦Evidence now exists for haploid gene expression during mammalian spermatogenesis. Because of this possibility, one must consider that a mouse heterozygous for two different alleles of the same gene which is known to be expressed during spermatogenesis could produce spermatozoa that are functionally different. Evidence for such a possibility has been demonstrated genetically where males heterozygous for the t-complex show transmission ratio distortion by transmitting the t-containing chromosome to 99% of its progeny. This is the only case in mice where such a phenomenon has been observed. The reason that other haploid-specific effects have not been detected may be due to the fact that all spermatids derived from a single spermatogonial cell are connected by intercellular bridges as a result of incomplete cytokinesis. Thus, it has been proposed that gamete equivalence is ensured due to the sharing of cytoplasmic constitutents. However, until the report listed above, no evidence for this possibility existed. Braun and co-workers generated male mice hemizygous for a chimeric transgene consisting of mouse protamine transcriptional regulatory sequences fused to a human growth hormone structural gene. Expression of the transgene was detected exclusively in haploid spermatids and not in spermatogonial cells or in spermatocytes. Nonetheless, visualization of the protein showed that the majority of the spermatids within a tubule of a hemizygous mouse were immunopositive for the protein. When quantitated, 91% of all sperm analyzed from hemizygous animals were immunopositive for the transgene product, despite the fact that the gene was transmitted to only 50% of the offspring. These results provide evidence against functional differences between sperm arising from postmeitoitc gene expression. They do not, however, address whether RNA and/or protein are being transmitted. The reasons that the t-complex demonstrates transmission ratio distortion despite cytoplasmic continuity are not understood at the present time. *Terry Magnuson*

Homeobox Genes 8

INTRODUCTION

Interest in the developmental roles of homeobox-containing genes continued unabated during the past year. Additional genes have been identified, interesting distributions have been described, and specific functions have been assigned. Other DNA sequences related to the homeobox sequence have also been identified.

The most excitement has been generated by the observation that homeobox-containing genes can serve as DNA-binding proteins that can modulate transcriptional activity. Such functional activity, like homeobox-containing proteins themselves, exists in many different biological species and is not restricted to *Drosophila* alone. In fact, functional homology is as exact as structural homology because homeobox-containing transcriptional factors from one species can effect transcription when introduced into another species.

Homeobox-containing genes were first associated with segmentation in *Drosophila* and have since been associated with pattern formation events during development. Gradient distributions of several homeobox-containing genes have been demonstrated. Such distributions suggest that these proteins, functioning as specific transcription factors, can effect gene expression in a way that would directly lead to the formation of patterns in developing tissues.

Other specific DNA sequences (e.g.., POU domain) have been identified, which share some functional and structural similarity with the homeobox domain. Their specific roles during development require additional exploration. Continued investigation of such domains will certainly expand our understanding of why they have remained so well conserved during evolution.

Progressively Restricted Expression of a Homeo Box Gene within the Aboral Ectoderm of Developing Sea Urchin Embryos
L. M. Angerer, G. J. Dolecki, M. L. Gagnon, R. Lum, G. Wang, Q. Yang, T. Humphreys, and R. C. Angerer
Genes Dev., 3, 370—383, 1989   8-1

Genes bearing homeo box sequences encode nuclear proteins having important regulatory functions in a wide range of organisms. There also is evidence that homeo proteins act as transcription factors regulating the expression of other genes. A homeo box-containing gene, Hbox1, is expressed in a highly conserved spatial pattern in the aboral ectoderm of two sea urchin embryos, *Tripneustes gratilla* and *Strongylocentrotus purpuratus*. On other molecular and cytologic criteria aboral ectoderm is a uniform region containing a single cell type.

Hybridization studies showed that the mRNA accumulates initially throughout the aboral ectoderm. Between the blastula and pluteus stages, however, the region containing Hbox1 mRNA gradually decreased in extent until finally only a small area about the vertex was labeled. In *S purpuratus*, the gene product probably is not involved in initial specification of cell fate, since the message is not significantly prevalent until nearly the hatching blastula stage — well after aboral ectoderm cells have begun a tissue-specific program of gene expression. RNA blot and RNAse protection studies showed low level of Hbox 1 mRNA in all adult tissues. The message was not detected in mature eggs.

The polarity observed does not conform precisely to either the animal-vegetal or oral-aboral embryonic axis, nor does it align closely with the anterior-posterior axis (Figure 8-1). It may be a clonal phenomenon wherein clusters of ectoderm cells derived from individual early blastomeres stop expressing Hbox 1 mRNA at different times. Alternately, cell-cell interactions might generate a wave of inactivation of Hbox1 gene expression that gradually moves through the tissue as a function of distance from some signal at either end.

The homeobox genes of these two sea urchin species also are highly conserved in sequences outside the homeodomain. As in several vertebrate homeo proteins, a short conserved sequence is encoded by an exon upstream of than endoding the homeodomain. These proteins also share a large region of high serine and proline content.

♦Although proteins containing homeobox domains are found in a wide variety of embryos and may well be universal, it is not clear that the specific developmental roles of these (putative) regulatory proteins are the same in diverse organisms. Analysis of homeobox genes in sea urchin embryos is of particular interest because they are expressed in two distinct stages of the life cycle, neither of which is segmented, i.e., the bilaterally symmetric pluteus larva and the radially symmetric adult (see references in this paper). This work documents an unusual and conserved pattern of expression for one homeobox gene, Hbox1, in embryos of two species of sea urchin. As is the case for other deuterostome (e.g., frog and mouse) embryos, the distribution of Hbox1 delimits a "region" that does not conform to histological borders defined by morphology or by the expres-

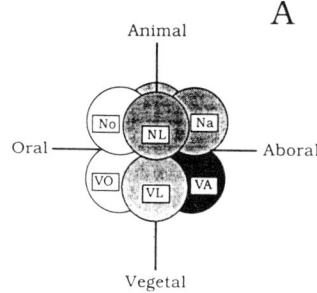

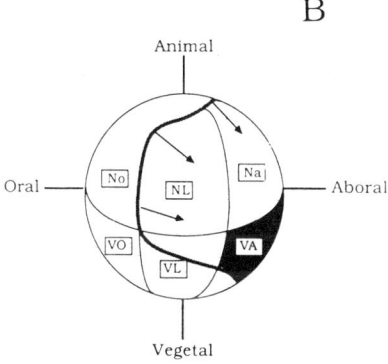

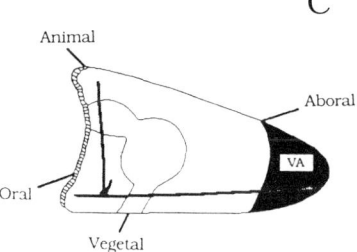

FIGURE 8-1. Relationship of spatial pattern of Hbox1 expression to early cleavage planes and embryonic axes. (A) The eight-cell embryo os shown with the blastomeres labeled according to the terminology of Cameron et al. (1987). The four blastomeres of the animal pole give rise exclusively to ectoderm, whereas the four blastomeres in the vegetal hemisphere contribute about half their volume to ectoderm and half to endoderm and mesenchyme. Six blastomeres contribute to aboral ectoderm, but unequally. The plane of the future oral-aboral axis lies at a 45° angle to the first two meridional cleavage planes, bisecting the **Na** and **VA** (and **No** and **VO**) blastomeres, which give rise only to aboral ectoderm. The four lateral blastomeres (NL asnd VL on the future right and left sides, respectively) yields both oral and aboral ectoderm. (B) Schematic representation of the blastula-stage embryo at the time of first appearance of Hbox1 mRNA. The regions contributed by individual eight-cell blastomeres are labeled. The thick line represents the approximate position of the border between presumptive oral and aboral ectoderm, based on in situ hybridizations carried out in our laboratory witrh probes specific for aboral ectoderm. In the pluteus, this border is defined by the ciliated band, Arrows indicate the direction in which the region of aboral ectoderm hybridizing with the Hbox1 probe retracts after blastula stage. (C) Schematic representation of the pluteus larva. The original animal-vegetal axis is bent as a result of expansion of the ectoderm. The ciliated band outlines the oral face (left). The skeletal rods, of which only the left is shown, meet at the vertex, which is the region covered by descendents of the VA blastomere (black). (From Angerer et al., *Genes Dev.*, 3, 370—383, 1989. ©Cold Spring Harbor Laboratory.)

sion of other tissue-specific genes, and this region becomes more spatially restricted as development progresses. Because SpHbox1 mRNA accumulates only after aboral ectoderm cells have begun to transcribe tissue specific mRNAs, the authors argue that the SpHbox1 gene product is not involved in initial specification of aboral ectoderm cells. This raises two interesting questions: (1) assuming that SpHbox 1 encodes a nuclear regulatory protein, what genes and what process is it regulating? (2) What information is arrayed along the oral-aboral axis which controls the pattern of Hbox1 expression? Because *Drosophia* homeo domain proteins are known to comprise a set of interactive and combinatorial regulators that control the architecture of the embryo, and are expected to do so in vertebrates as well, it will be interesting to determine spatial patterns for other Hbox mRNAs expressed in sea urchin embryos. Robert Angerer

---

The *C. elegans* Cell Lineage and Differentiation Gene *unc-86* Encodes a Protein with a Homeodomain and Extended Similarity to Transcription Factors
M. Finney, G. Ruvkun, and H. R. Horvitz
Cell, 55, 757—769, 198

Mutations in the gene *unc-86* alter development of the nematode C elegans by affecting cell lineages and cell differentiation. This gene likely acts in conjunction with other genes to specify cell fate. The authors molecularly cloned *unc-86* by chromosomal walking from linked polymorphic genetic loci, and identified the gene by locating polymorphisms specific for *unc-86* alleles.

A transcript containing a 467-amino acid open reading frame was inferred from the DNA sequence of a genomic clone. The *unc-86* transcript encodes a protein containing a 158-amino acid sequence, the pou domain, that very closely resembles sequences found in 3 mammalian transcription factors. Within the conserved region is a homeodomain related to, but distinct from homeodomains previously found in Drosophila and other organisms.

It appears that *unc-86* encodes a transcription factor, and that the related mammalian transcription factors may act to control cell differentiation and cell fate. The *unc-86* homeodomain probably functions in DNA binding. The protein presumably directly regulates transcription to control transitions from one cell state to another, in some instances in association with cell division.

♦ The *unc-86* gene of *C. elegans* has two similar kinds of effects. One effect is on division of certain blast cells that normally produce two daughter cells with different fates (both different from their mother).

Mutant analysis indicates that *unc-86* is necessary for one daughter cell of these divisions to become different from its mother (Chalfie et al., *Cell*, 24, 59-—69, 1981). Since a variety of blast cell types are affected by *unc-86* mutations, *unc-86* must specify the fate of each affected daughter by acting in combination with other genes. *unc-86* mutations also affect the differentiation of at least one dividing cell: the precursor to the HSN neuron fails to differentiate into the proper cell type (Finney, Ph.D. thesis, MIT, 1987; Desai-et al., *Nature,* 336, 638—646, 1989).

This paper is important because it shows that the *unc-86* protein predicted from DNA sequence analysis has extensive sequence similarity to three mammalian transcription factors (1) the rat pituitary transcription factor Pit-1 (Ingraham et al., *Cell,* 55, 519—529, 1988), (2) the human ubiquitous transcription factor Oct-1 (Sturm et al., *Genes Dev.,* 1988), and (3) the human lymphoid transcription factor Oct-2 (Clerc et al., *Genes Dev.*,1988). The region of homology called the *p o u* domain (for p*i*t, oct, and *u*nc-86) has three regions of strong conservation, one of which is a homeo-domain. By analogy to other homeo-domain-containg genes including Pit-1, Oct-1 and Oct-2, "the *unc-86* protein probably directly regulates transcription to control transitions from one cell state to another." In some cases this change in cell states is associated with a cell division and in the case of HSN it is associated with a nondividing cell. One of the big effects of *unc-86* mutations is to alter the subsequent lineages of daughters from affected cell divisions. In these cases we may view the change in "cell state" induced by *unc-86* as one that normally results in an altered pattern of cell divisions and cell fates (e.g., one normal function of *unc-86* may be to set into motion a sublineage program). In this respect we can think of a sublineage program as another form of differentiation. *Joseph G. Culotti*

The Sequence Specificity of Homeodomain-DNA Interaction
C. Desplan, J. Theis, and P. H. O'Farrell
Cell, 54, 1081—1090, 1988

Transcriptional regulators of gene activity are thought to have an important role in directing embryonic development. A group of Drosophila developmental genes are among the prime candidates for such regulators. The developmental gene *engrailed* encodes a sequence-specific DNA binding activity. One of the goals of this study was to define the domain of the *engrailed* encoded protein (En protein) that specifies DNA binding.

The use of deletion constructs expressed as fusion proteins in *E. coli* localized the sequence-specific DNA binding activity to the conserved homeodomain (HD). The binding site consensus TCAATTAAAT was found in clusters in the *engrailed* regulatory region. Initially weak binding of the En HD to one copy of a synthetic consensus was enhanced by adjacent copies. The distantly related HD encoded by *fushi tarazu* bound to the same sites as the En HD but differed in its preference for related sites. Both HDs bound a second type of sequence that was a repeat of TAA.

Within families of DNA binding proteins, close relatives can be expected to exhibit similar sequence specificies. Competition among related regulatory proteins may determine which occupies a given binding site, and accordingly the ultimate effect of *cis*-acting regulatory sites. Since a number of eukaryotic regulators have been found to share overlapping binding specificities, it may be that evolutionary duplication and divergence have created families of regulators having varying degrees of functional homology. Many HD-containing proteins guide embryonic pattern formation. At each step regulators are expressed in overlapping spatial distributions which act in combinational codes to control the spatial pattern of expression of subsequent regulators.

An *Ultrabithorax* Protein Binds Sequences Near its Own and the *Antennapedia* P1 Promoters
P. A. Beachy, M. A. Krasnow, E. R. Gavis, and D. S. Hogness
Cell, 55, 1069—1081, 1988   8-4

Ultrabithorax (Ubx) is a homeotic gene within the bithorax complex of Drosophila which specifies primarily the distinguishing features of parasegments 5 and 6. The gene encodes a family of closely related proteins that direct the developmental fates of posterior thoracic and anterior abdominal metameres. The authors examined the biochemical properties of *Ubx* proteins in order to elucidate the molecular basis of the regulatory interactions involved.

A member of the *Ubx* protein family was purified from an overproducing *E. coli* strain and found to be a sequence-specific DNA-binding protein. The protein binds tightly to sequences near its own promoter and also near the P1 promoter of *Antennapedia (Antp)*, another homeotic gene which genetic studies have shown is repressed by *Ubx* (Figure 8-4). The binding sites occur in clusters downstream from the transcription start sites, and far upstream at *Antp* P1. The sites contain tandem repeats of the trinucleotide TAA or of the related hexanucleotide TAA-TCG.

It appears that the regulatory functions of *Ubx* are mediated by binding of *Ubx* proteins to sequences in the promoter region. The importance of the *Ubx* binding site sequences is shown by their requirement for ectodermal expression in embryos of *Ubx* promoter-*lacZ* fusion genes, and also by their conservation in another Drosophila species, *D. funebris*.

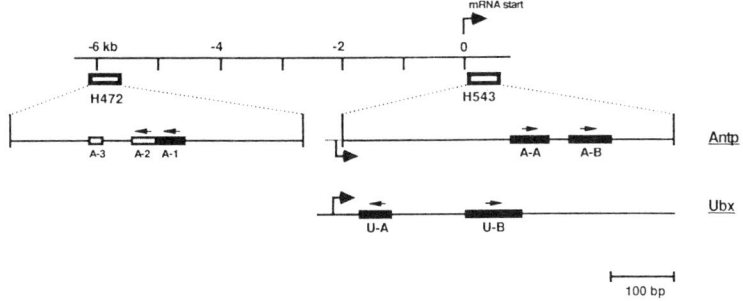

FIGURE 8-4. Locations of UBX Ib binding sites near the *Ubx* and *Antp* P1 promoters. The open boxes in the top portion of the figure show the locations of fragments H472 and H543 with respect to the *Antp* P1 transcription start site at coordinate 0. In the expanded diagrams below each open box, the solid and stippled boxes denote the locations of sequences fully or partially protected from DNAase I cleavage by UBX Ib. In the bottom half of the figure, two protected reagions downstream of the *Ubx* transcription start site are indicated. The *Antp* and *Ubx* transcription start sites are indicated by large arrows at coordinate 0 and the small arrows show the predominant orientation (5' to 3') of the TAA repeats within each region. (From Beachy et al., *Cell,* 55, 1069—1081, 1988. ©Cell Press.)

Activation and Repression of Transcription by Homeodomain-Containing Proteins That Bind a Common Site
J. B. Jaynes and P. H. O'Farrell
Nature, 336, 744—749, 1988   8-5

Many of the regulatory genes of Drosophila encode a homeodomain and in some instances the homeodomain proteins control transcription. The enhancer activity of a DNA fragment of the *ftz (fushi tarazu)* gene, a homeodomain-containing member of the pair-rule class of segmentation genes, suggests that the *ftz* gene product activates its own expression in the embryo.

In this study, the product of the *ftz* gene was shown to be a site-dependent activator of transcription. Defined binding sites acted as ftz-dependent enhancers in cultured cells. Strong ftz activation was always dependent on the presence of homeodomain-binding sites, but much smaller effects of ftz sometimes were found in the absence of such sites. Another homeodomain-containing protein, the *engrailed* gene product, competed for homoedomain-binding sites and repressed ftz activation. A combination of the *engrailed* and *ftz* gene products had an effect on transcription that was qualitatively distinct from that of ftz alone.

The product of the Drosophila gene *ftz* is a powerful site-dependent activator of transcription. It is likely that ftz acts by binding to the inserted sites to activate the linked promoter. Competition for binding sites could markedly influence the action of homeodomain proteins during embry-

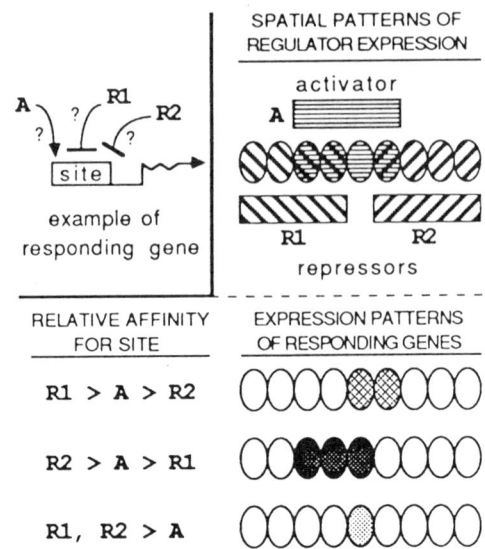

FIGURE 8-5. Multiple patterns of gene expression can be generated by competition among regulators for binding to related but different cis-acting sites. Beginning with overlapping patterns of expression of three regulatory proteins with related DNA-binding specificities, three novel patterns are produced by differential competition. A is an activator, whereas R1 and R2 are repressors. Related cis-acting sites in the responding genes have different relative affinities for the regulators. The first responding gene has a binding site with relative affinities R1 > A > R2, so that it is activated by A and repressed by R1, but not by R2. The second responding gene has a site with relative affinities R2 > A > R1, whereas the site in the third gene has a higher affinity for both R1 and R2 than for A. Such a scheme could operate in the embryo to refine spatial patterns of regulatory gene expression and to subdivide domains of expression. This simple example assumes that regulator concentrations are equal and constant in space and time. Similar competitive interactions could also generate multiple domains of responder expression from a gradient of regular concentration. (From Jaynes/O'Farrell, *Nature*, 336, 744—749, 1988.)

onic pattern formation. Binding sites are exposed to a mixture of homeodomain proteins that varies in a spatiotemporal manner. Each of the related binding sites may undergo transitions in occupancy at different relative concentrations of homeodomain proteins (Figure 8-5). Although direct evidence is lacking, the prevalence of a highly conserved DNA-binding domain among developmental regulators suggests that competitive interactions have a general role in the regulatory mechanisms governing space- and tissue-specific transcription.

Synergistic Activation and Repression of Transcription by Drosophila Homeobox Proteins
K. Han, M. S. Levin, and J. L. Manley

Cell, 56, 573—583, 1989   8-6

Combinational or competitive interactions between the products of regulatory genes may have a critical role in differentiation of the embryo. Many of the genes involved appear to encode transcriptional regulatory proteins. In this study, a cotransfection assay based on Drosophila tissue culture cells was used to examine the potential for homeobox gene proteins to function as transcriptional regulators. The transient expression assay utilizes Schneider L2 tissue culture cells.

A 96-bp fragment of the promoter region of the segment polarity gene *engrailed* — previously found to contain five copies of a 10-bp consensus binding site for these proteins — enhanced transcription in the presence of several homeobox protein expression vectors. No such effect was noted in the absence of vectors. Cotransfection with varying combinations of expression vectors encoding the homeobox proteins *fushi tarazu*, *paired*, and *zen* led to synergistic increases in expression. Activation took place at the RNA level. On the other hand, the products of the *even-skipped* and *engrailed* genes repressed the activation induced by the other proteins. Both wild-type and mutant homeobox proteins were able to quench activation.

Drosophila homeobox gene proteins can active transcription in tissue culture cells. Combinations of these proteins are able to function together to enhance transcription in a synergistic manner. The action of homeobox proteins to repress the activation of transcription induced by other proteins likely involves direct competition between active and inactive proteins for DNA binding sites. The authors envision a "multiswitch" model according to which the activity of a given target gene depends on the interactions of different homeobox proteins with multiple copies of a common binding site.

Transcriptional Activation by the *Antennapedia* and *fushi tarazu* Proteins in Cultured Drosophila Cells
G. M. Winslow, S. Hayashi, M. Krasnow, D. S. Hogness, and
M. P. Scott
Cell, 57, 1017—1030, 1989   8-7

Drosophila homeodomain proteins bind to specific DNA sequences *in vitro* and may regulate the transcription of other genes during development. The present study employed a cotransfection assay based on cultured Drosophila cells to determine whether homeodomain proteins encoded by certain genes can activate transcription from specific promoters. The proteins tested were those encoded by the homeotic gene *Antennapedia (Antp)* and the segmentation gene *fushi tarazu (ftz)*, as well as a hybrid homeodomain protein, *Antp-Ubx (Ultrabithorax)*.

All three of the proteins studied did activate transcription from specific promoters in cultured cells. Sequences downstream of the *Antp* P1 and *Ubx* transcription start sites mediate the activation. A TAA-richDNA sequence to which *Antp* protein binds *in vitro* can confer regulatory activity on a heterologous promoter. The *ftz* protein is able to activate the *Ubx* promoter. A *Ubx* homeodomain can substitute for the homeodomain of the *Antp* protein.

Homeodomain proteins are transcriptional regulators and, in cultured cells, different homeodomain-containing proteins can act on a common sequence to modulate gene transcription. It may be that subtle differences in the affinity of homeodomain proteins for target sequences are critical in this process. If so, competitive or cooperative interactions among different homeodomain proteins for the same *cis*-acting sequences could determine the level of transcriptional activity of the target genes. Interactions with nonhomeodomain proteins could also change the specificity of the protein-DNA interactions.

Transcriptional Activation and Repression by *Ultrabithorax* Proteins in Cultured Drosophila Cells
M. A. Krasnow, E. E. Saffman, K. Kornfeld, and D. S. Hogness
Cell, 57, 1031—1043, 1989

Homeotic genes of Drosophila such as *Ultrabithorax (Ubx)* and *Antennapedia (Antp)* are thought to select metameric identity during development through controlling the expression of various target genes. In this study a cotransfection assay using cultured Drosphila cells was used to assess the transcriptional regulatory activity of UBX and other proteins involved in Drosophila development.

UBX was able to repress an *Antp* promoter fusion and activate a *Ubx* promoter fusion. UBX proteins regulated the level of initiated *Antp* P1 and *Ubx* transcripts. Activation of the *Ubx* promoter required a downstream group of UBX binding sites. Binding site sequences served to confer regulation on a heterologous promoter regardless of their orientation or exact position.

It appears that UBX proteins can activate and repress transcription through binding to promoter-region sequences. All members of the UBX protein family have similar regulatory ability, but these proteins may have a modular design resembling other transcriptional regulators. An important question is how a single UBX protein can both activate and repress target genes. This may be a more general question for DNA binding transcriptional regulators.

Drosophila Homoeotic Genes Encode Transcriptional Activators Similar to Mammalian OTF-2
M. Thali, M. M. Muller, M. DeLorenzi, P. Matthias, and M. Bienz
Nature, 336, 598—601, 1988   8-9

Drosophila homeotic genes control the morphogenesis of segment-specific features. The proteins encoded by these genes all contain a conserved homeodomain that can bind DNA and may be involved in transcriptional regulation. To test the transcriptional activation ability of homeodomains, two Drosophila homeogenes, Ubx and Abd-B, were cotransfected into HeLa cells along with target sequences derived from the Ubx or engrailed genes in OVEC reporter plasmids.

The reporter gene without the trans-activator had little or no expression. Cotransfection with Ubx or Abd-B induced significant reporter gene expression. Therefore, these Drosophila homeoproteins appear to be transcriptional activators that can function in the absence of other Drosophila proteins. The homeoproteins were also capable of activating via the octamer binding site Octal, which is the target sequence for the mammalian transcription factor OTF-2. OTF-2 was also able to activate transcription from Ubx and Abd-B target sequences.

The Drosophila homeoproteins Ubx and Abd-B functioned as transcriptional activators in mammalian cells. The mammalian transcriptional activator OTF-2, which has been shown to contain a homeodomain was functionally equivalent to these two homeoproteins in this assay.

The *Drosophila* fushi tarazu Polypeptide is a DNA-Binding Transcriptional Activator in Yeast Cells
V. D. Fitzpatrick and C. J. Ingles
Nature, 337, 666—668, 1989   8-10

The Drosophila *fushi tarazu* (ftz) gene contains a homeobox region. To determine whether the ftz protein is a sequence-specific DNA-binding transcriptional activator, the ftz polypeptide was used to replace the acidic transcriptional activation domains of the GAL4 protein in yeast cells.

The GAL4/ftz fusion protein was almost as active as intact GAL4 protein in activating reporter gene transcription in yest cells. Deletion analysis of the ftz portion of the fusion gene demonstrated that transcriptional activation was destroyed by removal of the N-terminal 272 amino acids or the C-terminal 192 amino acids. To determine whether ftz had sequence-specific DNA-binding activity in yeast, a reporter gene containing six tandem repeats of a Drosophila ftz-binding sequence was used. This sequence functioned as a ftz-dependent upstream activating sequence in yeast. An intact homeobox was necessary for ftz transcriptional activation.

The Drosophila homeobox-containing protein, ftz, can function as a transcriptional activator in yeast. Therefore, the role of ftz during Drosophila embryogenesis may depend on its transcriptional activation properties.

Gene Activation and DNA Binding by Drosophila Ubx and abd-A Proteins
M.-L. Samson, L. Jackson-Grusby, and R. Brent
Cell, 57, 1045—1052, 1989   8-11

The Ubx and abd-A genes are homeobox-containing regulators of Drosophila embryonic development. To assess their DNA-binding and transcriptional regulatory function, LexA-Ubx and LexA-abdA fusion proteins were expressed in yeast along with reporter genes.

Both fusion proteins activated gene expression from target genes containing LexA operators or Ubx target sequences. Fusion proteins containing carboxy-terminal deletions were used to assess the role of the homeobox region in transcriptional activation. Homeobox deletions eliminated Ubx target activation, but not LexA target activation. This suggests that the homeobox region is involved in DNA sequence recognition, but not gene activation.

The properties of the yeast Ubx and abd-A fusion proteins suggest that these proteins function as DNA-binding transcriptional activators. The homeodomain appears to be involved in specific DNA binding, but not in dimerization, nuclear localization, or transcriptional activation. As binding of either Ubx or abd-A appeared to activate target sequences in yeast, the repression of genes that occurs after Ubx or abd-A binding in Drosophila probably requires auxillary proteins.

♦ Establishment of the Drosophila embryonic body plan is directed by approximately 40 maternally or zygotically expressed genes. Many of these genes have been ordered into a defined hierarchical pattern. More recently, molecular genetic experiments have begun to identify specific cross- and autoregulatory gene interactions as part of the ntework that leads to the establishment of embryonic pattern. However, so far it had not been possible to unambiguously demonstrate whether the observed gene interactions were direct or indirect.

A good number of these developmental genes have been cloned and nearly two thirds of them encode proteins containing a conserved 60-amino acid sequence termed homeodomain. There is good evidence to suggest that the homeodomain mediates binding of the protein to the DNA, possibly to regulate transcription.

The articles summarized here provide direct evidence not only that homeobox proteins can bind to promoter elements of pattern-forming genes, but also that they can function directly as transcriptional regulators.

The first two articles (series A) characterize in depth the interaction *in vitro* between homeoproteins and the promoter elements to which they bind. Article A1 by Desplan et al. characterizes the interaction with *engrailed* (en) promoter elements, of fusion proteins containing wild type or mutant forms of the *en* and *fushi tarazu (ftz)* homeodomains. One major finding is the demonstration that the homeodomain alone is sufficient to confer DNA sequence specificity of protein binding. A second major finding is that homeodomains from two different proteins (*en* and *ftz*) can recognize the same DNA sequence. These data are important to the understanding of gene interactions (competition among different regulatory proteins for the same binding site) that lead to the establishment of the embryonic body pattern. A third major finding is that the same homeodomain can recognize apparently unrelated DNA sequences, a finding that has important implications for the understanding of gene networks. Article A2 by Beachy et al. shows that the *Ultrabithorax (Ubx)* protein binds to DNA sequences residing within its own and within the *Antennapedia (Antp)* promoters. This study lends support to the hypothesis that the *Ubx* gene product stimulates expression of itself and represses expression of *Antp* by direct interaction with the corresponding promoters.

The articles above (and also previously published work) demonstrate that homeoproteins can bind promoters of developmentally important genes. The articles in series B and C go a step further in making the additional important point that homeoproteins can not only bind to promoters, but they actually regulate transcription of the target genes. The general approach in these studies was to cotransfect cultured cells with a producer and a responder plasmid. The producer plasmid expresses the homeoprotein from a strong constitutive promoter. The responder plasmid contains a reporter gene driven by the promoter under study. Homeoprotein-dependent expression of the reporter gene demonstrates the transcriptional regulatory nature of the homeoprotein. Deletion mapping of homeoprotein coding regions and also of various promoter elements identified protein and DNA sequences involved in the regulatory interactions.

In most cases the findings confirmed the results predicted from previous molecular genetic experiments. As expected, the homeodomain proved to be essential for DNA sequence recognition, but not for gene activation. For instance, when the binding function of the homeodomain was provided by alternate protein domains, the protein could still function as an transcriptional regulator. It was also shown that transcription can be modulated by competitive binding of the same promoter site by two homeoproteins that have antagonistic effects on transcription. In other cases homeoprotein combinations show large synergistic transcriptional effects when bound to nearby sites on the same promoter. Drosophila homeoproteins were even able to replace mammalian homeodomain-containing transcription factors that recognize octamer motifs when assayed in cultured HeLa cells (Article C1).

In some cases the results did not conform to previous expectations. For instance, the *even-skipped* protein has been shown to play an essential role in activating *engrailed* as well as its own expression in the embryo, while it had no or even an inhibitory effect in a cultured cell assay. While there is no detectable regulation of *Ubx* transcription by *Antp* protein in the embryo, this protein has a stimulatory effect in the cell culture assay. These unexpected observations can be variously ascribed to the lack of appropriate flanking sequences in the test constructs, to abnormally high levels of effector proteins in the cells, or to the absence of coregulators that are normally present in the embryo but not in the cultured cells. Despite these limitations, the assay system has already proven to be very helpful. This approach promises to yield further insights into the intricate cross-regulatory developmental pathways that lead to the establishment of the Drosophila embryonic body plan. *Marcelo Jacobs-Lorena*

Pattern Formation in the Developing Eye of *Drosophila melanogaster* is Regulated by the Homeobox Gene, *rough*
R. Saint, B. Kalionis, T. J. Lockett, and A. Elizur
Nature, 334, 151—154, 1988   8-12

Genes containing homeobox regions play important roles in the regulation of Drosophila early embryogenesis. This report describes a Drosophila homeobox-containing gene that does not have a role in early embryogenesis.

A concensus homeobox oligonucleotide was used to screen a Drosophila genomic DNA library and detected a DNA fragment that hybridized to region 97D3 to 5 of the right arm of chromosome 3. Molecular analysis of mutations in this region identified the novel homeobox as part of the *rough* (ro) locus. *In situ* hybridization confirmed that expression of this gene was limited to the retina region of the eye-antennal disc. Expression was highest at the site of the morphogenetic furrow. No expression of ro was detected during early embryogenesis.

The function of the ro homeogene appears to be different from all other Drosophila homeogenes identified to date. Rough does not appear to be involved in the generation of regional identities of groups of cells in early embryogenesis. Instead, ro appears to be involved in the differentiation of individual photoreceptor cells in the developing eye imaginal disc.

rough, a Drosophila Homeobox Gene Required in Photoreceptors R2 and R5 for Inductive Interactions in the Developing Eye
A. Tomlinson, B. E. Kimmel, and G. M. Rubin
Cell, 55, 771—784, 1988   8-13

During the development of the Drosophila eye, the fate of ommatidial cells (Figure 8-13) appears to be directed by cues from adjacent cells. A phenotypic and molecular analysis of the ommatidial assembly gene, *rough* (ro), is presented.

When flies are mutant for ro, ommatidia are aberrant in cell number and positioning. Analysis at the light microscope level indicated that first R8 and then R3 and R4 differentiated normally in ro eyes (Figure 8-13). However, subsequent development was not normal in ro eyes. Somatic mutation was used to generate marked mosaic ro+/ro− ommatidia. Mosaic analysis revealed that R2 and R5 had to be ro+ for wild type development of the ommatidia, despite the fact that in mutant ommatidia R2 and R5 always developed normally.

A P-element-induced ro allele was used to clone the locus for sequence analysis. The DNA sequence revealed the presence of a homeobox within the ro gene.

Mosaic analysis demonstrated that the ro gene was required only in R2 and R5 for normal ommatidia development. However, in mutant ommatidia, these two photoreceptors always develop normally, while subsequent development is aberrant. Therefore, the ro defect may involve the failure of R2 and R5 to send appropriate developmental signals to neighboring cells. The presence of a homeobox in the ro gene suggests that it may be involved in transcriptional regulation of genes involved in this developmental signaling pathway.

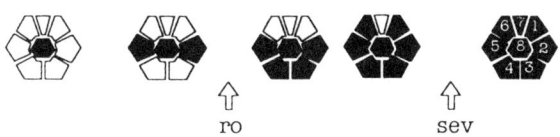

FIGURE 8-13. Schematic representation of the differentiation sequence of the photoreceptors in the ommatidium. The indivual stages displayed differ by approximately 3 to 4 h of developmental time. The *rough* mutant causes ommatidial assembly to go wrong between the onset of differentiation of the R2/R5 and R3/R4 cells, whereas *sevenless* causes a breakdown between the onset of differentiation of the R1/R6 and R7 cells. (From Tomlinson et al., *Cell,* 55, 771—784, 1988. ©Cell Press.)

♦ The Drosophila eye is composed of approximately 800 20-cell units called ommatidia, each ommatidium including 8 photoreceptor cells. Rather than cell lineage, cell-cell interactions are thought to play a key role in the differentiation of the photoreceptor cells. In other words, the pathway of differentiation of a given photoreceptor precursor is determined not by its ancestry, but by its neighbors. Mutations in the *rough* gene interfere with the differentiation of photoreceptor cells and lead to the formation of eyes containing disorganized ommatidia.

Saint et al. cloned the *rough* gene by virtue of its homeobox sequence. This is perhaps the first known Drosophila homeobox gene that is not expressed during embryonic development. The gene is expressed in the developing eye (as expected), but, surprisingly, also in embryonic brain. However, genetic analysis indicates no function for the gene outside the eye.

Tomlinson et al. undertook a careful analysis of the *rough* locus. Detailed morphological analysis indicated that *rough* acts relatively early in the pathway of ommatidium differentiation, probably at a time immediately following the differentiation of the first 3 photoreceptors (termed R8, R2, and R5). Significantly, analysis of genetic mosaics demonstrated that *rough* function is needed only in R2 and R5 for a normal ommatidium development. The results are entirely consistent with a model whereby expression of wild-type *rough* product is required only in cells R2 and R5. Interestingly, these two cells appear to develop normally in mutant ommatidia, suggesting that the role of the wild-type gene product is to promote the transmission of appropriate inductive signals to their neighbors. Because *rough* codes for a protein with a homeodomain, it is likely that the protein serves as a transcriptional regulator of other genes that produce the putative signal(s). The results contrast with the properties of *sevenless*, another well-characterized eye gene. As *rough*, *sevenless* also appears to be involved in cell communication, but unlike *rough*, it is involved in receiving rather than sending signals.

The exciting future prospects lie in understanding more precisely the nature of these signals and by what mechanism they act to direct cell differentiation. *Marcelo Jacobs-Lorena*

Differential Antero-Posterior Expression of Two Proteins Encoded by a Homeobox Gene in *Xenopus* and Mouse Embryos
G. Oliver, C. V. E. Wright, J. Hardwicke, and E. M. De Robertis
EMBO J., 7, 3199—3209, 1988   8-14

The *Xenopus laevis* homeobox gene 1 (XlHbox 1) was the first vertebrate homeobox to be isolated. It produces two transcripts from different promoters that generate products that differ by 82 amino acids. Antibodies specific for each predicted protein were generated and found to cross-react with the proteins from a mouse homeobox-containing gene with two promoters and two transcripts. The expression of these homeobox proteins was compared in these two species.

In the mouse embryo, the X1Hbox 1 proteins were localized in the central nervous system, the segmented mesoderm, and the visceral mesoderm of internal organs. The anterior border of expression in the mesoderm was more posterior than in the central nervous system. The short protein was expressed more anteriorly than the long protein. In Xenopus embryos the distribution of expression was similar. However, in Xenopus the anterior border of expression was the same in mesoderm and central nervous system.

Antibodies to Xenopus homeobox-containing proteins have been generated which allow expression to be compared between species during embryogenesis. These antibodies have been used to demonstrate that in both Xenopus and mouse the longer protein is expressed more posteriorly than the smaller homeobox protein. The two proteins appear to be under precise temporal and spatial regulation along the antero-posterior axis of the embryo. The difference in expression of these two proteins could have effects on embryonic cell determination.

A Gradient of Homeodomain Protein in Developing Forelimbs of Xenopus and Mouse Embryos
G. Oliver, C. V. E. Wright, J. Hardwicke, and E. M. De Robertis
Cell, 55, 1017—1024, 1988   8-15

Antibodies to the Xenopus homeobox-containing protein, X1Hbox 1, have been used to investigate the expression of this protein in the developing fore- and hindlimbs of frogs and mice.

In the forelimb bud mesoderm of both mouse and frog, there is an antero-posterior gradient of X1Hbox 1 expression, with highest levels of expression in the anterior region. However, there was no staining in hindlimb bud mesoderm. This represents an early molecular difference between forelimbs and hindlimbs. The innermost ectodermal layer stained for X1Hbox 1 protein throughout the forelimb and hindlimb buds. Therefore, expression of this protein in the ectoderm and mesoderm was not equivalent. In early Xenopus tadpoles, X1Hbox 1 protein was detected in the somtapleure in the region corresponding to the forelimb, but not hindlimb, presumptive field.

The expression of the homeodomain containing protein X1Hbox 1 has been analyzed in the developing limbs of mice and frogs. Differences in expression between hindlimbs and forelimbs and between ectoderm and mesoderm within the same limb were detected by antibody probes. The developing and regenerating limb system is useful in the study of pattern formation in vertebrates.

♦ An early and critical step in the development of embryos and tissues is the establishment and cellular interpretation of axial information, i.e., the direction from any given point of the anterior, posterior, dorsal, and ventral poles. Much progress has been made in explaining how this works in *Drosophila*, but in vertebrates the picture is considerably less clear. There are some clues, however, and the developing limb is one system with particular promise. A posterior-to-anterior gradient of retinoic acid, a demonstrated morphogen, exists in chick limb bud, and as these papers show, there is an oppositely oriented gradient in forelimb mesoderm of X1Hbox 1 expression. This homeobox pattern is similar in two distantly related vertebrates, frog and mouse. X1HBox 1 is expressed in earlier development in dorsal tissues, in a pattern that is also similar in these two organisms, although the alignments of its mesodermal and neural components differ somewhat. This conservation reinforces the belief that such expression is meaningfully related to the developmental organization of the relevant tissues. What is missing, of course, is a functional test of this hypothesis, but such experiments are becoming technically feasible and results are beginning to come in. See, for example, the papers by Altaba and Melton on Xhox3. *Thomas D. Sargent*

Bimodal and Graded Expression of the *Xenopus* Homeobox Gene *Xhox3* during Embryonic Development
A. R. I. Altaba and D. A. Melton
Development, 106, 173—183, 1989

A Drosophila *even skipped* homeobox was used to screen the *Xenopus* genome. This report describes the characterization of Xhox3, a novel frog homeobox-containing gene, that was isolated during this screen.

The Xhox3 mRNA was first detected at the midblastula transition and maximally expressed at the late gastrula and early neurula stages. During the gastrula and neurula stages, the highest level of expression occurred in the posterior mesoderm, and then it decreased along the anterior to posterior axis to form a gradient within the mesoderm. At the tailbud stage, Xhox3 was expressed in the central nervous system and the tail bud.

When embryos are treated with UV light, they become posteriorized. When teated with lithium ions, they become anteriorized. The expression opf Xhox3 was examined in these experimentally manipulated embryos. Anterior cell fate within the mesoderm was correlated with low levels of Xhox3 expression, while posterior cell fate within the mesoderm was correlated with high levels of expression of Xhox3.

These results suggest that the levels of expression of Xhox3 along the anterior to posterior axis of the Xenopus embryo are casually elated to mesodermal fate.

Involvement of the Xenopus Homeobox Gene Xhox3 in Pattern Formation along the Anterior-Posterior Axis
A. R. I. Altaba and D. A. Melton
Cell, 57, 317—326, 1989   8-17

During early Xenopus development, the Xhox3 gene is expressed in an anterior to posterior gradient within the mesoderm. To test the function of this gradient during development, Xhox3 mRNA was injected into the anterior end of the embryo, where Xhox3 levels are lowest, to obliterate the gradient.

Examination of injected embryos revealed that ectopic Xhox3 expression caused a graded series of axial defects (Table 8-17). In the most extreme cases head structures were completely missing. However, the

Table 8-17
Summary of Injection Results

A. Comparison of mRNAs Injected into Developing Embryos

mRNA	% Defective
Xhox3	70
Xhox3BSΔ	0
Xhox1A	0
Xβm	5

B. Dose-Dependent Effect of Xhox3 mRNA Injection

Concentration (μg/ml)	% Defective
50	58
100	75
200	78

C. Correlation of Localization of Defects and Site of Xhox3 mRNA Injection

Injection Site	% Defective
Equatorial region	70
Animal pole region	74
Vegetal pole region	47

Averages of defective embryos in anterior regions (%) were compiled from results obtained in several independent experiments. All head deficiencies as compared with control embryos were scored. More than 200 embryos were injected in each case. In (C) injections were at different sites in the early-cleavage embryo at the 2-, 4-, or 8-cell stage.

From Ruizi Altaba/Melton, Cell, 57, 317—326, 1989. ©Cell Press.)

timing and extent of anterior migration of mesoderm was normal in the injected embryos.

These results suggest that high levels of Xhox3 expression are incompatable with the development of anterior mesoderm. The level of Xhox3 expression in Xenopus axial mesoderm appears to be an important determinant of positional value along the anterior to posterior axis.

♦ Xhox3 expression has several interesting aspects: the gene turns on very early in frog development, it is expressed in multiple germ layers in the early embryo, and its RNA accumulates following a posterior-to-anterior gradient. If Xhox3 is involved in causing anterior-posterior patterning, then disturbing the expression pattern ought to affect the appearance of the embryo. The paper presents such results: synthetic Xhox3 mRNA injected into the fertilized egg or into early cleavage blastomeres survives long enough to obscure the endogenous Xhox3 mRNA gradient and clearly leads to the suppression of anterior development. The effect is not a homeotic transformation, however, in the sense that there does not appear to be a conversion of anterior to recognizably posterior structures. This head suppression is different from that caused by treatments that inhibit formation of the dorsalizing center, such as precleavage irradiation with UV light, in that the inward migration of dorsal mesoderm is not inhibited; gastrulation proceeds normally in RNA-injected embryos. Rather, the identity of cells in the anterior region seems to be altered from "anterior" to "something else". *Thomas D. Sargent*

Mix.1, a Homeobox mRNA Inducible by Mesoderm Inducers, Is Expressed Mostly in the Presumptive Endodermal Cells of Xenopus Embryos
F. M. Rosa
Cell, 57, 965—974, 1989

To examine Xenopus mesoderm induction, CDNAs corresponding to mRNAs synthesized in presumptive ectoderm in response to XTC-MIF were isolated. One of the cDNAs encoded a homeodomain containing protein, Mix.1.

Mix.1 synthesis was an immediate early response to XTC-MIF and other mesodermal inducing agents. Mix.1 mRNA was initially detected after midblastula transition, peaked at stage 10 and then decayed. By neurulation, Mix.1 was no longer detectable. In mid-blastula embryos, Mix.1 mRNA was uniformly expressed in the vegetal hemisphere, but was not expressed in the animal hemisphere (Figure 8-18A and B). At the late blastula stage, the marginal zone also expressed Mix.1 (Figures 5C and D). In early gastrula, expression was intense in the center and ventral side

Homeobox Genes   265

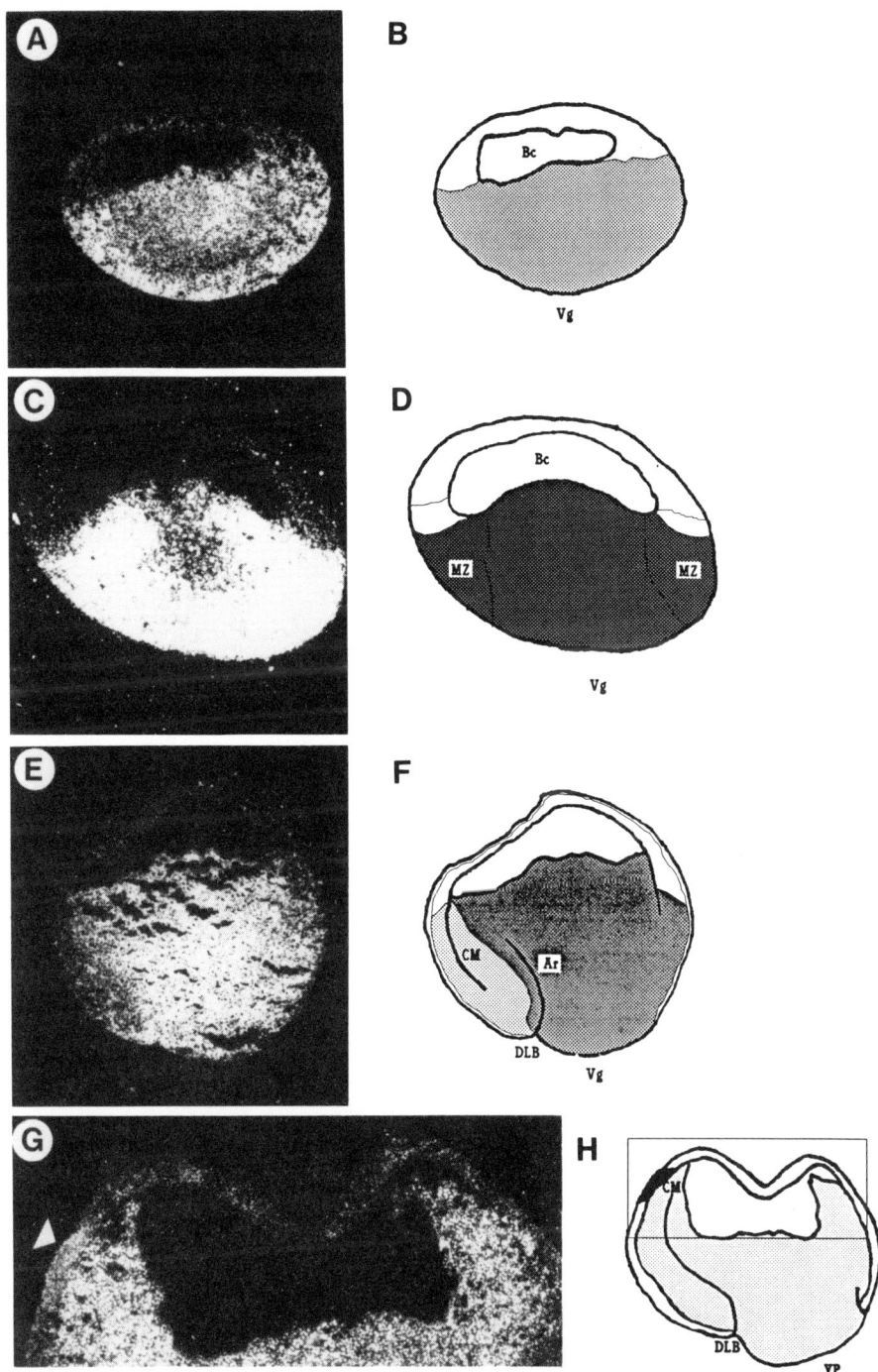

FIGURE 8-18. Localization of Mix.1 transcript in the early embryo by *in situ* hybridization. (From Rosa, *Cell,* 57, 965—974, 1989. ©Cell Press.)

of the vegetal hemisphere (Figure 8-18E and F). On the dorsal vegetal side, there was a gradient of decreasing expression from the center to the periphery. Therefore, Mix.1 appeared to be expressed in the prospective endoderm and also in the future mesoderm. During late gastrula, the ectodermal layers apposed to the tip of the involuting chordamesoderm/ anterior endoderm expressed Mix.1 (Figure 8-18G and H).

Synthesis of a Xenopus homeobox containing gene, Mix.1, is an immediate early response to mesoderm inducers. However, Mix.1 is expressed primarily in the presumptive endoderm.

♦ The discovery of Mix.1, an early-response, homeobox-containing gene activated by XTC-MIF in ectodermal cells, adds an intriguing new piece to the mesoderm induction puzzle. Mix.1 is turned on in ectoderm by a mesoderm inducer, so one would naively expect it to be normally expressed in mesoderm, but it isn't — Mix.1 (see Figure 8-18) is an endodermal marker with perhaps some expression in more vegetally derived mesoderm. Does this mean that XTC-MIF functions by turning part of the treated ectoderm into endoderm which then induces the remainder to become mesoderm? This possibility has been contemplated by several investigators in the field, and at least two opposing arguments can be brought to bear: first, ectoderm loses its competence to respond to endodermal tissue around the beginning of gastrulation (stage 10), a time when it is still responsive to XTC-MIF. Thus, the two-step model would require that ectodermal cells exposed to XTC-MIF at stage 10 be converted into endoderm and secrete mesodermal inducer almost instantaneously. This cannot be ruled out but seems unlikely. The other argument is that Mix.1 is not induced by concentrations of FGF that are effective in inducing mesoderm (muscle), so in such experiments mesoderm is forming without the induction of endoderm, at least judged by Mix.1 accumulation. *Thomas D. Sargent*

Duplicated Homeobox Genes in *Xenopus*
A. F. Fritz, K. W. Y. Cho, C. V. E. Wright, B. G. Jegalian, and
E. M. De Robertis
Dev. Biol., 131, 584—588, 1989   8-19

Two different X1Hbox two clones have been isolated from *X. laevis*. To determine whether these two clones represent two alleles of one gene or two distinct genes, *X. laevis/X. borealis* hybrids were created and analyzed. If the two clones represent alleles of one gene they should segregate independently in these crosses.

Southern blot analysis of 15 hybrids demonstrated that all carried a copy of each type of X1Hbox 2 clone. This lack of independent segregation indicates that the two clones represent independent loci.

The presence of two highly related X1Hbox 2 clones in *X. laevis* coupled with the tetraploid nature of the Xenopus genome, suggests that Xenopus laevis has two homeobox genes for each one present in other vertebrates. This gene duplication could be true for many other single copy genes in the Xenopus genome.

♦ The impressive evolutionary conservation of homeobox genes suggests that such genes serve important and necessary biological functions. In *Xenopus laevis*, which has a tetraploid genome, two types of clones have been identified for a particular homeobox gene, X1Hbox2. Such duplication has been reported for other *X. laevis* genes. In this paper, the authors show that the different types of clones are not due to the variations in four allelic copies of a single gene. By using interspecies hybrids and combining molecular genetic and classic genetic studies, they show that the different types of clones are due to two different genetic loci which presumably resulted from the duplication of the original gene. This study is consistent with recent studies on mammalian homeobox genes, which have shown some similar duplications. However, the possibility that all homeobox genes in *Xenopus* exist in duplicate has important, and potentially profound, evolutionary implications. *Joel M. Schindler*

The Pituitary-Specific Transcription Factor GHF-1 is a Homeobox-Containing Protein
M. Bodner, J.-L. Castrillo, L. E. Theill, T. Deerinck, M. Ellisman, and M. Karin
Cell, 55, 505—518, 1988   8-20

Growth hormone factor (GHF-1) is a pituitary-specific transcription factor that is involved in the expression of the growth hormone gene. The isolation and characterization of bovine and rat GHF-1 cDNA clones is described.

The GHF-1 protein was purified and a partial amino acid sequence was determined. Synthetic nucleotide probes based on this sequence were used to screen pituitary libraries to isolate GHF-1 cDNA clones. Sequence analysis detected a C-terminal sequence with significant homology to a homeobox concensus sequence. The GHF-1 homeobox region was expressed as a fusion protein in bacteria and functioned as the DNA binding region. Northern and immunofluorescence localization analysis demonstrated that expression of GHF-1 was restricted to somatotropic cells of the pituitary. These are the only cells which transcribe the growth hormone gene that is the GHF-1 transcriptional activation target.

GHF-1 is a homeobox containing gene which is directly responsible for controlling the transcription of a gene involved in the formation of the differentiated phenotype of somatotrope cells. Other mammalian homeobox containing genes may also be involved in the determination of cell type specificity through transcriptional activation of specific genes in specific cell types.

The B-Cell-Specific Oct-2 Protein Contains POU Box- and Homeobox-Type Domains
R. G. Clerc, L. M. Corcoran, J. H. LeBowitz, D. Baltimore, and P. A. Sharp
Genes Dev., 2, 1570—1581, 1988   8-21

Immunoglobin gene expression is restricted to B-cells and dependent on cis-acting elements. An octanucleotide sequence that is sufficient for B-cell specific expression is bound by a B-cell specific factor. A cDNA for this factor, oct-2, has been isolated from a human B-cell line cDNA library and characterized.

The oct-2 sequence possessed several conserved domains. A 60-amino acid region had 30% homology to a homeobox concensus. Fusion protein analysis indicated that this region functioned in DNA binding. The sequence also revealed a potential leucine zipper domain. A 75 nucleotide domain common to the NF-A2 family of proteins and termed a POU box was also detected in the oct-2 sequence. As several oct-2 cDNAs were isolated, oct-2 appears to encode multiple mRNAs with varying splicing patterns. The oct-2 cDNA also contained an overlapping reading frame of 278 residues.

The oct-2 protein is responsible for the specific transcription of immunoglobin genes in B cells. This gene has several conserved regions containing a homeobox concensus, a leucine zipper concensus and a POU-domain concensus sequence. It also has the potential to encode another overlapping gene with as yet unknown function.

A Tissue-Specific Transcription Factor Containing a Homeodomain Specifies a Pituitary Phenotype
H. A. Ingraham, R. Chen, H. J. Mangalam, H. P. Elsholtz, S. E. Flynn, C. R. Lin, D. M. Simmons, L. Swanson, and M. G. Rosenfeld
Cell, 55, 519—529, 1988   8-22

Cell-specific activation of the rat prolactin and growth hormone genes is controlled by multiple *cis*-active elements. These elements appear to bind a pituitary-specific positive transcription factor, Pit-1.

These *cis*-active DNA elements were used for DNA affinity chromatography to purify the binding protein. Expression vectors were then used to screen for the cDNA by affinity chromatography. The gene encodes a 33-

kDa protein with homology at the carboxy-terminus to homeodomains. Message for pit-1 was detected exclusively in the anterior pituitary in the somatotroph and lactotroph cell types, which synthesize growth hormone and prolactin. When pit-1 was expressed in HeLa cells, prolactin and growth hormone fusion genes were selectively activated. This suggests that pit-1 is sufficient to activate these genes in any cell type.

The protein required for the cell-specific activation of prolactin and growth hormone in the anterior pituitary has been identified. This protein is a member of the homeobox family. Other homeobox containing proteins may be involved in establishing the expression patterns of other differentiated mammalian cells.

A Human Protein Specific for the Immunoglobulin Octamer DNA Motif Contains a Functional Homeobox Domain
H.-S. Ko, P. Fast, W. McBride, and L. M. Staudt
Cell, 55, 135—144, 1988

The homeobox is a highly conserved 60 amino acid sequence found in several Drosophila proteins which are involved in the control of development. This sequence is involved in sequence-specific DNA binding. A human lymphoid-specific protein, oct-2, which binds to the immunoglobulin octamer regulatory motif, was cloned and sequenced. This protein also contains a homeobox.

A cDNA had been previously isolated from a lymphoid cell cDNA library and appeared to encode the lymphoid-specific immunoglobulin transcriptional activator. The cDNA was sequenced and was 33% homologous to Drosophila homeodomains (Table 8-23) over a 60 amino acid region. The oct-2 gene was mapped to chromosome 19, separate from all previously identified mammalian homeobox containing genes. Site-directed mutagenesis of the homeobox domain eliminated specific DNA binding.

To test whether homeobox containing genes have related DNA recognition and binding properties, oct-2 binding studies with oligonucleotides derived from the yeast mating locus, a Drosophila homeobox binding site, immunoglobulin binding sites and mutagenized sites were carried out. Mobility shift assays demonstrated that the best binding was to the wild type immunoglobulin octamer sequence, but some binding occurred between oct-2 and the yeast and Drosophila sequences.

The mammalian homeobox gene, oct-2, appears to encode the lymphoid-restricted immunoglobulin octamer binding protein. This protein can also bind yeast mating type sequences and Drosophila homeobox binding sites, indicating that all homeobox containing proteins may have closely related DNA binding characteristics.

Table 8-23
Percentage of Amino Acid Identity of the oct-2 Homeobox with other Homeoboxes

	Homeobox	% Amino Acid Identity	
		oct-2	c1
Human	oct-2	100	33
Human	c1	33	100
Human	c8	33	90
Human	c13	32	82
Human	Hu1	33	88
Human	Hu2	30	93
Mouse	Hox 1.1	33	93
Mouse	Hox 1.2	33	92
Mouse	Hox 1.3	33	88
Mouse	Hox 1.4	30	80
Mouse	Hox 1.5	28	72
Mouse	Hox 3	30	80
Mouse	En-1	27	52
Mouse	En-2	25	52
Sea urchin	HB1	28	87
Frog	MM3	33	97
Frog	AC1	33	90
Frog	Xhox-1	32	80
Drosophila	Antp	33	98
Drosophila	ftz	30	83
Drosophila	Ubx	32	88
Drosophila	Scr	33	90
Drosophila	Dfd	32	82
Drosophila	Eve	33	50
Drosophila	z1	27	62
Drosophila	z2	25	78
Drosophila	iab-7	33	60
Drosophila	cad	28	55
Drosophila	BSH4	28	38
Drosophila	BSH9	27	33
Drosophila	prd	27	37
Drosophila	bcd	25	42
Drosophila	en	23	50
Drosophila	inv	23	48
Honeybee	E60	23	50
Honeybee	E30	23	50
Yeast	MATa1 unspliced	33	23
Yeast	MATa1 spliced	32	25
Yeast	MATα2	22	28

References: c1 and c8: Simeone et al., 1987; c13: Mavilio et al., 1986; Hu1: Hauser et al., 1985; Hu2: Levine et al., 1984; Hox 1.1 Hox 1.2, Hox 1.4, Hox 1.5, Hox 3, and En-1: Martin, 1987 (references therein); Hox 1.3: Odenwald et al., 1987; En-2: Joyner and Martin, 1987; HB1: Dolecki et al., 1986; MM3: Muller et al., 1984; AC1: Carrasco et al., 1984; Xhox-1: Harvey et al., 1986; Antp, ftz, Ubx, Scr, Dfd, iab-7, cad, prd, bcd, en, and inv: Frigerio et al., 1986 (references therein); Eve: Frasch et al., 1987; z1 and z2: Rushlow et al., 1987; BSH4 and BSH9: Baumgartner et al., 1987; E60 and E30: Gehring, 1987; MATa1 and MATα2: Nasmyth et al., 1980.

From Ko et al., Cell, 135—144, 1988. ©Cell Press.)

A Cloned Octamer Transcription Factor Stimulates Transcription from Lymphoid-Specific Promoters in Non-B Cells
M. M. Muller, S. Ruppert, W. Schaffner, and P. Matthias
Nature, 336, 544—551, 1988   8-24

B-cell-specific transcription of immunoglobulin genes depends on *cis*-acting octamer motif elements. The isolation and characterization of a cDNA for a protein (OTF-2) which binds this octamer motif is reported.

A lymphoid expression library was screened for proteins that could bind oligonucleotides that contained the octamer motif. The binding ability of those clones isolated was confirmed by gel mobility shift assays and methylation interference assays with the octamer oligonucleotide. The distribution of this mRNA was assessed and was lymphoid specific. Sequencing of the cDNA revealed homology to Drosophila homeobox domains. Expression of the cloned sequence in HeLa cells was sufficient to activate transcription from immunoglobulin promoters.

A lymphocyte-specific transcription factor that binds to the octamer motif of immunoglobulin genes and activates transcription has been cloned. The OTF-2 sequence has homology to homeodomains.

A Human Lymphoid-Specific Transcription Factor that Activates Immunoglobulin Genes is a Homeobox Protein
C. Scheidereit, J. A. Cromlish, T. Gerster, K. Kawakami,
C.-G. Balmaceda, R. A. Curie, and R. G. Roeder
Nature, 336, 551—557, 1988   8-25

A conserved octamer sequence motif appears to be the chief determinant of B-cell specific immunoglobulin promoter expression. Octamer elements are recognized by the B-cell specific octamer-binding transcription factor, OTF-2. To understand the structure and function of this protein, cDNAs for this protein have been cloned.

Affinity chromatography was used to purify this protein. Purified OTF-2 was sequenced and oligonucleotides corresponding to this sequence were used to screen a cDNA library from a human Burkitt lymphoma line. Overlapping cDNA clones were sequenced to reveal a 1389 base pair open reading frame encoding a 463 amino acid protein of 49.5K. The region from amino acids 281 to 340 contained a homeodomain, while the region between amino acids 373 and 394 contained a potential leucine zipper. The OTF-2 message was expressed exclusively in B-cells.

Gel mobility shift assays with OTF-2 and octamer-containing oligonucleotides incubated with varying competitors demonstrated octamer-specific binding. Assays of truncated OTF-2 peptides indicated that the homeodomain was important for specific DNA binding. A Drosophila homeodomain oligonucleotide competed with octamer for OTF-2 binding, and OTF-2 specifically bound this oligonucleotide.

The lymphoid-specific transcription factor OTF-2 was cloned and characterized. It contained a homeobox and was capable of interacting with other homeobox binding sites in addition to the immunoglobulin octamer binding site.

The Ubiquitous Octamer-Binding Protein Oct-1 Contains a POU Domain with a Homeobox Subdomain
R. A. Sturm, G. Das, and W. Herr
Genes Dev., 2, 1582—1599, 1988   8-26

The octamer motif ATGCAAAT is recognized by two mammalian transcription factors, a lymphoid-specific factor, oct-2, and a ubiquitous factor, oct-1. Octamer-specific binding was assayed by the *in situ* filter method to detect oct-1 cDNA in an expression library.

One clone expressed a fusion protein with binding properties that were the same as HeLa oct-1. Oct-1 mRNA was detected in all human and mouse lines assayed with this clone. Polyclonal antiserum was raised to the cot-1 fusion protein and detected both oct-1 and oct-2. Deletion analysis of oct-1 revealed that DNA binding activity resides in a central highly charged domain of 160 amino acids. This POU domain was conserved between oct-1 and oct-2 and has also been detected in the pituitary transcription ofactor pit-1 and the *C. elegans* unc-86 cell lineage protein. This domain contains two conserved subregions, a POU-related homeobox and a POU-specific box.

The oct-1 protein is so far unique among homeobox containing proteins in that is is expressed ubiquitously. This suggests that the homeobox is a DNA-binding motif that is used by both general and cell-specific transcription factors to regulate gene expression by sequence-specific DNA binding.

The POU Domain Is a Bipartite DNA-Binding Structure
R. A. Sturm and W. Herr
Nature, 336, 601—604, 1988   8-27

The POU domain is a highly charged approximately 160 amino acid region found within three mammalian transcription factors and the nematode unc-86 protein. The POU domain is comprised of two subdomains, a C-terminal homeodomain and an N-terminal POU-specific region, which are separated by a short nonconserved linker. The POU domain is sufficient for sequence specific DNA binding. Homeodomains are involved in DNA binding and have been shown to be important in the specific DNA binding of the POU-containing proteins. Therefore, the function of the POU-specific region of the POU domain was investigated.

A gel retardation assay was used to monitor the effect of progressive deletion of the POU-containing protein, oct-1, on sequence-specific DNA binding. Deletion of either the homeobox or thePOU-specific box eliminated specific binding. To monitor the effect of more subtle mutations on specific binding, triple alanine substitutions were created at positions that are identical in all four POU proteins. These results also demonstrated that both the homeodomain and the POU-specific region were required for specific binding. However, linker substitutions had no effect on specific binding.

Therefore, the POU-domain can be described as a bipartite sequence-recognition-and-binding domain with a substructure composed of one homeodomain and one POU-specific domain. Thus, the POU domain represents a new class of DNA-binding structure.

Expression of a Large Family of POU-Domain Regulatory Genes in Mammalian Brain Development
X. He, M. N. Treacy, D. M. Simmons, H. A. Ingraham, L. W. Swanson, and M. G. Rosenfield
Nature, 340, 35—42, 1989   8-28

The conserved POU-domain is found in the mammalian transcription factors oct-1, oct-2 and pit-1, as well as the product of the *C. elegans* unc-86 gene. This report describes the identification of four new mammalian POU-domain genes, Brain-1 (Brn-1), Brn-2, Bn-3 and Testes-1 (Tst-1).

Degenerate oligonucleotides representing all possible codons for two 9-amino acid regions conserved among the four POU-proteins were used to prime the PCR with cDNA complementary to human and rat brain and rat testes mRNA serving as templates. Four novel POU-containing proteins were isolated: Brn-1, -2, -3 and Tst-1. All of these proteins were highly related to the previous members of the POU-domain family.

Bn-1, -2, -3 and Tst-1 were all expressed in the brain with varying patterns of expression. Brn-1 and -2 were the most widely expressed within the brain. Brn-3 was the only one of these proteins expressed in sensory ganglion cells. All four of these proteins was expressed in the neural tube and in the ventricular proliferative zone which gives rise to the CNS. The pattern of expression during development tended to reflect the adult loci of expression.

Pit-1 transcripts were detected in the neural plate and tube on embryonic days 10 to 13. On day 16 expression of pit-1 reappeared in the pituitary. Oct-2 was expressed in the neural tube and in certain structures of the adult brain. Oct-1 was also expressed in the neural tube, but at low levels. In the adult brain its expression was highly restricted. Oct-1 and Tst-1 were highly expressed in the egg cylinder.

The identification of new POU-domain genes confirms the idea that there is a large family of POU-domain proteins. All of the mammalian proteins have distinct patterns of expression during the development of the brain, suggesting that they play a role in the development of neuronal phenotypes.

♦ Three mammalian transcription factors (GHF-1/Pit-1, Oct-1, and Oct-2), four newly identified genes (*Brn-1, -2, -3,* and *Tst-1*) that are expressed in the nervous system and, in some cases, a limited number of other organs, and the product of the *unc-86* gene of *C. elegans* share a 160 amino acid region of sequence similarity called the POU domain. Located within this domain are a homeobox-related subdomain as well as a POU-specific subdomain. All the known POU-containing genes are expressed during brain development with widespread expression in the neural tube and different patterns of subsequent restriction. In the adult, GHF-1/Pit-1 is a pituitary-specific transcription factor that regulates expression of prolactin and growth hormone. Oct-1 and -2 are differentially expressed human promoter-binding proteins that recognize the same octameric sequence which is a *cis*-regulatory element of ubiquitously expressed (small nuclear RNA and histone H2B) and cell-specific immuno-globulin genes. Oct-1 and -2 were also expressed in adult brain. Loss of *unc-86* function results in altered neuroblast lineages and failure to differentiate at least one class of neuron. The carboxy terminus of the POU domain is distantly related to the *Drosophila Antp* homeobox. A second POU-specific subdomain, a 75- to 82 amino acid region that is highly conserved among the different proteins, is found just upstream from the homeo-subdomain. Based on sequence conservation, the POU-domain family can be divided into four subclasses: POU-1 (GHF-1/Pit-1), POU-II (Oct-1, -2), POU-III (*Brn-1, -2, Tst-1*), and POU-IV (*Brn-3, unc-86*). Questions that need to be addressed in future experiments center on determining whether the two POU subregions can be functionally dissociated. In addition, it is not known whether all the POU-containing proteins control cell fates during development as does the *unc-86* product or whether *unc-86* encodes a transcription factor as do the Pit and Oct proteins. *Terry Magnuson*

The Murine and Drosophila Homeobox Gene Complexes Have Common Features of Organization and Expression
A. Graham, N. Papalopulu, and R. Krumlauf

Drosophila genes involved in segment identity contain a conserved sequence element, the homeobox. Many of these genes are clustered into two adjacent complexes. The physical order of genes within the complex is identical to the order in which these genes are expessed along the anteroposterior embryo axis. The murine genome contains many homeogenes, which are organized into four clusters. One of these clusters, Hox-2, was analyzed for the relationship between sequence, organization and pattern of expression in the mouse embryo.

Gene-specific probes were generated for each member of the Hox-2 gene cluster. *In situ* hybridization demonstrated that every member of the cluster was expressed during embryogenesis. All members were expressed in embryonic spinal cord, with overlapping, but distinct, domains of expression for each gene. The position of a gene within the Hox-2 cluster reflected its relative area of expression along the embryo central nervous system anteroposterior axis.

In each of the four identified murine homeobox clusters the relative position of each subfamily member and the physical order and gene spacing is similar. Therefore, the Hox gene clusters can be aligned as an evolutionarily related complex (Figure 8-29).

In both mouse and fly, anteroposterior expression in the CNS correlates with the relative position of the genes within the complexes. The homeobox sequence of the most posteriorly expressed Hox-2 member, Hox-2.5, is most closely related to the most posteriorly expressed member of BX-C, Abd-B. This correlation is true for other genes in both mouse and Drosophila clusters. Therefore, there is a relationship between relative position within the gene cluster, sequence identity, and domain of expression along the anteroposterior axis for both mouse and fly homeobox

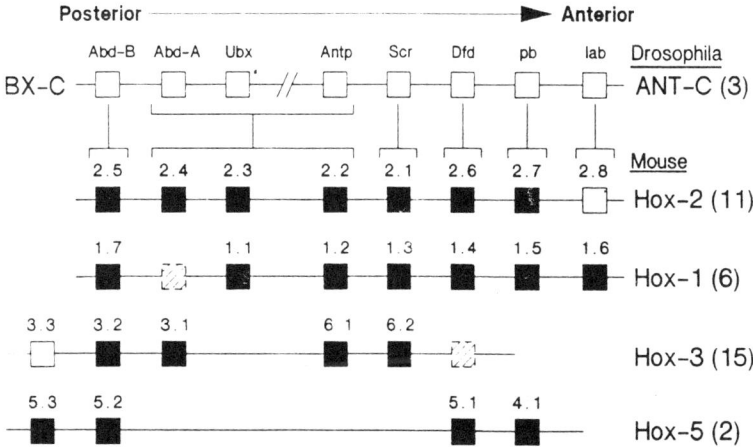

FIGURE 8-29. Diagrammatic representation of the relationships between Drosophila and murine homeobox gene clusters. (From Graham et al., *Cell,* 57, 367—378, 1989. ©Cell Press.)

gene clusters. This indicates a common ancestor for all of these complexes.

The Structural and Functional Organization of the Murine HOX Gene Family Resembles that of *Drosophila* Homeotic Genes
D. Duboule and P. Dolle
EMBO J., 8, 1497—1505, 1989   8-30

This report describes the cloning and characterization of the fourth murine homeogene complex, HOX-5.

Overlapping cosmid clones from the HOX-5 locus were screened by Southern blot analysis with a battery of homeobox probes. Two additional genes, Hox 5.2 and Hox .5.3, were detected at this locus. To test whether position within the complex reflects domain of expression along the anteroposterior axis, the expression of these two new genes during embryogenesis was examined. These two genes are at the 5' end of the complex and this was reflected by very posterior expression domains within the CNS. The expression of Hox 1.6, which is at the 3' end of the HOX-1 locus was also examined. The expression of this gene was at the anterior of the CNS.

The murine HOX-5 gene complex was characterized. Expression patterns from htis and other murine homeobox loci confirm that relative position within the complex reflects domains of expression along the anteroposterior axis. An alignment of Drosophila and vertebrate complexes with respective anterior expression boundaries along the CNS of the fly and mouse is proposed (Figure 8-30). An ancestral cluster of homeobox genes may have been present in the common ancestor of all these organisms.

♦During the last 2 to 3 years numerous reports have been published regarding the identification and characterization of expression of mouse homeobox-containing genes during development. It now appears that there are at least four homeobox-gene complexes in the mouse: *Hox-1* (chromosome 6), *Hox-2* (chromosome 11), *Hox-3* (chromosome 15), and *Hox-5* (chromosome 2). Data from the above listed reports as well as others cited by the authors indicate that members of each of these complexes are evolutionarily related not only to the prototypic *Drosophila Antp* homeodomain but also to one another. For example, *Hox-2.5, 1.7, 3.2, 5.2* and the *Drosophila Adb-B* genes show substantial identity in several regions of their predicted protein sequences, including the homeodomains, possibly forming a subfamily arising through a process of duplication and divergence of a common ancient ancestral gene. These investigators have divided the mouse homeobox genes into eight such

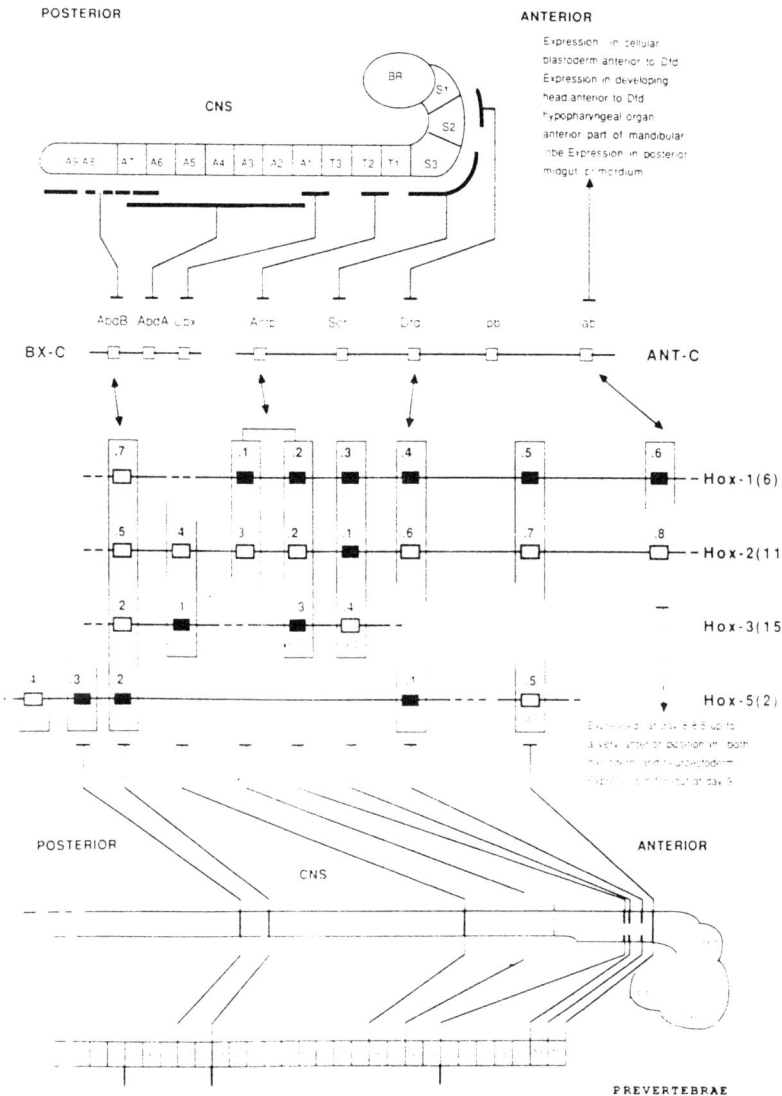

FIGURE 8-30. Schematic representation of the possible correlation between the Drosophila homeotic gene complexes and the murine (vertebrate) HOX gene netwrok. The upper part represents the domains of Drosophila homeotic genes, members of either BX-C or ANT-C, in the embryonic CNS. Other nonhomeotic genes (zen, bic, ftz) located within the ANT-C are not indicated for clarity. The central part represents the HOX complexes with, in filled boxes, genes which have been studies by comparative *in situ* hybridization experiments and whose expression domains (or, at least, the position of their anterior boundaries) have been defined. These boundaries are probably representative of those of all genes belonging to the same subfamily (indicated by the vertical open or closed rectangles). Thus, genes within the same rectangle are expected to have comparable expression boundaries or, at least, comparable antero-posterior relative positions within their respective complexes. The bottom part schematically represents the antero-posterior boundaries of expression of these genes along the fetal CNS and prevertebral column. In both structures a unique boundary is given for each gene subfamily without considering slight variations which might occur within a particular subfamily. (From Duboule/Dollé, *EMBO J.*, 8(5), 1497—1505, 1989. ©IRL Press.)

subfamilies. The putative role of mouse homeobox genes as regulators involved in establishing positional information is based solely on observed patterns of overlapping expression in the developing embryo. These genes are expressed primarily in derivatives of ectoderm, including the central and peripheral nervous system, and in mesoderm (lung, somites, gut, and kidney) but generally not in endoderm. The above reports also show that the phsycial order of the homeobox genes within the *Hox-2* and possibly the *Hox-1* cluster is identical to the order in which these genes are expressed in the anteroposterior axis. These results are similar to the *Drosophila* homeotic genes where the relative position of a gene within the Antennapedia and Bithorax cluster correlates with its domain of expression as well as its effects on specification of structures along the anteroposterior axis. This raises the question regarding the significance of conservation of expression and organization of these genes in mouse and flies. The authors suggest that it may be a consequence of *cis*-acting regulatory elements that are distributed along a cluster with each gene controlled by multiple elements. The normal pattern of gene expression would then result from additive effects of these elements. *Terry Magnuson*

Morphogenesis and Pattern Formation 9

INTRODUCTION

Biological hierarchy progresses from cells to tissues and from tissues to organs, ultimately culminating in the morphogenesis of a complete organism. The blueprint for the body plan of that organism can be established long before its overt expression. Pattern formation requires the coordinated integration of many types of signals in order to ensure the proper morphogenetic outcome.

Because pattern formation is the aggregate result of many different levels of control, abnomal morphogenesis could result, for example, from inappropriate cell movements, cell interactions, or cell determinations. A genetic defect whose phenotype was originally identified because of a morphological aberration could be the result of a mutation that affects an earlier developmental process, but ultimately expresses itself as a defect in morphogenesis.

As additional information about pattern formation accumulates, it is likely that we will begin to discover how biological processes that express themselves earlier in development can directly effect morphogenesis. Our awareness of the interrelationships between separate biological processes will continue to increase. Those events which are most important for normal morphogenesis to occur will become more obvious and the molecules responsible for those events will be identified. As a result, it should be possible to identify specific genes or gene families that have essential roles in morphogenesis, despite being directly relevant to an earlier developmental event.

A New Anatomy of the Prestalk Zone in *Dictyostelium*
K. A. Jermyn, K. T. I. Duffy, and J. G. Williams
Nature, 340, 144—146, 1989   9-1

The front prestalk zone of the Dictyostelium migratory slug has been considered to be composed of homogenous cells. Expression of the specific stalk-cell inducer, DIF-1, triggers expression of the extracellular matrix proteins, pDd56 and pDd63, in this zone. To examine their expression, the promoter of these two genes were fused to a reporter gene whose product could be detected immunologically.

Immunological staining of slug whole mounts revealed that cells expressing the pDd63 fusion gene, PstA cells, were restricted to the front 10% of slug length, approximately half the length of the prestalk cell region. Cells expressing the pDd56 fusion gene, PstB cells, were contained in a funnel-shaped central zone (Figure 9-1a and b). These fusion proteins accumulated for up to three days of slug migration. Anterior prestalk cells which did not express either marker were termed Pst0 cells (Figure 9-1c).

Multiple classes of anterior prestalk cells have been observed in the Dictyostelium migratory slug. These cell types are spatially separated and appear to differ in expression of extracellular matrix proteins. Dictyostelium pattern formation appears to be more complex than previously believed.

♦ The classical view of the multicellular assembly, or slug, of *Dictyosteli-*

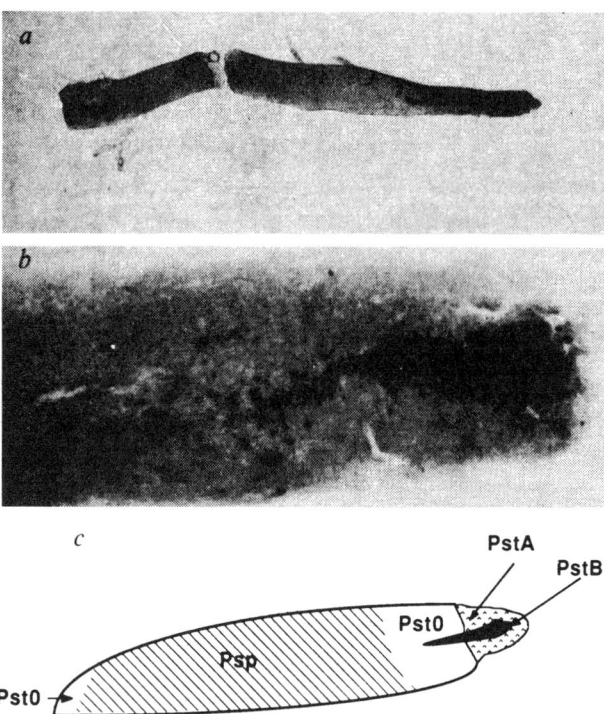

FIGURE 9-1. Analysis of the distribution within the slug of cells expressing the pDd56-Tag-cat fusion gene. a, A representative whole mount slug; b, the anterior regions of a representative whole mount slug; c, a diagrammatic representation of a slug showing cells expressing pDd63 (pstA), pDd56 (pstB), neither marker (pst0), and prespore (psp) cells. (From Jermyn et al., Nature, 340, 144—146, 1989. With permission.)

um was that it has two distinct cell types. The front 20% were the prestalk cells while the remaining 80% were prestalk. Some "anteriorlike" cells were dispersed throughout the posterior prespore zone and the most posterior basal disk cells were also prestalk. The data in this paper now show that the situation is considerably more complex and that at least three anatomical zones exist in the anterior zone. The authors have made two stable transformed cell lines that each express a different pretalk DIF-inducible gene that encodes an extracellular matrix protein. The genes are fused to a simian virus 40 (SV40) nuclear localization signal and antibodies to the SV40 are used to stain the nuclei of cells expressing these genes.

One of the genes pDd63 is expressed only in the anterior 10% of the slug although the anterior 20% of the same slug stains with neutral red which defines the entire prestalk zone. The second gene, pDd56, is expressed in an even more restricted area as a central funnel running through the entire prestalk zone. It is unclear whether the most anterior cells expressing pDd56 are also expressing pDd63. The remaining prestalk cells express neither of the genes. Thus the data indicate that there are at least three different prestalk zones defined by the differential expression of DIF-inducible prestalk genes. Previously published data indicate that only 90% of DIF-induced prestalk cells require cAMP to differentiate into stalk cells which also suggests the existence of multiple forms of prestalk cells. Overall, this report indicates a new complexity for pattern formation in *Dictyostelium* and the mechanisms by which diffusible morphogens act to define this pattern. *Stephen Alexander*

Posterior Pattern Formation in *C. elegans* Involves Position-Specific Expression of a Gene Containing a Homeobox
M. Costa, M. Weir, A. Coulson, J. Sulston, and C. Kenyon
Cell, 55, 747—756, 1988
9-2

In *C. elegans* the mab-5 gene controls developmental decisions and cell migration in the posterior region. To understand the activity of the mab-5 product, the gene was cloned by transposan tagging, sequenced, and its expression characterized.

The sequence of an mab-5 cDNA was determined and revealed a homeobox of the antennapedia class near the predicted carboxy terminus. To localize the expression of this gene, *in situ* hybirdization was performed on the first several larval stages. Expression of the mab-5 mRNA was detected only in the posterior region in cells that require mab-5 activity.

The mab-5 gene, required for normal posterior development in *C. elegans,* has been cloned and characterized. The sequence of mab-5 revealed a homeobox, which suggests that this gene exerts its effects through regulation of gene expression in the posterior region, where it is both expressed and active. This demonstrates that homeogenes influence pattern formation in organisms other than Drosophila.

♦ Groups of homologous cells are generated throughout the development of *C. elegans*. Many of these sets of homologs diversify, so that by the adult stage the body pattern is quite heterogeneous. *mab-5* is a gene that was previously shown to be involved in the diversification of sets of ectodermal (P,V, and Q) blast cells and even some mesodermal (M cell) cells in the posterior region of the body. In *mab-5* mutant, cells in the posterior of the animal are prevented from adopting fates that normally give this region of the body its identity, these cells instead adopt the fates of their more anterior homologs (Kenyon, *Cell,* 46, 477—487, 1986). *mab-5* acts cell autonomously, suggesting that it is required for initiation of specific modes of differentiation in response to posterior-localized positional information. The molecular characterization of *mab-5* reported here is consistent with this role for *mab-5*, both in terms of the spatial localization of its RNA product, which accumulates only in the posterior body region, and in terms of its primary sequence, which contains a homeobox similar to that of the *Drosophila antennapedia* gene. Two genes, *lin-22* (Horvitz et al., *Cold Spring Harbor Symp. Quant. Biol.,* 48, 453—463, 1983) and *pal-1*, have been shown to affect the spatial localization of *mab-5*.

These results are important because they show that homeobox-containing genes can play an important role in pattern formation in organisms other than *Drosophila*. The homology to *Antennapedia* is most intriguing since mutations in *mab-5* and *Antennapedia* each cause cells located together to undergo homeotic transformation and each is subject to spatial regulation at the RNA level. Yet there are great differences in the early development of *Drosophila* and *C. elegans*, and in the patterned expression of homeobox genes. In *Drosophila* spatial patterns of RNA expression can occur in the syncytial blastoderm, while nematodes do not have a syncytial development. This suggests that "... the experimental accessibility of *C. elegans* may make it possible to learn in detail how similar pattern forming mechanisms can be modified by evolution to produce strikingly different animal morphologies." *Joseph G. Culotti*

The *sqt-1* Gene of *C. elegans* Encodes a Collagen Critical for Organismal Morphogenesis
J. M. Kramer, J. J. Johnson, R. S. Edgar, C. Basch, and S. Roberts
Cell, 55, 555—565, 1988

Mutations of the *C. elegans* sqt-1 gene can lengthen, shorten or twist the body. The isolation, sequence and characterization of the sqt-1 gene is described.

Physical mapping of chromosomal deficiencies combined with transposan mutangenesis was used to identify and isolate the sqt-1 gene. The gene was sequenced and encoded a 32 kd member of the *C. elegans* collagen multigene family.

The transposan insertion mutant originally used to identify the sqt-1 gene was a reversion to wild type phenotype in a left roller mutant background. This revertant was analyzed for sqt-1 expression and had no detectable sqt-1 transcripts. Therefore, a sqt-1 null allele has a wild type phenotype.

The sqt-1 gene has been cloned and encodes a collagen. Therefore, collagen mutations can have dramatic effects on morphology in *C. elegans*. As seen in other multigene families, null mutations of sqt-1 are wild type. The morphological defects seen in sqt-1 mutants are probably due to incorporation of defective sqt-1 collagen into the cuticle.

---

dpy-13: **A Nematode Collagen Gene that Affects Body Shape**
N. von Mende, D. M. Bird, P. S. Albert, and D. L. Riddle
Cell, 55, 567—576, 1988   9-4

Mutations in the *C. elegans* dumpy-13 (dpy-13) gene result in a short chunky body. This gene was cloned and sequenced.

The dpy-13 gene was cloned by transposan tagging and chromosomal walking. The transcript was 1.2 kb and moderately abundant. Sequence analysis revealed that dpy-13 is a member of the *C. elegans* collagen multigene family. The predicted polypeptide contains 302 amino acids. The region between amino acids 56 and 103 is unique.

The dpy-13 gene was cloned and sequenced and is a member of the collagen multigene family. Therefore, specific body shape defects are associated with defective collagen, which is the primary constituent of the *C. elegans* cuticle.

♦ Unusual phenotypes that display morphological abnormalities should ultimately be explained by specific genotypic aberrations. In the case of pattern formation mutants that effect body plan, the likely candidates for genetic mutations are structural proteins, including collagens. In the nematode, collagen genes comprise a large gene family consisting of between 50 and 150 individual members. The specific papers highlighted identify two distinct genes in this gene family and demonstrate that aberrations in either one of them can result in the presentation of an abnormal body plan. These observations are important because they directly correlate the cause and effect of a particular aberrant gene or gene product with an abnormal phenotype. In addition, they identify the function of at least two members of this large multigene family and suggest that the functions of these related genes may both overlap and be distinct depending on certain structural motifs. *Joel M. Schindler*

Region-Specific Alleles of the *Drosophila* Segmentation Gene *hairy*
K. Howard, P. Ingham, and C. Rushlow
Genes Dev., 2, 1037—1046, 1988   9-5

During the formation of the Drosophila segmental pattern, the first evidence of periodicity is the expression of the primary pair-rule genes, *hairy, runt,* and *even-skipped.* These primary pair-rule genes may be responding to the cues of a prepattern established by maternal genes and gap genes. The signals in different parts of the blastoderm should be different; therefore, it should be possible to isolate pair-rule gene regulatory mutations that affect the response to some cues but not others. This paper describes four such hairy mutations: h-k1, h-m8, h-m7, and h-m3.

At the blastoderm stage of Drosophila development each of these mutants expressed part of the normal *hairy* pattern. There were corresponding gaps in ftz expression and deletions in the larval cuticle in the areas where hairy expression was missing. This indicates that hairy functions where it is produced. The lesions associated with these alleles were all mapped to the 5′ regulatory region of the hairy gene, with h-k1, the strongest allele, retaining the least amount of the 5′ region intact.

This suggests a model in which a prepattern is formed through the action of the maternal coordinate genes and the gap genes. This prepattern is decoded by primary pair-rule genes, such as hairy. The decoding appears to occur at the level of transcription in the case of the hairy gene. Several distinct 5′ regulatory elements appear to be involved in the control of hairy transcription in response to environmental cues in the blastoderm.

Early and Late Periodic Patterns of *even skipped* Expression Are Controlled by Distinct Regulatory Elements that Respond to Different Spatial Cues
T. Goto, P. Macdonald, and T. Maniatis
Cell, 57, 413—422, 1989   9-6

All Drosophila pair-rule genes are expressed as a characteristic pattern of seven transverse stripes. The expression of the primary pair-rule gene, even skipped (eve), was analyzed to reveal distinct regulatory programs that first establish and then refine the pattern of eve expression.

An eve/lacZ fusion gene was generated that reproduced part of the eve expression pattern. In the precellular blastoderm, it was expressed in stripes 2, 3, and 7. At gastrulation, expression in stripes 1, 4, 5, and 6 was detected, to generate seven regular stripes by the end of germ band extension. Therefore, eve expression has distinct early and late phases.

Deletion analysis of the 5′ regulatory region of the fusion gene was carried out. The seven-stripe pattern detected in the late phase of eve expression was controlled by a single element between −5.45 kb and −4.65 kb. A deletion from −3.8 kb to −2.6 kb eliminated expression from stripe 3. Stripe 2 required a region from −1.65 to −1.15 kb, while stripe 7 required a region between −2.6 and −0.3 kb. Therefore, a specific regulatory region appears to be required for the expression of each early eve stripe, while all late expression is controlled by a single small 5′ element.

The expression of eve was investigated in gap gene and pair-rule gene mutant embryos. The early eve expression patttern was altered by gap gene mutations but not by pair-rule mutations. The late period of eve expression was also altered in pair-rule mutants.

An analysis of the transcriptional control of the primary pair-rule gene eve has revealed two distinct programs of periodic expression. Several promoter elements are involved in the early response to environmental cues that established the periodicity of eve expression. In the late phase of eve expression, one region of the eve promoter responds to periodic cues from the primary pair-rule gene products to refine the periodic pattern and define parasegmental boundaries (Figure 9-6).

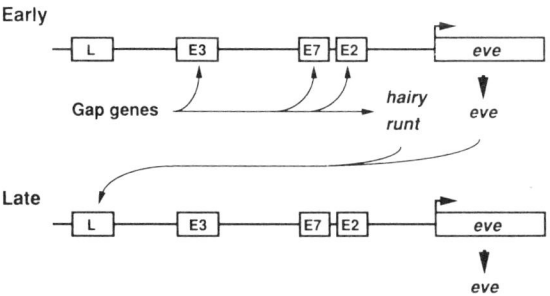

FIGURE 9-6. Model for two-step regulatory program of *even skipped* gene expression. In the first step, gap gene products specifically bind to individual regulatory elements (E3, E7, and E2), each of which specifies *eve* expression in a different stripe. It is likely that each element interacts with a different combination of gap gene products and possibly with maternal gene products. Although eve is expressed in seven stripes, the 7.3 kb 5′ flanking DNA fragment analyzed in this study contains only three of the stripe elements. The location of E7 has not been precisely determined. In the second step the products of the *hairy* and *runt* genes, and eve, which was expressed in the first step, interact with a single regulatory region (L) that specifies eve expression in all seven stripes. (From Goto et al., *Cell*, 57, 413–422, 1989. ©Cell Press.)

Autoregulatory and Gap Gene Response Elements of the *even-skipped* Promoter of *Drosophila*
K. Harding, T. Hoey, R. Warrior, and M. Levine
EMBO J., 8, 1205—1212, 1989

The primary pair-rule gene eve is involved in the establishment of the Drosophila segmentation pattern. P-element mediated transformation and eve promoter fusions were used to identify cis-active elements that regulate the characteristic 7 stripe eve expression pattern.

Various eve promoter fragments were attached to a reporter gene and integrated into the Drosophila genome via P-element mediated germ line transformation. A region between –5.9 and –5.2 kb upstream from the transcription start site allowed normal expression of all seven stripes in an eve+, but not an eve–, background. To confirm the autoregulatory response of this element, this DNA fragment was attached to an hsp70 promoter. In wild-type embryos, this region was sufficient to direct expression in seven stripes. In eve– embryos, this expression was abolished. A region required for stripe 3 was located between –4.7 and –3kb, while a region required for stripes 2 and 7 was located between –1.7 and –0.4 kb.

Several distinct cis-regulatory elements have been identified in the 5′ eve promoter region. Different stripes of expression appear to have separate cis-regulatory regions, while a distal region appears to be involved in autoregulation.

♦ *fushi tarazu (ftz)* is one of the best-characterized promoters of Drosophila genes acting in early development. It is composed of a "zebra element" that directs expression in stripes of a neurogenic element that is required only for expression in the nervous system and of an enhancer element that functions in an autoregulatory fashion by binding to the *ftz* protein itself.

The article by Howard et al. is a good example of the power of Drosophila as a system to anaylze mechanisms of development. These authors present a detailed study of four *hairy (h)* regulatory mutants. Each of these mutants is defective in only a portion of pattern that is disrupted in more extreme *h* alleles. Analysis of the mutant DNAs localized the mutations to distinct regions 5′ of the transcription initiation site, strongly suggesting that the promoter is a composite of regulatory elements each responsible for *h* expression in specific areas of the embryo.

The articles by Goto et al. and Harding et al. show that the *even skipped (eve)* promoter is composed of individual modules that independently regulate *eve* expression. Two early embryonic elements were identified, one that promotes *eve* expression in stripes 2 and 7, and the other that promotes expression in stripe 3. In late embryos, a single element seems to direct expression in all seven stripes. This element appears to be autoregulated by the *eve* protein itself.

The promoter organization of the three genes is quite different. While *ftz* appears to respond to the same cues at each of the segments where it is expressed or repressed, the *h* and *eve* promoters appear to respond to different regulatory signals in different areas of the embryo. This may be related to the fact that *h* and *eve* are considered to be "primary" pair rule genes because they act very early and their expression is independent of other pair rule genes. By contrast, *ftz* acts later and its expression is dependent on other pair rule genes. Thus, it is possible that the primary genes *h* and *eve* act to decode the positional information presented for instance by the "gap" class of genes and perhaps also by maternal genes. Since such genes are localized to specific areas of the embryo, different promoter elements would be needed to interpret positional cues in different parts of the embryo. The pattern established by the primary pair rule genes may then be transmitted to the subordinate genes. Obviously much more work will be required to substantiate such a model. *Marcelo Jacobs-Lorena*

Drosophila **Nurse Cells Produce a Posterior Signal Required for Embryonic Segmentation and Polarity**
K. Sander and R. Lehmann
Nature, 335, 68—70, 1988   9-8

The segmentation pattern in Drososphila depends on morphogenetic signals at the egg poles which are controlled by maternal genes. These maternal activities have been detected during oogenesis.

Cytoplasm from stage 10 nurse cells and oocytes was transplanted in the posterior region of osk mutant embryos, which lack abdominal structures, and into the anterior region of bcd mutant embryos, which lack head and thorax structures. The abdominal phenotype of osk was rescued with cytoplasm from these cells. However, no rescuing activity for anterior development was detected at this stage. This activity was first detected in stage 14 anterior pole cytoplasm.

Therefore, the anterior bcd-dependent and posterior osk-dependent factors become active at different times. Although bcd RNA is expressed in stage 10 nurse cells, bcd activity cannot be detected in these cells. It becomes active only in the anterior pole of the mature oocyte. On the other hand, posterior activity is distributed throughout the stage 10 nurse cell complex and becomes localized to the posterior oocyte pole by the end of oogenesis.

♦Gradients of specific molecules that establish morphogenetic patterns have been well characterized for the anterior-posterior axis of the fruit fly. These different molecules accumulate at each egg pole and interact in a manner that leads to opposite morphogenetic gradients. The current study demonstrates that the posterior signal, but not the anterior signal, can be recovered from the oocyte-nurse cell complex. This observation indicates that the posterior signal can function independently of position while the anterior signal must be localized and subsequently activated before it becomes functional. These observations are important because they demonstrate that the opposite morphogenetic gradients that establish the anterior-posterior axis develop differently. In addition, one of them (the anterior) requires some form of activation (e.g., physical localization, biochemical modification, or both) before becoming fully functional. *Joel M. Schindler*

A Mutation that Changes Cell Movement and Cell Fate in the Zebrafish Embryo
C. B. Kimmel, D. A. Kane, C. Walker, R. M. Warga, and M. B. Rothman
Nature, 337, 358—362, 1989

This report describes a zebrafish zygotic lethal mutation, spt-1, that changes the fate of embryo cells. The mutation was initially identified by changes in embryonic appearance.

The phenotype is zygotic recessive, with mutants surviving less than one week of development. The mutation segregates as a fully penetrant single locus. Initially development appears normal, but at 7 to 8 h cells accumulate abnormally at the approximate region of the prospective head/trunk boundary. As epiboly is completed at 10 h, cells accumulate abnormally at the prospective trunk/tail boundary. These cells appear to enter the tailbud, which becomes enlarged. During tail bud stages, paraxial mesoderm is deficient in the trunk region, while trunk and head region structures appear normal. Unusual cell death is not observed.

To examine the possibility that trunk mesoderm is deficient due to incorrect migration of cells, tracer dye was injected into cells from the lateral part of the germ ring at the gastrula stage. In haploid spt-1 mutants, these cells failed to converge dorsally, instead moving posteriorly. These cells sometimes adopted unusual fates.

The spt-1 gene affects cell fate in the early zebrafish embryo. Its function and pattern of expression are not yet known. It is possible that this gene has a direct effect on cell migrations within the developing embryo.

♦ Pattern formation in developing embryos results from the appropriate expression of multiple cell phenotypes, all of which are located in presumably the correct locations. However, the exact relationship between cell fate, cell movement, and pattern formation is not well understood. For example, do cells determined to be a particular fate move to the appropriate location in order to contribute to the proper pattern or do cells, having moved to a particular location, then become determined to a particular fate and pattern? Research with a mutant strain of zebrafish suggests that location influences fate. Cells which exhibit aberrant movement reach an incorrect location and express a new cell fate. However, the incorrect movement may reflect mutation in the determined cell fate, resulting in relocation. Thus, the current studies fail to clearly distinguish between the two possibilities. Further studies with chimeric embryos should facilitate establishing which mechanism operates during this phase of zebrafish development. *Joel M. Schindler*

Lithium-Induced Teratogenesis in Frog Embryos Prevented by a Polyphosphoinositide Cycle Intermediate or a Diacylglycerol Analog
W. B. Busa and R. L. Gimlich
Dev. Biol., 132, 315—324, 1989   9-10

Lithium ion is teratogenic to many types of embryos. In Xenopus embryos, microinjection of Li+ into prospective ventral cells at the early blastula stage is associated with duplication of dorsoanterior structures at the expense of posterior structures. The biochemical basis of this action of lithium ion remains to be elucidated. However, it is known that lithium inhibits the polyphosphoinositide (PI) cycle. To investigate the linkage between the teratogenic effects of lithium and inhibition of the PI cycle, lithium was microinjected with and without PI intermediates into Xenopus embryos.

Coinjection of lithium and equimolar myo-inositol, a PI cycle intermediate, prevented the teratogenic effects of lithium. Coinjection of epi-inositol, a nonbiologically active positional isomer of inositol, did not rescue the lithium phenotype. When stage 7 embryos were treated with phorbol myristate acetate, an analog of the PI cycle second messenger diacylglycerol, Li dorsoanterior duplication was prevented. However, treatment with phorbol myristate acetate-4-O-methyl ether, a nontransforming analog, did not prevent duplication. Administration of myo-inositol or phorbol myristate acetate in the absence of lithium had no effect on Xenopus embryogenesis.

Li-selective microelectrode measurements demonstrated that intracellular lithium ion levels were identical in cells microinjected with lithium in the presence or absence of myo-inositol. Clonal analysis demonstrated that blastomeres coinjected with lithium and myo-inositol contributed normally to structures in the embryo.

These results suggest that the effect of lithium ions in microinjected Xenopus embryos depends on lithium inhibition of the PI cycle. It is possible that PI cycle activity modulation may be involved in normal dorsoanterior specification.

Dorsalization of Mesoderm Induction by Lithium
K. R. Kao and R. P. Elinson
Dev. Biol., 132, 81—90, 1989

Lithium treatment of Xenopus 32-cell stage embryos causes dorsalization and can be used to rescue UV-irradiated dorsonaterior-deficient embryos. The effect of lithium on mesoderm induction was studied, to identify the lithium target cell.

The 32-cell Xenopus embryo is composed of four tiers of eight cells. Dorsal mesoderm is derived primarily from tier 3 cells. UV-irradiated, DAI grade 0 embryos, which lack all dorsal development, were microinjected with lithium in either a single tier 3 or tier 4 cell. Dorsal development was rescued in either case. To determine if the lithium injected cell contributed progeny to the rescued dorsal mesoderm, lithium was coinjected with a tracer (FDA). The microinjected-tier 4 cell did not contribute to dorsal mesoderm, while the microinjected-tier 3 cell contributed progeny to a variety of dorsal embryo structures (Figure 9-11).

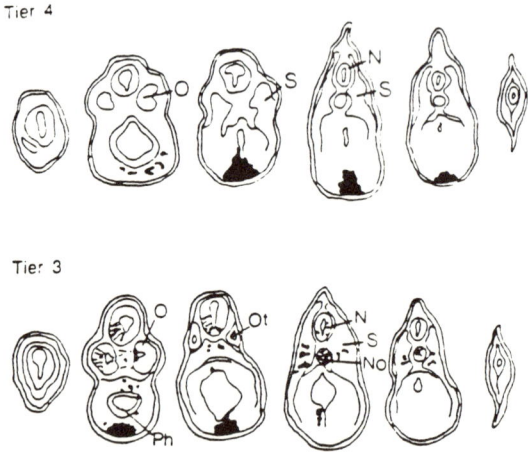

FIGURE 9-11. Lineage tracer distribution in tier 3 and tier 4 lithium-injected embryos. UV-irradiated embryos were microinjected with lineage tracer and lithium at the 32-cell stage and sectioned to determine the progeny of the injected cells. These maps represent composites of four tier 4 (upper row)-injected and five tier 3 (lower row)-injected embryos which had been scored as having normal development. (N) neural tube, (No) notochord, (O) optic vesicle, (Ph) pharynx, (Ot) otic vesicle, (S) somite. (From Kao/Elinson, *Dev. Biol.*, 132, 81—90, 1989. ©Academic Press.)

To determine whether lithium is a mesoderm inducer, animal explants from UV-irradiated fertilized eggs were cultured with or without lithium treatment. In the absence of lithium these explants did not elongate and appeared to form structures similar to ventral mesoderm. If the embryos were treated with lithium prior to explanting, there was no development of dorsal mesoderm. Therefore, lithium does not appear to induce dorsal mesoderm directly.

To determine if lithium modifies mesoderm induction to produce dorsonanterior mesoderm, ventral vegetal cells were sandwiched between lithium-treated animal caps. Of these explants, two-thirds elongated and of these, 82% differentiated into somites. If vegetal cells from lithium-treated embryos were sandwiched between untreated animal caps, the results were similar. Therefore, lithium modifies mesoderm induction so that dorsal mesoderm is always produced.

To determine if the vegetal signal or the animal cell response is modified by lithium, the requirement for a vegetal signal was bypassed by inducing lithium-treated animal caps with low doses of the primarily ventral-mesoderm inducer, fibroblast-growth factor (FGF). Lithium-treated FGF-induced animal caps underwent extensive elongation and differentiated into muscle. A lower concentration of FGF was required in lithium-treated embryos for mesoderm induction than in untreated embryos, suggesting that lithium raises the sensitivity of cells to mesoderm induction.

Therefore, although lithium is not a direct mesoderm inducer in Xenopus, it dorsalizes the response of animal cells to mesoderm induction signals. The lithium target cell appears to be the tier 3 cells of the 32-cell stage embryo.

Inductive Effects of Fibroblast Growth Factor and Lithium Ion on *Xenopus* Blastula Ectoderm
J. M. W. Slack, H. V. Isaacs, and B. G. Darlington
Development, 103, 581—590, 1988   9-12

Basic fibroblast growth factor (bFGF), acidic FGF (aFGF) and embryonal carcinoma derived growth factor (ECDGF) induce isolated Xenopus ectoderm to form mesoderm. As FGF appears to be involved in Xenopus mesoderm induction, the response of Xenopus blastula ectoderm to FGF and to lithium was assessed.

Both a- and bFGF had a 50% mesoderm induction level at 1 to 2 ng/ml, with increased muscle formation up to 100-200 ng/ml. Explant elongation had a similar dose-response relationship. A minimum exposure of 90 minutes was required for induction. The ectoderm was competent to respond from midblastula to early gastrula stage. The timing of the response to FGF was similar to that for the natural vegetal inducer.

Lithium treatment of isolated ectoderm was without effect. In the presence of FGF, lithium enhanced FGF-induced elongation and muscle formation. Ventral marginal explants were capable of responding to lithium without FGF treatment. This suggests that ventral marginal explants have received a treatment similar to low-dose FGF within the embryo.

Therefore, an endogenous Xenopus FGF may be responsible for induction of ventral mesoderm. In the dorsal region, high concentrations of FGF or a synergistic molecule may be present to promote development of somitic muscle.

♦Slack and colleagues have shown previously that in frog embryos, "ventral" mesoderm can be induced to differentiate into more "dorsal" tissue by a signal produced by extreme dorsal mesoderm. The identity of this signal is still not known, but it presumably is important in setting up the dorsal/ventral organization of the embryo. The dorsalizing effects of lithium salts have been fascinating, if somewhat difficult to interpret, in this context.

One puzzling observation is that combined treatment of ectoderm with FGF and lithium enhances muscle differentiation but does not lead to notochord formation, and yet injection of lithium into ventralized embryos results in restoration of the entire dorsal axis, including notochord. Another interesting result is that the synergistic effects of lithium and FGF can be obtained even when the lithium is administered as much as 2 h either before or after FGF exposure.

Busa and Gimlich show that the lithium effect can be nullified by coinjecting myoinositol, which supports the idea that the response mechanism to dorsalizing induction utilizes phosphoinositides, probably IP3, for signal transduction. When more is learned about the receptors for FGF and TGFβ (and for other less-characterized inducers) these observations of lithium effects should be quite helpful in deciphering the downstream signaling pathways. *Thomas D. Sargent*

Reversal of Dorsoventral Polarity in *Xenopus laevis* Embryos by 180 Rotation of the Animal Micromeres at the Eight-Cell Stage
P. Cadellini
Dev. Biol., 128, 428—434, 1988   9-13

During the eight-cell stage of Xenopus development, a 180 degree rotation of the four animal cells with respect to the four vegetal cells reverses dorsoventral polarity (D/V polarity).

More than 300 operations were made, but only 130 of these embryos developed. After those performed at the 8-cell stage, 15 to 30 minutes after cell division, 11% developed with normal polarity, 75% had inverted D/V polarity and 14% produced twins with gastrulation occurring on both sides. At the 16-cell stage, normal polarity was present in 76% of the embryos, while 13% had inverted D/V polarity and 11% were twins. Polarity inversion could be obtained at high frequency up to approximately 2 min prior to the fourth cell division.

Rotation of the four animal cells by 180 at the eight-cell stage of Xenopus development causes inversion of the D/V polarity of the embryo (Figure 9-13). This indicates that at the beginning of the 8-cell stage, dorsal and ventral macromeres must be nearly equivalent. It is hypothesized that the dorsal micromeres contain an unidentified morphogenetic factor that instructs the underlying macromeres to become the center of mesoderm induction.

♦ It has been known for some time that from the 32-cell stage through late blastula the dorsal/ventral polarity of the frog embryo resides in the vegetal hemisphere. Cardellini shows, by means of elegant blastomere transplant experiments, the surprising result that this vegetal dominance does not extend back in time beyond the 16-cell stage. The animal quartet of an 8-cell stage donor imposed its dorsal polarity when grafted to an oppositely polarized vegetal quartet in about 75% of the cases (14% resulted in twins and 11% retained the vegetal polarity). The results were exactly opposite one cell division later. Cardellini suggests that a dorsalizing signal originates in the animal hemisphere and is communicated to the vegetal cells between the third and fourth cleavage. Judging from the length of the cell cycle at this time and the timing of the transplantations, this induction might take as little as 2 min, which would make it much faster than other known inductive interactions in *Xenopus*. Thomas D. Sargent

Spatial Distribution of Cellular Protein Binding to Retinoic Acid in the Chick Limb Bud
M. Maden, D. E. Ong, D. Summerbell, and F. Chytil
Nature, 335, 733—735, 1988   9-14

Retinoic acid may be the natural morphogen of the chick limb bud. A cellular retinoic acid-binding protein (CRABP) has been identified and may be involved in this process. To investigate the distribution of CRABP in the developing chick limb, antibodies were generated to this protein.

At stage 18, the limb bud emerges from the flank. At this stage, antibody to CRABP detected protein in the distal tip of the limb bud. At stage 24, the bud elongates and differentiation begins. At this stage, the progress zone of the distal tip was stained by antibody. In the region behind the progress zone where differentiation begins, there was less

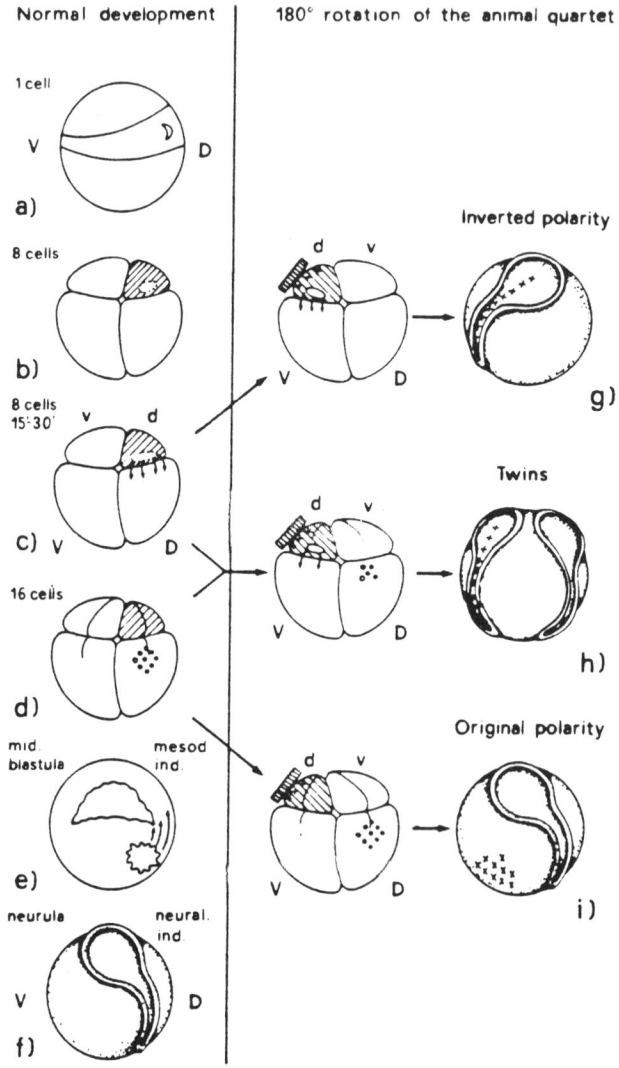

FIGURE 9-13. Dorso-ventral polarity induction. Left column, interpretation of normal development. (a) One-cell stage, cytoplasmic segregation of yolk-free material (moon shape) on dorsal side. (b) Eight-cell stage, progress of dividing furrow (0 to 15 min). (c) Eight-cell stage, second half of the cell cycle (15 to 30 min), first step of induction of D/V polarity by activation of dorsal endoderm to become future center of mesoderm induction. (d) Sixteen-cell stage, dorsal macromeres are activated, as denoted by a cluster of small circles. (e) Early blastula stage, mesoderm induction. (f) Neurula stage, morphological appearance fo dorsal structures. Right column, results of 180° rotation. (g) Operation at 8-cell stage, after 15 min, D/V polarity of the embryo is inverted. (h) Some cases of double axis formation in 8-cell stage and early 16-cell stage, induction is activated on both sides. (i) Operation at early 16-cell stage. Induction has already occurred in dorsal macromeres; rotation does not change original D/V polarity. The small spots indicate the macromeres' dorsalizing response to micromere induction. The small crosses indicate neutral red mark, and the small arrows indicate the dorsalizing inductive action. The small rectangle represents an agar block with neutral red. (From Cardellini, *Dev. Biol.*, 128, 428—434, 1988. ©Academic Press.)

CRABP staining. The developing humerus did not stain for CRABP, while peripheral muscle and connective tissue stained intensely for CRABP. The stage 22 progress zone was sectioned and stained with antibody to examine anterior to posterior staining intensity. There was intense staining at the anterior margin and light staining at the posterior margin of the progress zone.

During the development of the chick limb bud, the levels of CRABP are temporally and spatially regulated. If retinoic acid is the chick limb bud morphogen, CRABP may act by steepening the morphogen gradient to influence development of the chick limb.

Spatial and Temporal Expression of the Retinoic Acid Receptor in the Regenerating Amphibian Limbs
V. Giguere, E. S. Ong, R. M. Evans, and C. J. Tabin
Nature, 337, 566—569, 1989   9-15

Retinoic acid may be the natural morphogen of the vertebrate developing limb. To identify and study the expression of the retinoic acid receptor (RAR) in the newt, an early limb bud stage newt cDNA library was screened with human RAR cDNA.

A newt gene was identified, which was similar to the human beta-RAR. An expression vector assay was used to confirm functional binding of the product of this gene to retinoic acid. Retinoic acid perturbation of newt limb regrowth was most effective on day 7. Therefore, expression of RAR was assayed by *in situ* hybridization 7 days after limb amputation. Mesenchymal cells directly under the wound epidermis had increased RAR expression. No RAR anteroposterior nor proximal/distal gradient of expression was detected. RAR expression decreased by stage 25, when retinoic acid no longer effected the pattern of limb growth.

The expression of RAR in the regenerating mesenchymal blastema cells during the period when retinoic acid can exert an effect on limb development supports the establishment of limb positional information by retinoic acid acting through the RAR receptor.

♦Retinoic acid (RA) has been shown to function as a morphogen during limb development in at least two different classes of organisms, avian and amphibian. As a morphogen, this molecule functions by providing the positional information necessary to establish the anteroposterior axis of the limb. Several molecules have been identified that seem to play a role in the mechanism of action of RA as a morphogen. These include a cytoplasmic binding protein (CRABP) and a nuclear receptor (RAR). Therefore, it is possible that a functional morphogenetic gradient could be the result of differential distributions of not only RA, but also CRABP or

RAR. The papers by Maden et al. and Giguere et al. demonstrate that both spatial and temporal expression of CRABP and RAR exist in developing limbs. In addition, an activation step for preexisting RAR seems necessary before it becomes a functional receptor. These papers are important because they describe several different molecules, and the processing of those molecules, that could play central roles in RA-induced positional information. *Joel M. Schindler*

Segmental Patterns of Neuronal Development in the Chick Hindbrain
A. Lumsden and R. Keynes
Nature, 337, 424—428, 1989   9-16

In higher vertebrates, transient periodic swellings or rhombomeres are particularly distinct in the developing hindbrain. The development of these rhombomeres was examined in the chick hindbrain to determine if this development has a segmental basis.

Rhombomeres appeared between stages 9 and 12 and were detected up to stage 24. The first neurons appeared within alternate rhombomeres and were most numerous in rhombomeres 2, 4, and 6 at stage 13 and 14. At stage 16, motor axons grew in the even numbered rhombomeres. However, this alternating pattern of neuron density was lost beginning at stage 16.

Rhombomere boundaries could be detected by axon concentrations at stage 13. Neuronal cell bodies and axons were located on embrasures underlying the ventricular ridges. the intercellular spaces were large and has speckled immunoreactivity to laminin. N-CAM was located in the central region and Ng-CAM/L1 was expressed at rhombomere boundaries.

Branchiomotor nerves originated from the rhombomere containing the nerve root. Subsequent axons appeared in the posteriorly adjacent rhombomere. Each neuronal group spanned the anterior to posterior length of the rhombomere.

The branchial nerves entered the anterior region of their respective branchial arches, which were ventral to the rhombomeres in which the corresponding motor neuron cell bodies were located. Each of the three principle branchial arches was in register with its respective hindbrain segment pair.

The segmented morphological pattern of the vertebrate hindbrain is expressed at the cellular level. It is the neural crest, which has rhombomeric origins, which is responsible for patterning the skeletal and muscular components of the branchial arches. Formation of the head may depend on the early segmentation of the cranial neuroectoderm.

♦ Segmentation is an important feature in the development of invertebrates and some vertebrates. In invertebrates, segmentation is apparent in the nervous system which forms as a series of ganglia. The sensory and autonomic nervous systems of vertebrates also contains segmentally arranged ganglia. However, there has been little evidence of segmentation in the vertebrate central nervous system.

This report by Lumsden and Keynes provides a detailed documentation of the vertebrate hindbrain, which is composed of a segmental pattern of neuromeres or "rhombomeres". In the chicken embryo, the rhombomeres first become apparent on the second day of development as a periodic pattern of ridges dividing the hindbrain rudiment into seven rostrocaudal segments. The segmental pattern is reflected in the iterative pattern of neurogenesis, with even-numbered segments developing recognizable neurons (assayed by neurofilament immunoreactivity) before the odd-numbered segments. The rhombomere boundaries in chick are distinct from central portion of the rhombomeres: they contain speckled laminin immunoreactivity and Ng-CAM/L1 immunoreactivity which is absent from the central regions. The three principal branchial arches lie in register with rhombomeres 2 to 7, and the branchial motor nucleus is constituted by neurons from the two adjacent rhombomeres that correspond to the apropriate branchial arch.

These findings show that the hindbrain contains early segmented neuronal populations as well as neuronal processes which appear to navigate along segment boundaries. The mechanism of segmentation appears to involve a two-segment repetition. Similar "pair-rule" patterns have been described for early segmentation of the Drosophila embyo. This paper elegantly describes the segmental properties of the hindbrain which appear to be unique within the central nervous system of higher vertebrates. *Marianne Bronner-Fraser*

AUTHOR INDEX

A

Abraham, J. A., 89
Ackerman, L., 132
Acres, B., 103
Adams, T. H., 53, 54
Akam, M., 124
Akers, R., 91
Akeson, R., 46
Albert, P. S., 283
Alexander, S., 209
Alliegro, M. C., 146
Allison, J. P., 232
Alt, F. W., 16
Altaba, A. R. I., 262, 263
Ambros, V., 69
Amon-Böhn, E., 3
Amrein, H., 215
Anderson, D., 47, 103
Anderson, J. M., 228
Angerer, L. M., 149, 245
Angerer, R. C., 149, 245
Aramayo, R., 54
Arnheim, N., 29
Arthur, D. C., 228
Asarnow, D. M., 232
Ashburner, M., 134
Austin, J., 156
Ausubel, F. M., 11

B

Baird, S. M., 15, 47
Baker, B. S., 217, 218
Balling, R., 39
Balmaceda, C.-G., 271
Baltimore, D., 204, 219, 268
Bangham, J. W., 28
Barbel, S., 132
Barberis, A., 66
Baroffio, A., 202
Barthels, D., 176
Barton, S. C., 139, 141
Basch, C., 282
Basler, K., 196
Bauer, G., 144
Beach, D., 85
Beacy, P. A., 250
Beer-Romero, P., 240

Behringer, R. R., 242
Belin, D., 36
Bell, L. R., 215
Benfey, P. N., 77
Bennett, F., 108
Bhatt, S., 111
Bienz, M., 255
Bird, C. R., 12
Bird, D. M., 283
Bishop, C. E., 237
Blackwell, T. K., 16
Block, M. L., 181
Bloom, B. R., 233
Blow, J. J., 86
Bluthmann, H., 173, 174, 230
Bodner, M., 267
Bok, D., 45
Bonas, U., 76
Boned, A., 176
Bonner, J. T., 81, 82
Bonneville, M., 205
Bonyhadi, M., 232
Borkird, C., 76
Born, W. K., 234
Borrelli, E., 45, 47
Bowtell, D. D. L., 197
Brand, M., 218
Brandhorst, B. P., 61
Braun, R. E., 242
Brechner, T., 3
Brenner, C. A., 30, 78
Brenner, M. B., 233
Brenner, M., 205
Brent, R., 130, 256
Brinster, R. L., 75, 242
Britten, R. J., 59, 185
Brizuela, L., 85
Bronner-Fraser, M., 198
Brown, A., 42
Brown, D. D., 72
Brown, L. G., 240
Brown, S. S., 2
Browne, L., 209
Brunning, R. D., 228
Burdsal, C. A., 146
Burgoyne, P. S., 235
Burki, K., 105
Burtis, K. C., 218
Busa, W. B., 289

Busslinger, M., 66

C

Cacheiro, N. L. A., 28
Cadellini, P., 292
Calzone, F. J., 59
Campbell, R., 42
Campos, J., 74
Campos-Torres, J., 74
Capecchi, M. R., 22
Castrillo, J.-L., 267
Caudy, M., 218
Chan, K., 238
Chantry, D., 102
Chen, R., 268
Chiang, A., 82
Cho, K. W. Y., 266
Choi, J. H., 76
Chorneau, R., 76
Chua, N.-H., 77
Chytil, F., 293
Clappoff, S., 42
Clark, S. C., 108
Clarke, A. R., 24
Clarke, H. J., 140
Claussen, U., 49
Clerc, R. G., 268
Clevenger, W., 101
Cline, T. W., 215, 219
Coffin, J. M., 41
Coligan, J. E., 232
Conlon, P. J., 103
Corcoran, L. M., 268
Cosman, D., 101, 103
Costa, M., 281
Coulson, A., 7, 281
Craine, W. R., 67
Cromlish, J. A., 271
Cronmiller, C., 219
Cui, X., 29
Cunningham, B. A., 172, 178
Curie, R. A., 271

D

D'Andrea, A. D., 203
Dahm, L., 170
Dallas, A., 234
Darlington, B. G., 291
Das, G., 272
Davidson, E. H., 59, 185
Davis, L., 209

De Bruyn, J., 232
De Robertis, E. M., 260, 261, 266
Deerinck, T., 267
DeLorenzi, M., 255
Desai, C., 212
Desplan, C., 249
Deutsch, U., 39
Devreotes, P. N., 56, 57
Dickson, C., 109, 110
Dike, L. E., 67
Dingermann, T., 3
Disteche, C. M., 238
Dixon, M. S., 87
Dodd, J., 98
Doetchman, T., 24
Dolecki, G. J., 245
Dolle, P., 276
Donaldson, D. D., 107, 108
Dove, W. F., 34
Drager, B. J., 64
Driever, W., 118, 128
Duboule, D., 276
Duffy, K. T. I., 279
Dunphy, W. G., 85
Dupin, E., 202
Durston, A. J., 97
Dush, M. D., 31

E

Early, A. E., 208
Edelman, G. M., 172, 178
Edgar, B. A., 83
Edgar, R. S., 282
Edstrom, J.-E., 217
Egelhoff, T. T., 2
Eisenman, J., 103
Elinson, R. P., 95, 290
Elizur, A., 258
Ellisman, M., 267
Elsholtz, H. P., 268
Eplen, J. T., 235
Epperlein, H. H., 165
Erickson, H. P., 146
Erlich, H. A., 29
Ettensohn, C. A., 146
Evans, E. P., 235
Evans, G. A., 45
Evans, R. M., 45, 47, 295

F

Fast, P., 269

Feldmann, M., 102
Fenner, S., 41
Ferrier, P., 16
Finney, M., 248
Firtel, R. A., 57
Fitzpatrick, V. D., 255
Flynn, S. E., 268
Foster, J. W., 237
Frangoulis, B., 20
Franks, R. R., 185
Franz, G., 76
Fraser, S. E., 198
Fritz, A. F., 266
Frohman, M. A., 31
Fulop, G. M., 16
Fundele, R., 141
Furley, A. J. W., 16

G

Gagnon, M. L., 245
Gajl-Peczalska, K., 228
Gallin, W. G., 172
Garrett, N., 224, 225
Garriga, G., 212
Gautier, J., 86
Gavis, E. R., 250
Gearhart, P. J., 18
Gearing, D. P., 107
Geissler, E. N., 37
Gerhart, J. C., 96
Gerisch, G., 144
Germain, R. N., 205
Gerster, T., 271
Giguere, V., 295
Gillespie, L. L., 87, 90
Gillis, S., 101, 102
Gimlich, R. L., 289
Giusto, J., 195
Glover, J. C., 200
Goldberg, J. I., 172
Gollahon, K. A., 75
Gong, Z., 61
Goodfellow, P. N., 237
Goodwin, R. G., 101, 103
Gooley, A., 209
Gooley, A. A., 210
Goralski, T. J., 217
Goridis, C., 176
Gorman, M., 215
Gossler, A., 20
Goto, T., 284
Gough, N. M., 107

Grabstein, K. H., 103
Graham, A., 274
Grainger, R. M., 160
Graves, J. A. M., 237
Gray, D., 105
Gray, G. E., 200
Greengard, P., 178
Greenoak, G. E., 41
Greenwald, I., 152, 154, 192
Grierson, D., 12
Grills, G. S., 113
Gruss, P., 25, 39
Gulizia, R. J., 15
Gundersen, R., 57
Gurdon, J. B., 160, 161, 224
Gyllensten, U. B., 29

H

Haaparanta, T., 89
Hafen, E., 196
Hagman, J., 75
Han, K., 252
Handler, A. M., 9
Hanes, S. D., 130
Happ, M. P., 234
Har, D., 81
Hardin, J., 147
Harding, K., 286
Hardwicke, J., 260, 261
Harkey, M. A., 64
Haser, W., 205
Hatzopoulos, P., 76
Hay, B., 132, 133
He, X., 273
Heath, J. K., 87, 107
Heikkila, J. J., 71
Heimfeld, S., 99
Heinke, L. B., 16
Herr, W., 272
Herrmann, B., 34
Heyman, R., 45
Heyman, R. A., 47
Hill, R. L., 59
Hilton, D. J., 107
Hjerrild, K. J., 101
Hochstenbach, F., 205
Hoey, T., 286
Hofman, F. M., 233
Hogness, D. S., 250, 254
Holoshitz, J., 232
Hooper, M. L., 24
Horsthemke, B., 49

Horvitz, H. R., 189, 212, 248
Hough-Evans, B. R., 185
Housman, D. E., 37
Howard, K., 284
Howlett, S. K., 139
Hsi, M., 45
Huang, L. H., 122
Huarte, J., 36
Hudson, G. C., 210
Hulskamp, M., 123
Humphreys, T., 245
Hunsicker, P. R., 28
Hunt, T., 86
Hunter, D. D., 169, 227
Huppi, K., 18
Hurley, D. L., 149
Hyman, R., 47

I

Ingham, P., 284
Ingles, C. J., 255
Ingraham, H. A., 268, 273
Irish, V., 124
Isaacs, H. V., 88, 291
Ishida, I., 205
Ito, K., 205
Iwata, M., 64

J

Jackle, H., 123, 128
Jackson-Grusby, L., 256
Jacobson, A. G., 167
Jan, L. Y., 132, 133, 218
Jan, Y. N., 132, 133, 218
Janeway, C. A., Jr., 205
Janis, E. M., 233
Jaynes, J. B., 251
Jegalian, B. G., 266
Jenkinson, E. J., 106
Jermyn, K. A., 279
Jerzy, R., 101
Jessell, T. M., 98
Jim, Z.-H., 76
Johnson, J. J., 282
Johnson, R. L., 56
Jonas, E. A., 70
Joyner, A. L., 20, 27

K

Kalionis, B., 258
Kamboj, R. K., 144

Kane, D. A., 288
Kaneshima, H., 13
Kao, K. R., 290
Kappler, J. W., 231
Karin, M., 267
Kaufmann, S. H. E., 233
Kaur, S., 46
Kawakami, K., 271
Kay, R. R., 208
Keenan, T. W., 144
Kemler, I., 66
Kenyon, C., 281
Kesbeke, F., 5
Key, B., 46
Keynes, R., 296
Kiff, J., 7
Kimble, J., 156
Kimelman, D., 89
Kimmel, A. R., 56
Kimmel, B. E., 259
Kimmel, C. B., 288
King, M. L., 137
King, T. R., 34
Kingston, R., 106
Kintner, C., 164
Kirschner, M. W., 89
Kishi, H., 173
Kisielow, P., 173, 174, 230
Klein, P. S., 56
Klein, W. H., 63
Knochel, W., 135
Ko, H.-S., 269
Kohara, Y., 7
Koning, F., 232
Koopman, P., 235
Korhary, R., 42
Kornfeld, K., 254
Koster, C. H., 97
Koster, M., 135
Kozak, C. A., 238
Kraft, B., 3
Kramer, J. M., 282
Krasnow, M. A., 250, 254
Krone, P. H., 71
Kroschewski, R., 20
Kruisbeek, A. M., 175, 205
Krumlauf, R., 274
Kubo, R., 234
Kumagai, A., 57
Kuziel, W. A., 232

L

Lam, T. L., 144

Landel, C. P., 45
Landis, S. C., 226
Landmesser, L., 170
Lane, M. C., 148
Lasko, P. F., 134
Lau Y.-F., 238
Lebecque, S. G., 18
LeBien, T. W., 228
LeBowitz, J. H., 268
Leder, P., 74
LeDourin, N. M., 202
Lehmann, R., 124, 287
Lennarz, W. J., 63
Lesley, J., 47
Levin, M. S., 252
Levine, M., 286
Levy, N. S., 18
Li, H., 29
Liberman, M., 13
Liberty, J. A., 228
Lin, C. R., 268
Livingston, B. T., 187
Lockett, T. J., 258
Lofberg, J., 165
Lohka, M., 86
London, C., 91
Ludecke, H.-J., 49
Lum, R., 245
Lumsden, A., 296
Lumsden, A. G. S., 98
Lupton, S., 101, 103

M

Macdonald, P., 284
Macdonald, P. M., 120, 129
Mackie, K., 178
Maden, M., 293
Maeda, N., 24
Maine, E. M., 215
Majors, J., 200
Maller, J., 86
Malpiero, U. V., 18
Malynn, B. A., 16
Mangalam, H. J., 268
Maniatis, T., 284
Manley, J. L., 252
Mansour, S. L., 22, 111
Manstein, D. J., 2
March, C. J., 101
Mardon, G., 238, 239
Mark, D., 30, 78
Marrack, P., 231
Marschalek, R., 3

Martin, D. I. K., 203
Martin, G. M., 31
Martin, G. R., 111
Marusic-Galesic, S., 205
Matsuzaki, F., 172
Mattei, M. G., 237
Matthias, P., 255, 271
McBride, W., 269
McCaw, P. S., 219
McClay, D. R., 146
McCune, J. M., 13
McIntire, S. L., 212
McLaren, A., 235, 238
McMahon, A. P., 109, 111
McNeish, J. D., 40, 46
Mege, R.-M., 172
Melton, D. A., 138, 262, 263
Melton, D. W., 24
Merlie, J. P., 169, 227
Metcalf, D., 107
Miake-Lye, R., 57
Michot, J. L., 237
Minshull, J., 86
Mirabito, P. M., 53
Mochizuki, D., 103
Mochizuki, D. Y., 101
Modlin, R. L., 233
Mohum, T. J., 224
Mohun, T., 225
Moody, W. J., 181
Moran, C., 41
Moreau, J.-F., 108
Moreau, J., 107
Morris, P. C., 12
Morrissey, P. J., 103
Moser, A. R., 34
Mosher, R., 238
Mosier, D. E., 15
Mosley, B., 101
Moury, J. D., 167
Muller, M. M., 255, 271
Murre, C., 219
Mutter, G. L., 113

N

Nagamine, C. M., 238
Naim, A. C., 178
Nakanishi, N., 205
Namen, A. E., 101, 103
Namikawa, R., 13
Nemazee, D. A., 105
Newport, J., 85
Nicola, N. A., 107

Nishioka, Y., 238
Nisson, P. E., 67
Norbury, C., 86
Nordan, R. P., 103
Norris, M. L., 139, 141
Nothiger, R., 215, 221
Nurse, P., 86
Nussbaum, R., 240
Nussenzweig, 74
Nusslein-Volhard, C., 118, 128

O

O'Brien, R. L., 234
O'Brochta, D. A., 9
O'Farrell, P. H., 83, 249, 251
Oliver, G., 260, 261
Ong, D. E., 293
Ong, E. S., 295
Orkin, S. H., 203
Otte, A. P., 97
Owen, J. T., 106

P

Page, D. C., 237—240
Palacios, R., 102
Palisi, T. M., 89
Palmer, E., 234
Palmer, M., 237
Palmiter, R. D., 242
Papalopulu, N., 274
Pardoll, D. M., 233
Passage, E., 237
Paterno, G. D., 87, 90
Pease, S., 107
Pelegri, F., 76
Perris, R., 165
Peschon, J. J., 242
Peters, G., 109, 110
Peterson, A., 42
Pfeifle, C., 123
Phillips, C., 91
Phillips, C. R., 92
Phillips, R. A., 16
Pieler, T., 135
Pirmez, C., 233
Placzek, M., 98
Pondel, M. D., 137
Por, S. B., 209
Poting, A., 135
Potter, S. S., 40, 46
Pow, A. M., 24

Prideaux, V. R., 140
Pullen, A. M., 231
Pupillo, M., 57

R

Raff, R. A., 182
Rajewsky, K., 18
Rammensee, H.-G., 20
Rappolee, D. A., 30, 78
Rathbun, G. A., 16
Ray, J., 12
Rea, T. H., 233
Reed, S. G., 103
Reik, W., 141
Reinke, R., 157
Richman, D. D., 47
Rickles, R. J., 36
Riddle, D. L., 283
Rio, D. C., 8
Roberts, C., 237
Roberts, S., 282
Roeder, R. G., 271
Roes, J., 18
Rogers, D., 107
Rosa, F. M., 264
Rosenfeld, M. G., 268, 273
Rossant, J., 20, 27, 42, 140
Rothman, M. B., 288
Rowning, B., 95, 96
Rubin, G. M., 8, 197, 259
Ruoslahti, E., 114
Ruppert, S., 271
Rushlow, C., 284
Russell, L. B., 28
Russell, W. L., 28
Rutishauser, U., 170
Ruvkun, G., 195, 248
Ryan, M. A., 37

S

Sablitzky, F., 18
Saffman, E. E., 254
Saiki, R. K., 29
Saint, R., 258
Saito, T., 205
Samson, M.-L., 256
Sander, K., 287
Sanes, J. R., 169, 200, 227
Santoni, M. J., 176
Sargent, T. D., 70, 93
Sato, S. M., 93

Savage, R., 92
Saxe, C. L., III, 56
Schaffner, W., 271
Scharf, S. R., 96
Schedi, P., 215
Schedl, P., 219
Scheidereit, C., 271
Schmid, H., 221
Schmierer, A., 101
Schneider-Gadicke, A., 240
Schotzinger, R. J., 226
Schroder, C., 123, 128
Schuch, W., 12
Schultz, K., 170
Schultz, R., 30, 78
Schwartz, R. H., 233
Scott, B., 173, 174
Scott, M. P., 253
Scott, W. J., 40
Seifert, E., 128
Senger, G., 49
Seydoux, G., 192
Shah, V., 169, 227
Sharp, P. A., 268
Shaw, A. C., 74
Shedlovsky, A., 34
Shelby, M. D., 28
Shultz, L. D., 13
Silberman, T. L., 228
Simmons, D. M., 268, 273
Simon, M. A., 197
Simoncini, L., 181
Simpson, E., 235
Sims, J., 103
Sinclair, A. H., 237
Singer, A., 175
Sinn, E., 74
Siu, S.-H., 144
Skarnes, W. C., 20, 27
Skarvall, H., 105
Slack, J. M. W., 87, 88, 90, 291
Smith, A. G., 107
Smith, C. A., 106
Smith, C. J. S., 12
Smith, E., 209, 210
Smith, J. C., 160
Smith, R., 110
Smithies, O., 24
Snaar-Jagalska, B. E., 5
Snape, A. M., 70
Snoek, G. T., 97
Soll, D. R., 3
Solter, D., 139, 140

Solursh, M., 148
Sorkin, B. C., 178
Spangrude, G. J., 99
Spencer, J. A., 237
Spudich, J. A., 2
Stadler, J., 144
Staerz, U. D., 230
Stahl, M., 107
Staudt, L. M., 269
Steinbrech, D., 3
Steinmann-Zwicky, M., 221
Steinmetz, M., 230
Sternberg, P. W., 188, 189
Steward, R., 122
Stewart, C. L., 107
Stock, J., 46
Storb, U., 75
Stoye, J. P., 41
Strickland, S., 36
Strober, S., 232
Struh, K., 129
Struhl, G., 120, 124, 129
Studer, R., 235
Sturm, R. A., 271
Sucov, H. M., 185
Sulston, J., 7, 281
Summerbell, D., 293
Sun, T. J., 56
Sung, Z. R., 76
Superti-Furga, G., 66
Surani, M. A., 139, 141
Suthers, H. B., 81, 82
Svhedl, P., 122
Swanson, L., 268
Swanson, L. W., 273
Symes, K., 160

T

Tabin, C. J., 295
Takeda, S., 102
Tautz, D., 123, 128
Taylor, M., 225
Taylor, M. V., 224
Teh, H. S., 173, 174
Tessier-Lavigne, M., 98
Thali, M., 255
Theill, L. E., 267
Theis, J., 249
Theze, N., 59
Thiebaud, P., 59
Thomas, K. R., 22
Thompson, S., 24

Thomson, J. A., 139, 140
Tigelaar, R. E., 232
Timberlake, W. E., 53, 54
Tomlinson, A., 259
Tomlinson, C. R., 63
Tonegawa, S., 205
Torigian, V., 233
Treacy, M. N., 273
Treisman, R., 225
Tsai, S.-F., 203
Tucker, P. W., 232
Tuma, R., 218
Turner, M., 102

U

Uematsu, Y., 173
Urdal, D., 101
Uyemura, K., 233

V

Van Haastert, P. J. M., 5
Varmuza, S., 140
Vassalli, A., 36
Vassalli, J.-D., 36
Vassin, H., 218
Vitelli, L., 66
von Boehmer, H., 173, 174, 230
von Mende, N., 283
Vopper, G., 176
Voytas, D. F., 11

W

Wagner, E. F., 107
Walker, C., 288
Wang, G., 245
Warga, R. M., 288
Warrior, R., 286
Waterston, R., 7
Watson, C. F., 12
Weir, M., 281
Weissman, I. L., 13, 99

Weith, A., 237
Werb, Z., 30, 78
Wessel, G. M., 63
Weston, K., 152
Whiteley, A. H., 64
Wignall, J., 101
Wilffe, A. P., 72
Wilkinson, D. G., 109, 111
Wille, W., 176
Williams, G. T., 106
Williams, J. G., 208, 279
Williams, K. L., 209, 210
Williams, R. L., 107
Wilson, D. B., 15
Wilson, T. A., 107
Wilt, F. H., 187
Winoto, A., 204
Witek-Giannotti, J., 108
Wolgemuth, D. J., 113
Wong, G. G., 107, 108, 203
Wong, L. M., 144
Wormann, B., 228
Wray, G. A., 182
Wright, C. V. E., 260, 261, 266
Wu, M., 96

Y

Yagoob, M., 160
Yamaguchi, Y., 114
Yancopoulos, G. D., 16
Yang, M., 3
Yang, Q., 245
Yisraeli, J. K., 138
Yochem, J., 152, 154

Z

Zhang, W., 63
Zhao, J., 45
Zimmer, A., 25
Zipursky, S. L., 157
Zon, L. I., 203
Zusman, S. B., 122

SUBJECT INDEX

A

alpha-A-crystallin regulatory region, 46—47
α-actin
 gene expression, 225—226
 gene promoters, 71—72
abaA, 53—54
abd-A genes, 256—258
abd-A proteins, gene activation and DNA binding by, 256—258
Aboral ectoderm, 245—248
AC, see Anchor cell
Acetylcholinesterase, 226—227
achaete-scute complex, 218—219
achaete-scute proteins, 219—221
Actin mRNA, 67—69, 149—152
Actin RNA, 160—161
Adenylate cyclase, stimulation of in *D. discoideum*, 5—7
Adhesion molecules, 170—171
Adult chimeras, 139
Adult sea urchin, tube feet of, 66—67
Affinity chromatography, 271—272
AIDS, 15—16
Allelic exclusion, 74—75
beta-Amino-propionitrile (BAPN), 63—64
Ammonia (NH_3), 81—82
 thermotaxis and, 81—82
Amorphic alleles, 221—224
Amphibian induction, 160—161
Anastrepha suspensa, 9—11
Anchor cell (AC), 192—195
Androgenetic blastomeres, 139—140
Androgenetic cells, 139—140
Androgenetic cells, detected in embryo and yolk sac, 139
Androgenetic chimeras, 140
Aneroposterior expression, 274—276
Animal development, 161—164
Anopheles gambiae, 9—11
Antennapedia (Antp), 250—251, 253—254, 281—282,
Antennapedia, transcriptional activation by, 253—254
Antennapedia-class homeoproteins, 130—132
Anterior morphogen, 124—127

Anterior pituitary, growth hormone in, 268—269
Anterior prestalk cells, 279—281
Anterior-posterior axis, pattern formation along, 263—264
Anti-CD3 treatment, 106—107
Anti-CD4 monoclonal antibody, 175—176
γδ Antigen receptor genes, 232
Antigen-specific T cells, 174
Antisense RNA, 36—37
 inhibition, 12—13
Antp, see *Antennapedia*
Antp promoter fusion, 254
Antp-Ubx (Ultrabithorax), 253—254
Apoptosis, 106—107
Apoptosis, induced in immature T cells, 106—107
Apoptotic thymocytes, 106—107
Arabidopsis thaliana, 11—12
Archenteron elongation, 147—148
Ascidians
 accelerated development into adults, 182—185
 loss of larval stages, 182—185
Aspergillus development, 53—54
Aspergillus nidulans, 53—54
 growth and development in, 54—56
ATP-dependent helicases, 133—134
Autonomic nervous system, 226—227
Auxin 2,4-dichlorophenoxyacetic acid, 76—77
Axial defects, 263—264
Axolotl, 165—167
 neural plate and epidermis in, 167—168
Axon-axon interactions, 170—171
Axon-cell interactions, 170—171
Axon-matrix interactions, 170—171

B

B-adrenergic receptor, 56—57
BAPN, see beta-Amino-propionitrile
Basal lamina (BL), 169—170
bcd-dependent transcriptional activation, 129—130
bcd protein, 118—120, 130—132

homeodomain recognition helix of, 130—132, see also *bicoid*

B cell
  memory, 105—106
  ontogeny, 228—230
  tolerance, 105
B cells, in the presence of antigen, 105
Bicoid activator protein, 130—132
bicoid (bcd), 129—132, 219
  gradient, 118—120
  mRNA, 120—121
  mRNA localization, 120—121
  protein, 118, 128—129
Bipartite DNA-binding structure, 272—273
BL, see Basal lamina
Blastomeres, 160
B-lineage cells, molecular cloning and growth factor activity on, 101—102
BLIN-1 line, 228—230
B lymphocytes, 48—49, 75—76
boss/sev double mutant, 157—159
br1A, 53—54
Brachiopods
  accelerated development into adults, 182—185
  loss of larval stages, 182—185
bride-of-sevenless gene, 157—159, 197—198
Brittle stars
  accelerated development into adults, 182—185
  loss of larval stages, 182—185
Bryozoans
  accelerated development into adults, 182—185
  loss of larval stages, 182—185
Burkitt lymphoma line, 271—272

C

c-fos, 225—226
c-kit, 37—39
c-mos, 113—114
  mRNA, 113—114
  ranscript levels, 113—114
Caenorhabditis elegans, 7—8, 69—70, 152—153, 189—192, 195—196, 212—215, 221—224, 248—249
  cell fate decision in, 192—195
  cell fate determination in, 192—195
  dumpy-13 (dpy-13) gene, 283—284

genome, 153—155
germline, 156—157
glp-1, 189—192
  mutations of the, 282—283
  posterior pattern formation in, 281—282
sqt-1 gene of, 282—283
  temporal pattern in, 195—196
unc-86 gene of, 248—249
  vulval development in, 188—192
  vulval induction in, 188—192
Ca-independent neural (CAM), 172—173
Calcium currents lineage-specific, development of, 181—182
CAM cDNA transfection of, cells with, 172—173
CAM, see Ca-independent neural
cAMPn 279—281
  analogues, 208
  effects on chemotaxis and gene expression, 57—58
  receptor, 56—58
  sequencing of cDNA for, 56—57
  signal transduction mechanism, 5—7
CaMV 35S enhancer, constitutive expression of, 77—78
Candida albicans
  high-frequency switching systems in, 3
  switching involved in pathogenicity, 3—5
Carbohydrate modifications, 210—212
Cardiac actin
  expression, 89—90
  gene, muscle-specific transcription of, 224
CArG promoter sequence, 224
CAT
  anti-sense RNA probe, 186
  mRNA, 186
$cdc2^+$, 86
CD4⁻CD8⁻cells, 204—205
CD4⁺CD8⁺ thymocytes, 230—231
CD4⁻CD8⁻ T lymphocyte, 232—233
cdc2 protein, 85
CD3/T-cell receptor complex antibodies to, 106—107
CD4⁺T cells, 205—206
CD8⁺T cells, 205—206
CD4 thymocytes, 13—15
CD8 thymocytes, 13—15
CD4⁺8⁺ thymocytes
  cDNA

cloning of, 203—204
library, 101, 169—170
rapid production of full-length, 31—34
sequence, 144—145
Cell ablation experiments, 188—189
Cell adhesion, molecules, 172—173
Cell-binding domain, mapping of, 144
Cell-cell communication, 161—164
Cell-cell interactions, 259—260
Cell differentiation, underlying mechanisms of, 157—159
Cell interactions, 143—179
Cell lineage analysis, 198—204, 248—249
  developmental fate and, 181—206
  evolutionary modification of, 182—185
Cell-matrix interactions, 165—167
Cell migration, 164—165
Cell migratory response, 148—149
Cell morphology, long-range effects on, 172—173
Cell proliferation, 114—115
Cells, subpopulation of, 234—235
Cellular retinoic acid-binding protein (CRABP), spatial distribution of, 293—295
Central nervous system, signals involved in the induction of, 93—94
ChAT, see Choline acetyltransferase
Chemotropism, 98—99
Chick ciliary neurons, 169—170
Chick
  ciliary neurons, 169—170
  hindbrain
    branchiomotor nerves, 296—297
    neuronal development in, 296—297
  lumbosacral motor neurons, 170—171
  optic tectum, 200—202
Chicken cell, adhesion molecules, 172—173
Chicken optic tectum, clonally related cells in, 200—202
Chimeric genes, 71—72
Chimeric mice, production of containing embryonic stem cells, 25—27
Chimeric mouse embryos, 140—141
Chinese hamster ovary, 114—115
Chlorambucil, 28—29
4-{p[bis(2-Chloroethyl)amino]-phenyl}butyric acid, 28—29
CHO, see Chinese hamster ovary
Choline acetyltransferase (ChAT), 226—227

Cholinergic sympathetic target, 226—227
Cholinergic transmission, 226—227
Chordamesoderm, 167—168
Chromatin condensation, 106—107
Chromosomal deficiencies, physical maping of, 282—283
Chromosomal determinants, influence of, 138—139
Chymomyza proncemis, 9—11
alpha-Crystallin-synthesizing cells, targeted ablation of, 46—47
Clonal anergy, 20, 231
Clonal deletion, 231
Clone-forming ability of migratory neural crest cells, 202—203
C10-MJ2 cells, 108—109
Collagen, probed in *Stronglyocentrotus purpuratus,* 63—64
Complete somatic embryogenesis, 76—77
Concanavalin A-agarose, 169—170
Constitutive alleles, 221—224
Cortical cytokeratin, breakdown of, 137
Cortical tractor model, 167—168
cot-2 sequence, 268
CRABP, see Cellular retinoic acid-binding protein, 293—295
Cyclic AMP, 208
Cyclin mRNA, 86—87
  translation of, 86—87
Cyclin protein synthesis, 86—87
CyI, 67—69
CyIIa, 149—152
CyIIIA, 67—69
CyIIa actin, 63—64
CyIIIa actin, 63—64
CyIIIa gene, activation of, 59—61
Cytodifferentiation, 207—244
Cytoplasmic rotation, 95—96

D

da, see *daughterless*
DA cells, 108—109
da gene, single functional product of, 218—219, see also *daughterless*
daughterless (da), 218—219
  gene, 219—221
  maternal expression of, 219—221
  proteins, 219—221
DBL, see Dorsal blastopore lip tissue
dEC, see Dendritic epidermal cell

Decorin, 114—115
Deletion mutations, in mouse germ cells, 28—29
Dendraster excentricus, 64—66
Dendritic epidermal cell (dEC), 232
Developing embryos, pattern formation in, 288—289
Development of the HSN neurons, genetic pathway for, 214
Development of the mouse skeleton, mutation affecting, 39—40
Developmental cell biology, 81—115
Developmental expression, 156—157
Developmental gene expression, 53—79
Developmental genetics, 1—51
Developmental switch, larva-to-adult, 69—70
Developmentally regulated genes, 20—22
DIA, see Differentiation inhibitory activity
Diacylglycerol analog, 289—290
Dictyostelium, 144—145, 208—209
  discoideum, 81—83, 144—145
    contact site A glycoprotein of, 144—145
    development in, 56—57
    extensive size polymorphisms in, 210—212
    hygromycin resistance as a selectable marker in, 2—3
    integration of mobile genetic elements in, 3—5
    slug migration in, 209—210
    stimulation of adenylate cyclase in, 5—7
    strain AX2, 3—5
    in study of chemosensory mutants, 5—7
  G protein role in, 57—58
  high-frequency switching in, 3
  prestalk zone in, 279—281
  signal-transduction mutants of, 56—57
  spore maturation in, 208
  surface glycoprotein, 144
Dictyostelium fgd A mutants, 5—7
DIF, see Differentiation inhibitory factor
Differentiation inhibitory activity (DIA), 108—109
Differentiation inhibitory factor (DIF), 107, 208—209
Diptheria toxin, 47—49

District cell-cell, interactions in, 153—155
DNA
  affinity chromatography, 8—9, 268—269
  binding and dimerization motif, 219—221
  fingerprinting, 235—236
  hybridization, 153—155
  sequence recognition, 256—258
  sequences, analysis of in single human sperm and diploid cells, 29—30
  specificity, 130—132
DNase footprint, 66—67
Dominant-white spotting (w) locus, 37—39
Dormant mRNA activation, antisense RNA preventing, 36—37
Dorsal blastopore lip (DBL) tissue, 93—94
Dorsal explants, 98—99
dorsal (dl) gene, 122—123
dorsal RNA, 122—123
Dorsalventral polarity, 97
Dorsoventral (D/V) polarity, 292—293
  genes that influence, 83—85
  induction, 294
doublesex (dsx), 218
dpy-13, 283—284
Drosophila, 9—11
  Adb-B genes, 276—278
  Antp homeobox, 273—274
  Antp homeodomain, 276—278
  bicoid (bcd) gene, 118
  clusters, genes in, 274—276
  compound eye of, 196—197
  early development of, 281—282
  embryo, 39—40, 118, 124—127
    anterior development of, 128—129
    anterior localization of *bicoid* mRNA in, 120—121
    cell division patterns in, 83—85
    determined by *bicoid* protein, 118—120
    hunchback transcription in, 128—129
    posterior development in, 123—124
    posterior segmentation of, 123—124
    role of the posterior determinant system in, 124—127
  embryogenesis, 122—123
    temporal and spatial patterns of cell division during, 83—85
  even skipped

homeobox, 262—263
    promoter of, 286—287
  eye, 259—260
    ommatidia in, 197—198
  funebris, 250—251
  fushi tarazu polypeptide, 255—256
  gap segmentation gene, 128, 134—135, 153—155
  germ line of, 221—224
  hawaiiensis, 9—11
  homeobox gene complexes, 274—276
  homeobox proteins, 252—253
  homeo domain proteins, 245—248
  homeodomains, 269—270
  homeotic genes, 255, 276—278
  identification and purification of, 8—9
  K_c tissue culture cells, 8—9
  maternal dorsal mRNA in, 120—121
  melanogaster, 8—12
    pattern formation in the eye of, 258—259
  Notch, 152—153, 189—192
  nurse cells, 287—288
  polar granules, 132—134
  primary signal for sex determination, 219—221
  retina cell-cell, interaction in, 157—159
  segmentation gene, 284
  segmentation pattern in, 287—288
  simulans, 9—11
  somatic sex development in, 215
Drosophilids, P elements in, 9—11
dsx, see doublesex
D/V, see Dorsoventral polarity
dystonia musculoram (dt), 42—44

E

Early embryo, vegetal polarity in, 185—187
Early lineage restriction, 202—203
Early sea urchin embryo, blastomeres of, 149—152
EC, see Embryonal carcinoma, 108—109
Echinoderms, 64—66
Echinonectin (EN), 146—147
ECM disruption, 63—64, see also Extracellular matrix
Electrophoretic variants, 210—212
Embryo cDNA libraries, 134—135
Embryo, anterior structures of, 120—121
Embryonic body plan, establishment of along the dorsal-ventral axis, 122—123
Embryonic chick muscle, 170—171
Embryonic genes in plants, developmental regulation of, 76—77
Embryonic primary mesenchyme genes, expression of, 64—66
Embryonic segmentation and polarity, 287—288
Embryonic spinal cord, 274—276
Embryonic stem cells (ES), 20—24, 107
    differentiation, 107—108
    gene targeting in, 24—25
    production of chimeric mice containing, 25—27
Embryos, deciliation of, 61—62
EN, see Echinonectin
En-2 gene, production of a mutation in mouse, 27—28
engrailed, 249—250
    gene product, 251—252
    genes, 252—253
Enhancer trap construct, 20—22
5-Enolpyruvylshikimate-3-phosphate synthase (EPSP), 77—78
Epi 1, 92—93
    expression of, 91—92
Epidermal growth factor, 152—153
Epidermis-specific keratin expression, 70—71
Epithelial tissues, morphogenesis of, 147—148
EPSP, see 5-Enolpyruvylshikimate-3-phosphate synthase
ES, see Embryonic stem cells
Escherichia coli, 9—11
    full length bcd protein expressed in, 128—129
    hygromycin resistance gene of, 2—3
    strain rec A⁻, 3—5
    Xenopus bFGF produced in, 89—90
N-Ethyl-N-nitrosourea (EtNU), 28—29, 34—36
EtNU, see N-Ethyl-N-nitrosourea
Evasterias troschellii, 64—66
eve, see even-skipped
eve/lacZ fusion gene, 284—285
even skipped, 284
    expression, 284—285
    gene expression, 284—285
    genes, 252—253
    promoter, 286—287
    protein, 256—258

Extracellular matrix (ECM), 98—99, 114—115, 148—149, 165—167
  disruption of, 63—64
  influence of on gene expression during sea urchin development, 63—64
  molecules, 169—170
Extraembryonic matrix, 146—147
exu, 118—120
Exuperantia genes, 120—121

F

Feedback inhibition, 75—76
Fetal thymocyte clones, 102
fgd A mutants, 5—7
FGF, 90—91, see also Fibroblast growth factor
Fibroblast growth factor (FGF), 88—89, 110—111
  inductive effects of, 291—292
Fibronectin (FN), 146—147
Filopodial contraction, 147—148
Filopodial-directed migration, 148—149
Finger motif, 135—136
Flow cytometry, 76—76
FN, see Fibronectin
Foreign DNA, 42—44
Frog
  accelerated development into adults, 182—185
  egg
    dorsal/ventral polarity of, 292—293
    fibroblast growth factor in, 89—90
    lithium-induced teratogenesis in, 289—290
  homeobox pattern in, 261—262
  loss of larval stages, 182—185
ftz, see *fushi tarazu* gene
Function-perturbing antibodies, 170—171
Functional homeobox domain, 269—270
fushi tarazu (ftz) gene, 251—256, 286—287
  homeodomains, 256—258
  transcriptional activation by, 253—254
Fusion protein analysis, 268

G

Gα-common antiserum, 5—7
β-Galactosidase activity, 9—11
G proteins, 5—7, 56—57

G-protein subunit, 5—7
G-alpha protein subunits, regulation and function of, 57—58
G418 resistance, 22—24
GAL4/ftz fusion protein, 255—256
Gamete equivalence, 244
GANC-r, see Gancyclovir resistance
Gancyclovir resistance (GANC-r), 22—24
Gastrulating mouse embryos, 139—140
Gastrulation, 91—92
GDPβS, inhibition of cAMP binding by, 5—7
Gel mobility shift, 66—67
Gelatin agarose affinity chromatography, 146—147
Gene expression
  control of, 56—57
  developmental regulation of, 72—74
Gene isolation, tags for, 41
Gene rearrangement, feedback inhibition of, 75—76
Gene transplacement strategy, 54—56
Gene trap construct, 20—22
Gene vectors, in insect and noninsect systems, 9—11
Genes, regulation of, 67—69
Genome linking, 7—8
Germ cell sex determination, nonautonomy of, 221—224
Germ line transmission and expression, 24—25
GF-1, 203—204
GHF-1, see Growth hormone factor
Globin gene family, regulation of, 203—204
glp-1, 153—155
  transcript, analysis of, 156—157
N-Glycanase, 107—108
Glycogen synthase 3 (GSK-3), 178—179
Growth hormone, cell-specific activation of, 268—269Growth hormone factor (GHF-1), 267—268
Gα2 subunits, 57—58
GSK-3, see Glycogen synthase 3
GTP-binding regulatory protein, 5—7
GTPγS, inhibition of cAMP binding by, 5—7
Gynogenetic cells, 139—140

H

H-Y antigen, 235—236

H. erythrogramma, 182—185
H2A-2 genes, 66—67
H2B-2 genes, 66—67
hairless, 42—44
hairy, 284
Haploid gene expression, 244
HAT-R cells, 24—25
hb, see also *hunchback* gene
   expression, zygotic pattern of, 129—130
   gene, 124
   promoter, 129—132
HBGF, see Heparin-binding growth factor
Hbox1, 245—248
HD, see Homeodomain
Heat shock promoters, 71—72
Heavy chain loci, 75—76
Heliocidaris erythrogramma, 182—185
Heliocidaris tuberculata, 182—185
Helper T cells, 105
Hematolymphoid differentiation and function, 13—15
Hematopoiesis, 103—104
Heparin-binding growth factor (HBGF), 88—89
Hermaphrodites, 156—157
Hermaphrodite-specific neuron, see HSN
High-affinity surface receptors, 5—7
High-frequency switching, 3
Hindbrain, segmental properties of, 296—297
Homeobox concensus, 268
Homeobox containing gene, 42—45
Homeobox domains, 245—248
Homeobox gene, 245—248
   analysis of, 245—248
Homeobox sequences, 246
Homeodomain (HD), 249—250
   -DNA interaction, sequence specificity of, 249—250
   protein, gradient of, 261—262
   proteins, 251—252
   recognition helix, 130—132
Homeogene, overexpression of, 42—44
Homologous recombination, production of mutation by, 27—28
Hox 1.1 allele, mutated by homologous recombination, 25—27
Hox-1.4, 42—44
Hox-2 gene cluster, 274—276

HPRT gene, germ line transmission and expression of, 24—25
Hprt, see Hypoxanthine-quanine phosphoribosyltransferase
Hprt gene, 24
HSN (hermaphrodite-specific neuron), 212—215
   axonal outgrowth, 212—215
   defects, 212—215
   development, 212—215
   motor neurons, 212—215
   migration, 212—215
   serotonin expression, 212—215
hsp30/CAT, 71—72
hsp30 expression, 71—72
hsp30 promoter, 71—72
hsp70, 71—72
hsp70/CAT, 71—72
Human beta globin gene, 70—71
Human cDNA, 101—102
Human genome, cloning of by microdissection and enzymatic amplification, 48—51
Human hematopoietic cells, 13—15
Human immune system, 13—16
Human interleukin 7, 101—102
hunchback activity, 124—127
hunchback (hb) genes, 123—124, 130—132
Hya$^+$, 235—236
Hya, location of, 237
Hybridoma analysis, 76—76
Hygromycin resistance, in *Dictyostelium discoideum*, 2—3
Hypermutation mechanism, activated by V gene rearrangement, 18
Hypoxanthine-quanine phosphoribosyltransferase (Hprt), 24

I

Ig lambda-producing B cells, 75—76
IL, see also Interleukin
IL-1, 102—103
IL-7, 102—103
IL-7-induced proliferation, 102—103
Immature T cells, cell death in, 106—107
Immature thymocytes, 106—107
Immune system reconstitution, 13—15
Immunoglobin gene

activated by human lymphoid-specific transcription factor, 271—272
B-cell-specific transcription of, 271
expression, 219—221, 268
Immunoglobulin octamer DNA motif, 269—270
Immunoglobulin V-H genes, onset of somatic mutation in, 18—20
Imprinted genes, action of, 140—141
Inositol phospholipid turnover, 97—98
Insertional mutagenesis, 40—41
int-2, 31—34, 86—87, 109—110, 111—113
  cDNAs, 111
  disruption of in mouse embryo-derived stem cells, 22—24
  expression, 111
    spatial and temporal pattern of, 111—112
  FGF-related, 111—113
  gene, 22—24
  mRNAs, 111
Interaction between antigen and B cells, 105—106
Intercellular junctions, formation of, 172—173
Intercellular signaling pathways, 189—192
Interleukin 7, 101—103, see also IL
intersex gene, 218
Intestinal intraepithelial lymphocytes, 205—206
Intracellular calcium levels, 106—107
Intracellular signaling pathways, elucidation of, 108—109
Isolated chemosensory mutants, 5—7
Isolated ectoderm, lithium treatment of, 291—292

K

Kappa/lambda isotypic exclusion, 75—76
Keratin expression, 71—72
KFGF, 86—87

L

Lactotrope differentiation, 48—49
Lactotropes, 48—49
lacZ gene, 20—22
lacZ, 42

Laminin B1 gene, cloning and characterization of, 227—228
Laminin, 227—228
Laminin-like adhesive protein, 227—228
Laser ablation, 146—147
L-CAM, 172—173
Legless, 40—41
Leukemia inhibitory factor (LIF), 108—109
LiCl treatment, 187—188
LIF, see Leukemia inhibitory factor
lin-4, 69—70
$I\alpha\beta$ lineage, 204—205
lin-12, 152—155, 188—192
  activation of, 192—195
  cell autonomy of, 192—195
  genotype, 192—195
  transcript, analysis of, 156—157
lin-14, 69—70, 195—196
  analysis of mutations in, 195—196
  protein product of, 195—196
  temporal developmental switch, 195—196
lin-15 mutant, 188—189
lin-28, 69—70
lin-29, 69—70
Lineage specificity, 203—204
Lithium ion, inductive effects of on Xenopus blastula ectoderm, 291—292
Lithium, 187—188
  -injected embryos, lineage tracer distribution, 290—291
  -induced teratogenesis, 289—290
Low-affinity surface receptors, 5—7
LpS3, probed in *Lytechnius variegatus* and *L. pictus,* 63—64
Lymphocytes, 233
  expressing T cell receptor, 234—235
Lymphoid expression, 271
Lymphoid-specific promoters, 271
Lymphokines, 103—104
Lytechinus, 147—148
  pictus, 67—69
  variegatus, 63—64

M

M actin, 67—69
mab-5, molecular characterization of, 281—282
mab-5 gene, 281—282

Major histocompatibility antigens, 174–175
Major histocompatibility complex (MHC), 173–174
Mammalian brain development, POU-domain regulatory genes in, 273–274
Mammalian meiotic maturation, 113–114
Mammalian oogenesis, 113–114
Mammalian spermatogenesis, 244
Mammalian sympathetic neurons, 226–227
Mammalian transcription factors, 273–274
Mammalian central nervous system, 98–99
Mammary tumor formation, 111–113
Maternal chromosomes, 139
Maternal controls, cytoplasmic determinants, and imprinting, 117–142
Maternal *dorsal* mRNA, 120–121
Maternal gap genes, *grandchildless-knirps*, 133–134
Maternal genome contributions, 139–140
Maternal mRNA, 135–136
Mature T cells, 103–104
  generation of, 174–175
Meiotic recombinants, mapping to *T* to *H-2* region of mouse genome with, 34–36
Memory B cells, 105–106
Memory cell response, 105–106
Mesenchymal cells, int-2 expression detected in, 111–112
Mesenchyme blastula stage, 149–152
  experimental treatments leading to arrest at, 63–64
Mesoderm formation, 160
Mesoderm inducers, 264–266
Mesoderm induction, 160
  dorsalization of, 290–291
Mesoderm specification, 225–226
Mesoderm inducing factor (MIF), 160–164
Mesoderm-inducing activity, 88–89
Metallothionine, probed in *Lytechnius variegatus* and *L. pictus*, 63–64
Methylation interference assays, 66–67
MHC, see also Major histocompatibility complex
  class I, 174–175
  molecules, 174
Microcarrier, transplantation paradigm, 165–167
Microcarriers, 165–167
Microtubules, involvement of cytoplasmic rotation in, 95–96
Midblastula embryos, 264–266
Midgestation chimeras, 139
MIF, see Mesoderm inducing factor
Migrating slugs, phototaxis of, 82–83
Migratory neural crest cells, 202–203
Migratory patterns, 200–202
Migratory slug, 279–281
Mitosis, 85, 86
Mitotic patterns, 200–202
Mix.1, 264–266
Mix.1 mRNA, 264–266
Mobile genetic elements, 3–5
Molecular resolution, 34–36
Molecular tag, 42–44
Mollusks
  accelerated development into adults, 182–185
  loss of larval stages, 182–185
Monoclonal antibodies, 99–100, 169–170, 210–212
Monoclonal antibody 46F11, 132–133
Morphogenesis and pattern formation, 279–297
Morphogenetic patterns, gradients of specific molecules that establish, 287–288
Morphological abnormalities, unusual phenotypes displaying, 283–284
Mosaic analysis, 197–198
Mouse
  brain, NCAM cDNA in, 176–178
  chromosome mapping and expression, 238–239
  embryos, gradient of homeodomain protein in, 261–262
  fibroblasts, functionally interchangeable elements in, 225–226
  gastrulation and neurulation in, 109–110
  germ cells, deletion mutations in, 28–29
  hematopoietic, 99–100
  homeobox genes, putative role of, 276–278
  homeobox pattern in, 261–262

spermatogenesis in, 221—224
Y chromosome
  sex-determing region of, 238—240
  testis determination on, 235—236
  Y-finger, 238—239
MPF, 85
MUD50, 209—210
Multilocus deletions, radiation-induced, 28—29
Multiple RNAs, 110—111
Multiple segmentation genes, 39—40
Multipotency, revealed by cell lineage analysis, 198—200
Murine cDNA, 101—102
Murine genome, 274—276
Murine homeobox gene complexes, 274—276
Murine Hox gene family, structural and functional organization of, 276—278
Murine NCAM, 176—178
Murine primary B cells, 74—75
Murine thymocytes, 103—104
Muscle actin gene, 71—72
Muscle-lineage blastomeres, 181—182
Mutagens, role of endogenous retrovirus, 41
Mutant human IgM genes, 74—75
myc proteins, 219—221
Mycobacterium tuberculosis, 234—235
  primary immune response to, 233—234
Myeloid leukemia inhibitory factor (LIF), 107
MyoD, proteins, 219—221

N

N-CAM, 144—145, 164—165, 172—173, see also Neural cell adhesion molecule
nanos (nos, ns) gene, 123—124
nanos functions, 124
NC, see Neural crest
NCAM cDNA, 176—178
NCAM N-terminal domains of, 176—178
neo gene, 27—28
Neural adhesion molecule, altered expression of, 164—165
Neural cell adhesion molecule (N-CAM), 176—178
Neural crest (NC), 165—167, 198—200
Neural ectoderm, 91—92
Neural fold formation, 167—168

Neural tube, *Dystonia* locus expressed in, 42
Neurite guidance, 170—171
Neuromuscular junction, 169—170
  synaptic cleft of, 227—228
Neuronal development, segmental patterns of, 296—297
Neuronal populations, 296—297
Neurotransmitter plasticity, 226—227
NH_3, see Ammonia
Non-B cells, 271
Nondrosophilids, *P* elements in, 9—11
Nonselectable genes, general strategy for targeting mutations to, 22—24
Noradrenergic sympathetic neurons, 226—227
Normal chimeras, 140
nos, see *nanos* gene
Notch, 188—189

O

oct-1, 272
oct-2, 268
  cDNAs, 268
Octamer sequence motif, 271—272
Oligonucleotide primer, single gene-specific, 31—34
Oligonucleotide
  probe, 135—136
  screening, 219—221
Ommatidia development, 259—260
Oncogenic transformation, 87—88
Oocyte
  cDNA libraries, 86—87
  5S genes, 72—74
Oogenesis, 118—120
Oral-aboral axis, 59—61
Organismal morphogenesis, 282—283
OTF-2, 271—272
Ovarian cDNA expression library, 133—134
Overexpression, 164—165

P

P13-agarose, chromatography on, 85
Pair-rule patterns, 296—297
paired, 219
Paracrine VPC signaling, 189—192
Parallel microtubules, 95—96
Parthenogenetic blastomeres, 139—140

Parthenogenetic cells, 139—140
  detected in embryo and yolk sac, 139
  developmental potential of, 141—142
  systematic elimination of, 141—142
Parthenogenic chimeras, 140
Paternal genome contributions, 140
Pax 1
  function in murine development, 39—40
  point mutation in the paired box of, 39—40
PCR, see Polymerase chain reaction
pDE109, used to transform genes, 2—3
PDGF, 78—79, see also Platelet-derived growth factor
PDGF-A, 30—31
P-element mobility, 9—11
P element transposition, 9—11
PG expression, 12—13
Phenotypic events, timing of, 181—182
Phorbol myristate acetate-4-O-methyl ether, 289—290
Phototaxis, effect of NH_3 gas on, 82—83
PHT1-1 transgenic mice, 40—41
PI cycle, lithium inhibition of, 289—290
Pigment cell migration, 165—167
Pituitary gland, 48—49
Pituitary phenotype, 268—269
PKC, see Protein kinase C
Platelet-derived growth factor (PDGF), 30—31
Pluteus larvae tubulin, mRNA accumulation, 61—62
P-mediated transposition, 8—9
Pole cells, transplantation of, 221—224
Polygalacturonase (PG) gene, 12—13
Polygalacturonase gene expression
  antisense RNA inhibition in, 12—13
Polymerase chain reaction (PCR), 31—34, 78—79
Polymorphic self-antigens, 231
Polyphosphoinositide cycle intermediate, 289—290
Posterior structures, compensatory reduction of, 96—97
Postmeiotic gene expression, 242—244
Potassium currents, 181—182
p o u domain, 248—249
POU domain, 272—273
  regulatory genes, 273—274
Pr1 expression, 48—49
Pre-B cell growth factor, 103—104

Prestalk-specific genes, transcription of, 208—209
Primary immune response, 18—20
Primary mesenchyme cell lineage, 185—187
Primary mesenchyme cells, 148—149
  cell biology of, 64—66
Pro-B-lymphocyte, 102
Pro-T-lymphocyte, 102
Program mutants, 5—7
Promoter elements, binding of, 256—258
Protein kinase C (PKC) translocation, 97—98
Proteins, transcriptional activation and repression by, 254
Proviral insertion, 41
P transposable element, inverted repeats of, 8—9
P transposase, coding for, 9—11
Purification and characterization of stem cells, 99—100
Purified polypeptides, 107—108
Purkinje cell differentiation *int-2*, transcripts found in, 111—113
Putative, diffusible neural inducer, 167—168

R

R-cells, 157—159
RA, see Retinoic acid
RACE, see Rapid amplification of cDNA ends
Radial arrangement, 198—200
Rapid amplification of cDNA ends (RACE), 31—34
RAR, see Retinoic acid receptor
Rat prolactin, cell-specific activation of, 268—269
Receptor-mediated signal transduction, 97—98
γδ Receptors
Recessive insertional mutation, 42
Recombinant interleukin 7, 103—104
Recombinant ricin, lens-specific expression of, 45
Region-specific alleles, 284
Retinoic acid (RA), 295—296
Retinoic acid receptor (RAR), 293—296
Rhombomeres, 296—297
RNA binding proteins, 215
RNA-injected embryos, gastrulation in, 263—264

RNA polymerase III, 72—74
RNase protection assays, 110—111
ro, see *rough*
rough (ro) locus, 258—259
rough gene, mutations in, 259—260
runt, 284

S

s-laminin, 169—170, 227—228
s-laminin cDNAs, 169—170
S. droebachiensis, 64—66
Saccharomyces cerevisiae, tRNA genes in, 3—5
Salmonella, high-frequency switching in, 3
SCID-hu mouse, 13—15
Sea cucumbers
　accelerated development into adults, 182—185
　loss of larval stages, 182—185
Sea urchin
　blastomeres, 149—152
　development, 66—67
　embryo, 59—61
　　cells, 149—152
　　early cleavages, 187—188
　　aboral ectoderm of, 245—248
　　additional mechanisms required in, 61—62
　　animal blastomeres of, 187—188
　　embryological studies, 182—185
　　expression, tubulin gene expression during, 61—62
　　extraembryonic matrix of, 146—147
　　gastrulation, 147—148
　　genome, 61—62
　　primary mesenchyme cells of, 185—187
　　transcription factor of, 66—67
Secondary mesenchyme cells (SMC), 147—148
Segment identity, Drosophila genes involved in, 274—276
Segmentation, 296—297
Sequence conservation, 111—13
Sequence motif, 135—136
Serum response factor, SRF, 225—226
Serum-responsive (SRE) promoter, 225—226
sevenless (sev), 157—159, 196—197, 259—260
　cDNA, 196—197

ommatidia, 196—197
　mutation, 197—198
　ubiquitous expression of, 196—197
Sever combined immunodeficiency (SCID) 15—16
sev[+] ommatidia, 196—197
Sex differentiation, control of, 217
Sex-determination signal, 240—242
Sex-determining (Sxr) region, 238
Sex-lethal (Sxl), 215
Sex-reversed mice, 240—242
Sex reversed (Sxr) region, 235—237
Sex-specific polypeptides, 218
Sex-specific RNA splicing, 215
Side branch formation, N-CAM antibodies decreased by, 170—171
Signal transduction, 5—7
Simian virus 40 (SV40), 279—281
Skeletal matrix protein gene, lineage-specific expression of, 185—187
SMC, see Secondary mesenchyme cells
Solid reaggreagates, 161—164
Somatic 5S genes, 72—74
Somatic mutation, 18—20
Somatotropes, 48—49
Spec 1, probed in *Stronglyocentrotus purpuratus*, 63—64
Spec 2, probed in *Stronglyocentrotus purpuratus*, 63—64
Specificity, temporal and spatial, in *Aspergillus* development, 53—54
Spermatids, 242—244
Spermatogenesis, 113—114
　syncytial nature of, 243
Spicule growth, biomineralization process of, 185—187
Spicule matrix protein (SM50), 187—188
Spleen colonies, 99—100
Spore formation, 54—56
Spore maturation, 208
Sporogenous mutants, 208—209
spt-1, 288—289
SRE, see Serum responsive promoter
SRF, see Serum response factor
5S ribosomal RNA genes, developmental regulation of, 72—74
Stage III oocytes, 138—139
Starfishes
　accelerated development into adults, 182—185
　loss of larval stages, 182—185
stg gene, 83—85

Subject Index

Stronglylocentrotus purpuratus, 64—69, 245—248
   skeletogenic system of, expression of embryonic genes in, 64—66
Substrate adhesion protein, 146—147
Suicide vector, creation of, 45—46
Surface cAMP receptors, 5—7, 5—7
SV40, see Simian virus 40
swa embryos, 118—120
Swallow, 120—121
Switching activity, blocking of, 69—70
Sxl, 221—224, see also *Sex-lethal*
   alleles of, 221—224
Sxr, 235—237, 240—242, see also Sex reversed region

T

Targeted mutagenesis, 20—22
T-cell antigen receptor, 204—205
T cell clones, 231
T cell receptor (TCR), 173—175, 205—206, 233
$\alpha\beta$ T-cell receptor, 174—175
   heterodimers, 205
$\gamma\delta$ T-cell receptor, 233—234
   heterodimers, 205
T-cell repertoire, 231
T-cell-receptor transgenic mice, 230—231
T cells, 105, 174
   lineages of, 204—205
$\gamma\delta$ T cells, 233—234
T lymphocytes, 48—49, 230—231
t-Pa, see Tissue plasminogen activator
$\alpha\beta$ TCR, 230—231, see also T cell receptor
TDF, see Testis-determining factor gene
Tdy, 235—236
   location of, 237
Temporal developmental switch, 195—196
Teratocarcinoma endoderm cells, 111—113
Testis-determining factor (TDF) gene, 238—242
Testis-determining genes, 240—242
TFG-beta, 102—103
TFG-beta genes, 78—79
TFIIA, binding of, 72—74
TFIIIA, 72—74
TFIIIB, 72—74
TFIIIC, 72—74

TGF, see Transforming growth factor
TGF-alpha, 78—79
TGF-beta, 78—79, 90—91, 136—137
TGF-beta-2, 88—89
TGFβ2, 138—139
TGFβ-like polypeptide, 138—139
Thoracic duct lymphocytes, 105—106
Thy-1⁺ dendritic epidermal cells
Thy-1⁺ dendritic epidermal cells, 234—235
Thymic MHC antigens, 173—174
Thymidine-kinase obliteration, 47
Thymidine kinase obliteration system (TKO), 48—49
Thymocyte selection and maturation, 175—176
Tissue plasminogen activator (t-Pa), 36—37
Tissue-specific activator protein (TSAP), 66—67
Tissue-specific expression, of XK8IA1, 70—71
Tissue-specific transcription factor, 268—269
Toxic gene, expression of, 45—46
Toxic phenotype, in animal cells, 45—46
Toxotrepana curvicauda, 9—11
tra-2, 215—217
tra expression, regulation of, 219—221
tra gene, 219—221
TRA007, 54—56
Transcription
   activation and repression of, 251—252
   synergistic activation and repression of, 252—253
Transcriptional activators, 255
Transcriptional regulation, 70—71
   of genes, 259—260
Transcriptional regulators, 249—250
Transfer RNA (tRNA), 3—5
   genes, 3—5
transformer-2 (tra-2), 217
Transforming growth factor (TGF)-α, 30—31
Transforming growth factor (TGF)-β1, 30—31
Transgene, insertion of, 42—44
Transgenic mice, 105
   creation of with controlled immune deficiency, 47
   developmental defects in eyes of, 45
   lens-deficient eyes in, 46—47
   with inducible dwarfism, 48—49

Transgenic parthenogenic embryos, 140—141
Transgenic plants, regulated genes in, 77—78
Transmembrane signal transduction, 5—7
Transposable elements, 11—12, 41
Transposan insertion mutant, 282—283
Tripneustes
  esculentus, 147—148
  gratilla, 245—248
Trophectoderm, 139—140
Trophectoderm-derived tissue, 139—140
Trypanosoma, 144—145
TSAP, see Tissue-specific activator protein
Tubulin, 61—62
  gene transcription, 61—62
  mRNA stability, autoregulation of, 61—62
  synthesis, 61—62
Tunicamycin treatment, 136—137

U

Ubiquitin, probed in *Lytechnius variegatus* and *L. pictus*, 63—64
Ubx genes, 256—258
Ubx promoter, 254, see also *Ultrabithorax*
  -lacZ fusion genes, 250—251
  transcription, 256—258
Ultrabithorax (Ubx), 250—251, 254
  protein, 256—258
unc-86, 248—249
  mutations in, 248—249
undulated, 39—40

V

V gene rearrangement, 18
V regions, 18
Vasa mutations, females homozygous for, 124—127
vasa , 133—135
  flies, 133—134
  gene, 134—135
Vβ element, 231
VDJ recombinase mechanism, affected by *scid* defect, 16—18
VDJ region, 18—20
VδM23, 234—235
Vegetal blastomeres, 187—188
Vegetal cortex, 136—137

Vegetal signal, modified by lithium, 290—291
Vegetal-specific gene products, 187—188
Vegetal-specific molecules, 187—188
Ventral structures, compensatory reduction of, 96—97
Ventral uterine (VU), 192—195
Vg1, 136—137
Vg1 mRNA cortical vegetal localization, 138—139
Vg1 RNA, 138—139
VPCs, see Vulval precursor cells
VU, see Ventral uterine
Vulval precursor cells (VPCs), 188—192

W

wetA, 53—54
White axolotl mutant, 165—167
Wild-type alleles, 221—224
W locus, 37—39
W^{44} mutation, 37—39

X

X-chromosome/autosome (X:A) ratio, 221—224
X. laevis/X. borealis hybrids, 266—267
X1Hbox 1 protein, 261—262
X:A, see X-chromosome/autosome
Xenopus
  blastula
    animal regions, 161—164
    ectoderm, 291—292
  cell-free system, methionine incorporated in, 86—87
  duplicated homeobox genes in, 266—267
  eggs, treated with D20, 96—97
  ectodermal tissue, 160—161
  embryogenesis, 289—290
  embryos, 90—91, 160
    cardiac actin gene in, 224
    dorsalization in, 92—93
    early neural development in, 164—165
    FGF cell surface receptors of, 90—91
    functionally interchangeable elements in, 225—226
    lithium treatment of, 290—291
    neural inducing capacity in, 93—94

presumptive endodermal cells of, 264—266
FGF, 291—292
genomic library, 135—136
homeobox containing gene, synthesis of, 264—266
homeobox gene in, 260—261
homolog, 86
mesoderm in, 25—226
oocyte, 86—87, 89—90, 136—137
  finger, motif proteins identified in, 135—136
  maternal mRNA localized following injection into, 138—139
  tissue differentiation, 70—71
Xenopus laevis, 72—74, 136—137, 260—261
  embryos, 71—72
    epidermis-specific marker in, 91—92
    reversal of dorsoventral polarity in, 292—293
  mesoderm induction in, 160
  neural induction, 97—98
  nonepidermal pathway of differentiation in, 92—93
  oocytes, 135—136
Xhox3, 263—264

biomodal and graded expression of, 262—263
expression, 263—264

Y

Y-borne zinc-finger (ZFY) protein, 237—238
YAC, see Yeast artificial chromosome
YAC maps, 7—8
Yeast artificial chromosome (YAC), genome linking with, 7—8

Z

Zaprionis tuberculatus, 9—11
Zebrafish
  embryo, cell movement and fate in, 288—289
  zygotic lethal mutation, 288—289
ZFX, 238—242
ZFY (Y-borne zinc-finger), 237—238, 240—242
  marsupials in, 237—238
Zfy-1, 239—240
Zfy-2, 239—240
Zinc finger, 135—136